INTERFACIAL PHENOMENA IN CHROMATOGRAPHY

SURFACTANT SCIENCE SERIES

1. Nonionic Surfactants, edited by *Martin J. Schick* (see also Volumes 19, 23, and 60)
2. Solvent Properties of Surfactant Solutions, *edited by Kozo Shinoda* (see Volume 55)
3. Surfactant Biodegradation, *R. D. Swisher* (see Volume 18)
4. Cationic Surfactants, *edited by Eric Jungermann* (see also Volumes 34, 37, and 53)
5. Detergency: Theory and Test Methods (in three parts), *edited by W. G. Cutler and R. C. Davis* (see also Volume 20)
6. Emulsions and Emulsion Technology (in three parts), *edited by Kenneth J. Lissant*
7. Anionic Surfactants (in two parts), *edited by Warner M. Linfield* (see Volume 56)
8. Anionic Surfactants: Chemical Analysis, *edited by John Cross*
9. Stabilization of Colloidal Dispersions by Polymer Adsorption, *Tatsuo Sato and Richard Ruch*
10. Anionic Surfactants: Biochemistry, Toxicology, Dermatology, *edited by Christian Gloxhuber* (see Volume 43)
11. Anionic Surfactants: Physical Chemistry of Surfactant Action, *edited by E. H. Lucassen-Reynders*
12. Amphoteric Surfactants, *edited by B. R. Bluestein and Clifford L. Hilton* (see Volume 59)
13. Demulsification: Industrial Applications, *Kenneth J. Lissant*
14. Surfactants in Textile Processing, *Arved Datyner*
15. Electrical Phenomena at Interfaces: Fundamentals, Measurements, and Applications, *edited by Ayao Kitahara and Akira Watanabe*
16. Surfactants in Cosmetics, *edited by Martin M. Rieger* (see Volume 68)
17. Interfacial Phenomena: Equilibrium and Dynamic Effects, *Clarence A. Miller and P. Neogi*
18. Surfactant Biodegradation: Second Edition, Revised and Expanded, *R. D. Swisher*
19. Nonionic Surfactants: Chemical Analysis, *edited by John Cross*
20. Detergency: Theory and Technology, *edited by W. Gale Cutler and Erik Kissa*
21. Interfacial Phenomena in Apolar Media, *edited by Hans-Friedrich Eicke and Geoffrey D. Parfitt*
22. Surfactant Solutions: New Methods of Investigation, *edited by Raoul Zana*
23. Nonionic Surfactants: Physical Chemistry, *edited by Martin J. Schick*
24. Microemulsion Systems, *edited by Henri L. Rosano and Marc Clausse*
25. Biosurfactants and Biotechnology, *edited by Naim Kosaric, W. L. Cairns, and Neil C. C. Gray*
26. Surfactants in Emerging Technologies, *edited by Milton J. Rosen*
27. Reagents in Mineral Technology, *edited by P. Somasundaran and Brij M. Moudgil*
28. Surfactants in Chemical/Process Engineering, *edited by Darsh T. Wasan, Martin E. Ginn, and Dinesh O. Shah*
29. Thin Liquid Films, *edited by I. B. Ivanov*
30. Microemulsions and Related Systems: Formulation, Solvency, and Physical Properties, *edited by Maurice Bourrel and Robert S. Schechter*
31. Crystallization and Polymorphism of Fats and Fatty Acids, *edited by Nissim Garti and Kiyotaka Sato*

32. Interfacial Phenomena in Coal Technology, *edited by Gregory D. Botsaris and Yuli M. Glazman*
33. Surfactant-Based Separation Processes, *edited by John F. Scamehorn and Jeffrey H. Harwell*
34. Cationic Surfactants: Organic Chemistry, *edited by James M. Richmond*
35. Alkylene Oxides and Their Polymers, *F. E. Bailey, Jr., and Joseph V. Koleske*
36. Interfacial Phenomena in Petroleum Recovery, *edited by Norman R. Morrow*
37. Cationic Surfactants: Physical Chemistry, *edited by Donn N. Rubingh and Paul M. Holland*
38. Kinetics and Catalysis in Microheterogeneous Systems, *edited by M. Grätzel and K. Kalyanasundaram*
39. Interfacial Phenomena in Biological Systems, *edited by Max Bender*
40. Analysis of Surfactants, *Thomas M. Schmitt*
41. Light Scattering by Liquid Surfaces and Complementary Techniques, *edited by Dominique Langevin*
42. Polymeric Surfactants, *Irja Piirma*
43. Anionic Surfactants: Biochemistry, Toxicology, Dermatology. Second Edition, Revised and Expanded, *edited by Christian Gloxhuber and Klaus Künstler*
44. Organized Solutions: Surfactants in Science and Technology, *edited by Stig E. Friberg and Björn Lindman*
45. Defoaming: Theory and Industrial Applications, *edited by P. R. Garrett*
46. Mixed Surfactant Systems, *edited by Keizo Ogino and Masahiko Abe*
47. Coagulation and Flocculation: Theory and Applications, *edited by Bohuslav Dobiáš*
48. Biosurfactants: Production • Properties • Applications, *edited by Naim Kosaric*
49. Wettability, *edited by John C. Berg*
50. Fluorinated Surfactants: Synthesis • Properties • Applications, *Erik Kissa*
51. Surface and Colloid Chemistry in Advanced Ceramics Processing, *edited by Robert J. Pugh and Lennart Bergström*
52. Technological Applications of Dispersions, *edited by Robert B. McKay*
53. Cationic Surfactants: Analytical and Biological Evaluation, *edited by John Cross and Edward J. Singer*
54. Surfactants in Agrochemicals, *Tharwat F. Tadros*
55. Solubilization in Surfactant Aggregates, *edited by Sherril D. Christian and John F. Scamehorn*
56. Anionic Surfactants: Organic Chemistry, *edited by Helmut W. Stache*
57. Foams: Theory, Measurements, and Applications, *edited by Robert K. Prud'homme and Saad A. Khan*
58. The Preparation of Dispersions in Liquids, *H. N. Stein*
59. Amphoteric Surfactants: Second Edition, *edited by Eric G. Lomax*
60. Nonionic Surfactants: Polyoxyalkylene Block Copolymers, *edited by Vaughn M. Nace*
61. Emulsions and Emulsion Stability, *edited by Johan Sjöblom*
62. Vesicles, *edited by Morton Rosoff*
63. Applied Surface Thermodynamics, *edited by A. W. Neumann and Jan K. Spelt*
64. Surfactants in Solution, *edited by Arun K. Chattopadhyay and K. L. Mittal*
65. Detergents in the Environment, *edited by Milan Johann Schwuger*

66. Industrial Applications of Microemulsions, *edited by Conxita Solans and Hironobu Kunieda*
67. Liquid Detergents, *edited by Kuo-Yann Lai*
68. Surfactants in Cosmetics: Second Edition, Revised and Expanded, *edited by Martin M. Rieger and Linda D. Rhein*
69. Enzymes in Detergency, *edited by Jan H. van Ee, Onno Misset, and Erik J. Baas*
70. Structure–Performance Relationships in Surfactants, *edited by Kunio Esumi and Minoru Ueno*
71. Powdered Detergents, *edited by Michael S. Showell*
72. Nonionic Surfactants: Organic Chemistry, *edited by Nico M. van Os*
73. Anionic Surfactants: Analytical Chemistry, Second Edition, Revised and Expanded, *edited by John Cross*
74. Novel Surfactants: Preparation, Applications, and Biodegradability, *edited by Krister Holmberg*
75. Biopolymers at Interfaces, *edited by Martin Malmsten*
76. Electrical Phenomena at Interfaces: Fundamentals, Measurements, and Applications, Second Edition, Revised and Expanded, *edited by Hiroyuki Ohshima and Kunio Furusawa*
77. Polymer-Surfactant Systems, *edited by Jan C. T. Kwak*
78. Surfaces of Nanoparticles and Porous Materials, *edited by James A. Schwarz and Cristian I. Contescu*
79. Surface Chemistry and Electrochemistry of Membranes, *edited by Torben Smith Sørensen*
80. Interfacial Phenomena in Chromatography, *edited by Emile Pefferkorn*
81. Solid–Liquid Dispersions, *Bohuslav Dobiáš, Xueping Qiu, and Wolfgang von Rybinski*

ADDITIONAL VOLUMES IN PREPARATION

Modern Characterization Methods of Surfactant Systems, *edited by Bernard P. Binks*

Handbook of Detergents, *editor in chief: Uri Zoller*
    Part A, Properties, *edited by Guy Broze*

Interfacial Forces and Fields, *edited by Jyh-Ping Hsu*

Silicone Surfactants, *edited by Randal M. Hill*

Surface Characterization Methods: Principles, Techniques, and Applications, *edited by Andrew J. Milling*

Dispersions: Characterization, Testing, and Measurement, *Erik Kissa*

INTERFACIAL PHENOMENA IN CHROMATOGRAPHY

edited by
Emile Pefferkorn

Institut Charles Sadron
Strasbourg, France

CRC Press
Taylor & Francis Group
Boca Raton London New York

CRC Press is an imprint of the
Taylor & Francis Group, an **informa** business

First published 1999 by Marcel Dekker

Published 2019 by CRC Press
Taylor & Francis Group
6000 Broken Sound Parkway NW, Suite 300
Boca Raton, FL 33487-2742

© 1999 by Taylor & Francis Group, LLC
CRC Press is an imprint of Taylor & Francis Group, an Informa business

First issued in paperback 2019

No claim to original U.S. Government works

ISBN 13: 978-0-367-45568-2 (pbk)
ISBN 13: 978-0-8247-1947-0 (hbk)

Visit the Taylor & Francis Web site at
http://www.taylorandfrancis.com

and the CRC Press Web site at
http://www.crcpress.com

Preface

Chromatography is largely employed in characterization and separation methods in the biomedical, industrial, and environmental domains, and the technical development has called for diversified detection and analysis procedures. Concerning principles, the major efforts to improve separation efficiency have been directed to (1) the establishment of specific interactions between groups of the stationary phase and groups belonging to molecules present in the mobile phase and (2) the elimination of all specific interactions between constituents of the stationary phase and those of the mobile phase. A great number of books and reviews have been devoted to the improvement of chromatographic stationary phases.

The present book is aimed at the presentation of different chromatography methods in which interactions between the stationary and mobile phases are employed to enhance and/or control separation selectivity. Obviously, separation requires the stationary phase to be well characterized from the physicochemical and structural viewpoints.

The principles of gas chromatography are presented first to show the multiplicity of information on the gas–solid interaction provided by the use of the viral coefficient theory. Since efficient separation requires the best adhesion (adsorption, retention, etc.) of the solute in the stationary phase, inverse gas chromatography has been directed toward the optimization of the properties of fibers and fillers for reinforced materials, but it may serve as a convenient technique for characterization of stationary phases used in liquid chromatography. Inverse gas chromatography has been applied to fibers and fillers in order to determine the different components of their surface free energy. It has been employed to determine the modifications resulting from chemical surface treatments of natural and synthetic fibers as well to detect the particularities of amorphous and crystalline powders. General and particular results are presented and discussed in Part I.

Interaction phenomena in liquid chromatography have been found to be efficient in the separation of colloids and solutes as a function of size and chemical nature. Interfacial interactions between silica and solutes have been suppressed through adsorption and grafting methods to allow for the use of silica beads in size exclusion chromatography. The separation efficiency of the inverse method, which is based on the development of strong or solvent-modulated solid–solute interactions, has been tested. In Part II, as in Part I, general and particular aspects of the different methods are presented.

The book presents results related to the less explored possibilities offered by establishing and/or designing interfacial phenomena in chromatography. Separation processes in industrial and biomedical applications are expected to benefit greatly from these new methods, and environmental areas requiring better determination of the multiple selective interactions of gaseous and liquid species with soil constituents should also benefit.

Emile Pefferkorn

Contents

Preface   iii
Contributors   vii

Part I   Stationary Phase Characteristics

1. Henry's Law Behavior in Gas–Solid Chromatography: A Virial
    Approach   1
    Thomas R. Rybolt and Howard E. Thomas

2. Inverse Gas Chromatography as a Tool to Characterize Dispersive and
    Acid–Base Properties of the Surface of Fibers and Powders   41
    Mohamed Naceur Belgacem and Alessandro Gandini

3. Interactions in Cellulose-Polyethylene Papers as Obtained Through
    Inverse Gas Chromatography   125
    Bernard Riedl and Halim Chtourou

4. Inverse Gas Chromatography: A Method for the Evaluation of the
    Interaction Potential of Solid Surfaces   145
    Eugène Papirer and Henri Balard

Part II   Characterization and Separation Methods

5. Chromatography of Colloidal Inorganic Nanoparticles   173
    Christian-Herbert Fischer

v

6. Chromatographic Behavior and Retention Models of Polyaromatic
   Hydrocarbons in HPLC   227
   Yu. S. Nikitin and S. N. Lanin

7. Polymer-Modified Silica Resins for Aqueous Size Exclusion
   Chromatography   263
   Yoram Cohen, Ron S. Faibish, and Montserrat Rovira-Bru

8. Polycation-Modified Siliceous Surfaces for Protein Separations   311
   Yingfan Wang and Paul L. Dubin

9. Adsorption Processes in Surface Area Exclusion Chromatography   329
   Emile Pefferkorn, Abdelhamid Elaissari, and Clarisse Huguenard

10. Separation of Polymer Blends by Interaction Chromatography   387
    Harald Pasch

Index   435

Contributors

Henri Balard   Institute of Chemistry of Surfaces and Interfaces, French National Center of Scientific Research, ICSI–CNRS, Mulhouse, France

Mohamed Naceur Belgacem   Department of Paper Science and Technology, University of Beira Interior, Covilhã, Portugal

Halim Chtourou*   Department of Wood Science, CERSIM, Laval University, Quebec City, Quebec, Canada

Yoram Cohen   Department of Chemical Engineering, University of California, Los Angeles, Los Angeles, California

Paul L. Dubin   Department of Chemistry, Indiana University—Purdue University, Indianapolis, Indiana

Abdelhamid Elaissari   Chemistry of Supports, CNRS–Biomérieux (UMR-103), Lyon, France

Ron S. Faibish   Department of Chemical Engineering, University of California, Los Angeles, Los Angeles, California

---

* *Current affiliation*: Rétec Inc., Granby, Quebec, Canada.

Christian-Herbert Fischer Department of Solid State Physics, Hahn-Meitner-Institut Berlin, Berlin, Germany

Alessandro Gandini Department of Polymeric Materials, Ecole Française de Papeterie et des Industries Graphiques, St. Martin d'Hères, France

Clarisse Huguenard Laboratoire RMN et Chimie du Solide, Université Louis Pasteur, Strasbourg, France

S. N. Lanin Department of Chemistry, M. V. Lomonosov State University of Moscow, Moscow, Russia

Yu. S. Nikitin Department of Chemistry, M. V. Lomonosov State University of Moscow, Moscow, Russia

Eugène Papirer Institute of Chemistry of Surfaces and Interfaces, French National Center of Scientific Research, ICSI–CNRS, Mulhouse, France

Harald Pasch Department of Polymer Analysis, German Plastics Institute, Darmstadt, Germany

Emile Pefferkorn Institut Charles Sadron, Strasbourg, France

Bernard Riedl Department of Wood Science, CERSIM, Laval University, Quebec City, Quebec, Canada

Montserrat Rovira-Bru Department of Chemical Engineering, University of California, Los Angeles, Los Angeles, California

Thomas R. Rybolt Department of Chemistry, University of Tennessee at Chattanooga, Chattanooga, Tennessee

Howard E. Thomas Department of Chemistry, Erskine College, Due West, South Carolina

Yingfan Wang Department of Chemistry, Indiana University—Purdue University, Indianapolis, Indiana

1
Henry's Law Behavior in Gas–Solid Chromatography: A Virial Approach

THOMAS R. RYBOLT   Department of Chemistry, University of Tennessee at Chattanooga, Chattanooga, Tennessee

HOWARD E. THOMAS   Department of Chemistry, Erskine College, Due West, South Carolina

I.	Introduction	2
II.	Theoretical Background	3
	A. Adsorption and the second gas–solid virial coefficient	3
	B. Gas–solid chromatography	4
	C. Flat single-surface model	6
	D. Flat two-surface model	7
	E. Parallel plate model	8
	F. Cavity model	9
III.	Experimental	12
	A. Apparatus	12
	B. Experimental considerations	13
	C. Data collection	15
	D. Conversion to B_{2s}	16
IV.	Analysis and Discussion	18
	A. Single-surface approach	18
	B. Two-surface approach	20
	C. Cavity model	26
	D. Comparison of flat, parallel plate, and cavity models	30
	E. Correlations with other physical properties	33
	References	37

I. INTRODUCTION

Changes in the distribution of molecules between the gas and adsorbed phases as a function of pressure and temperature have been used for many years to study surface area, surface structure, and gas–solid interactions. Gas–solid chromatography provides a useful alternative to conventional adsorption experiments and can serve as a method to study these aspects of physical adsorption.

In the virial coefficient treatment of physical adsorption, the moles of gas adsorbed per gram of adsorbent, n_a, are related to the second gas–solid virial coefficient, B_{2s}, which is a measure of the interaction of isolated gas molecules with a solid surface. Adsorption isotherms or gas chromatographic retention times measured in the Henry's law region of low adsorption can be used to provide values of second gas–solid virial coefficients. In this chapter we will explore the theoretical basis of the second gas–solid virial coefficient, how these values are determined from gas–solid chromatography experiments, and how an analysis of these values and their temperature dependence for various adsorbate gases can provide adsorbent structural information, solid surface areas, and measures of gas–surface interactions. In addition, we will examine how retention times and B_{2s} values can be correlated with other physical or structural properties.

In the pulse flow technique, a gas or vapor is injected into a flowing carrier gas. The adsorbate flows through a powder-packed column, and molecules are distributed between an adsorbed form and the gas phase. The retention time depends on the magnitude of the equilibrium between the adsorbed and free forms of the adsorbate. As we shall see in the theory section, the second gas–solid virial coefficient depends on the corrected flow rate of the carrier gas through the column, the retention time of the sample gas measured relative to a noninteracting reference gas, and the mass of powder in the column.

The virial coefficient treatment of physical adsorption was developed primarily by Steele and Halsey [1] and others [2–4], discussed by Steele [5], and thoroughly reviewed by Pierotti and Thomas [6] who covered the exact statistical thermodynamic basis of this approach. The importance of a Henry's law approach to studies of surface heterogenity has been discussed by Bakaev and Chelnokova [7]. An excellent review of the historical background and approaches used in a virial analysis of adsorption is provided by Rudzinski et al. [8]. Although most early work used volumetric or gravimetric adsorption techniques, chromatographic studies included work by Boucher and Everett [9], Rudzinski et al. [10], and Ross et al. [11].

Since the introduction of the virial coefficient theory [1–4], continued attention has been given to physical adsorption and the application of the virial coefficient theory [8,12–14]. Adsorption in micropores has been studied at various pressures and temperatures [15,16], and the size and shape of the micropores

[17–20], as well as their formation and structure [21,22], have also been studied. Virial coefficients have been used to analyze experimental data [23,24], and calculations have been carried out to compare the adsorption in slit-like pores to flat surfaces and to find surface areas and volumes of the solids [25]. Other studies have been conducted to analyze Henry's law constant [7] and the energy aspects of adsorption [24,26,27].

A variety of studies have used gas–solid chromatography to determine second or second and third virial coefficients for gases interacting with solid surfaces [10,28–37]. Other studies have examined the theoretical basis of the virial approach or used traditional adsorption experiments to determine virial coefficients or Henry's law constants [8,24,38–52]. The use of gas chromatography to determine equilibrium properties and second virial coefficients for gases was reviewed by Conder [45]. Gas–solid chromatography has been used by Rybolt and Thomas to determine B_{2s} values, find powder surface areas, specify the gas–solid interaction energies, determine structural parameters for microporous solids, determine the relative amounts of higher and lower energy surfaces on two-surface solids, and correlate B_{2s} values or adsorption energies obtained from B_{2s} values with other molecular physical properties [28,29,32,37,53].

II. THEORETICAL BACKGROUND

A. Adsorption and the Second Gas–Solid Virial Coefficient

A virial approach to physical adsorption and gas–solid chromatography is based on the following equation:

$$n_a = \sum_{i=1} B_{i+1,\,s}(f/RT)^i \tag{1}$$

where n_a is the moles of gas (adsorbate) adsorbed per gram of solid adsorbent, f is the fugacity, R is the gas constant, T is the temperature, and $B_{i+1,\,s}$ is the $(i + 1)$th gas–solid virial coefficient [6]. B_{2s} represents the interaction of one adsorbate molecule with the surface, B_{3s} represents a pair of gas molecules interacting with each other and the surface, B_{4s} represents a triplet of adsorbate molecules, and so forth. In the Henry's law region of adsorption where only a small portion of the surface is covered with adsorbate, the adsorbate–adsorbate interaction is negligible and the higher order terms drop out of the power series in the previous equation.

As the pressure, P, approaches zero, Eq. (1) may be simplified and written as [6]:

$$n_a = B_{2s}(P/RT) \tag{2}$$

B_{2s} values vary with temperature and contain information about the structure of the solid as well as the strength and nature of the gas–solid interactions. The probably more familiar Henry's law constant, K_H, relates pressure to moles adsorbed as $n_a = K_H P$ where $B_{2s}/RT = K_H$. As we shall show in this chapter, B_{2s} values have an exact definition based on statistical thermodynamic considerations that make them useful in analyzing gas–solid interactions.

In order to extract the maximum amount of information about a solid, it is necessary to have adsorption data or gas chromatographic data for several different adsorbates over a range of temperature. Sets of B_{2s}–temperature pairs for varied gases on a particular adsorbent can be generated from careful experimental measurements of corrected retention times, corrected flow rates, and the mass of the powder in the packed column.

Experimental values of B_{2s} are related to the surface structure and gas–solid interaction through the integral expression:

$$B_{2s} = \int_V [\exp(-u_{1s}/kT) - 1] dV \qquad (3)$$

u_{1s} is the gas–solid interaction potential, k is the Boltzmann constant, and dV is the volume element in the gas phase [5]. Note that B_{2s} values may be calculated from Henry's law constants but have the useful feature of being related to the gas–solid interact potential by an exact statistical thermodynamic derivation [5,6]. Approximations enter into the equation only as one develops specific equations to represent u_{1s}.

Given the functional form of u_{1s} and the appropriate parameters to characterize the solid surface and gas–solid interaction, one can calculate B_{2s} values and predict adsorption and chromatographic data. However, the approach discussed below is based on the availability of experimental B_{2s} data and the desire to extract gas–solid molecular parameters from these data. In the following sections, we will show how B_{2s} is related to chromatographic retention times and focus on four different models that have been used to represent the gas–solid interaction. These models are by necessity simplified versions of the surface structure but can nevertheless provide useful data and insights into unique methods to calculate surface area, surface structural parameters, and gas–solid interaction parameters. The theoretical equations for the flat single-surface, flat two-surface, cavity, and parallel plate models are presented in this section while applications are presented later, in the analysis section.

B. Gas–Solid Chromatography

The diffusion of an adsorbate pulse transported through an adsorbent packed column by an inert carrier gas stream is given by [11,54–56]:

$$n_{gc}/n_{ac} = L/tv \qquad (4)$$

where n_{gc} is the moles of the unadsorbed gas in the column, n_{ac} is the moles of adsorbed gas in the column, L is the length of the powder packed column, v is the interstitial gas velocity, and t is the retention time of the adsorbate pulse. Eberly and Spencer [56] showed that the previous equation is the solution to the linear partial differential equation that describes the mass conservation between the adsorbed and desorbed or free phases in a pulse flow system where the Gaussian distribution is approximated by an infinitesimally narrow pulse. Ross et al. [11] showed that Eq. (4) may be rewritten as:

$$n_{gc}/n_{ac} = V_g/tF_1 \qquad (5)$$

where F_1 is the actual flow through the column and V_g is the interstitial volume in the column. Using the ideal equation of state to model the adsorbate gas gives:

$$V_g = n_{gc}RT_1/P_1 \qquad (6)$$

where P_1 is the average pressure of the sample gas in the packed column, R is the gas constant, and T_1 is the column temperature.

In the Henry's law region, the virial equation of state for adsorption from Eq. (2) may be written as:

$$n_{ac} = mB_{2s}(P_1/RT_1) \qquad (7)$$

since $n_{ac} = mn_a$ where m is the mass of powder in the packed column and n_a is the moles of sample gas (adsorbate) adsorbed per gram of solid adsorbent. Substituting Eq. (6) and Eq. (7) into Eq. (5) gives:

$$B_{2s} = tF_1/m \qquad (8)$$

The adsorbate retention time is given by

$$t = t_s - t_m \qquad (9)$$

where t_m is the time from injection to detection of pulse maximum of suitable marker gas with negligible gas–solid interaction and t_s is the time from injection to detection of pulse maximum of sample gas. This subtraction corrects the sample pulse time to the residence time, t, that arises due to sample gas adsorption in the column.

Since the measured flow rate, F_m, is determined outside the column, it is necessary to calculate F_1 from F_m. The corrections to the measured flow rate must take into account the pressure drop across the column, outlet and atmosphere pressure difference in the flow meter, and the temperature of the column compared to the flow meter temperature. The corrected column flow rate, F_1, is given by:

$$F_1 = 1.5\, F_m(T_1/T_f)(P_o/P_a)[(P_i/P_o)^2 - 1]/[(P_i/P_o)^3 - 1] \qquad (10)$$

where P_i is the inlet pressure, P_o is the outlet pressure, P_a is the atmospheric pressure, and T_f is the temperature of the flow meter [57–59].

Equations (8)–(10) can be used in conjunction with experimental measurements to obtain values of the second gas–solid virial coefficient for different adsorbates over a range of temperature for a variety of solid adsorbents. In the remainder of this section, we will show how these values of B_{2s} may be expressed in terms of gas–solid energetic and solid structural parameters.

C. Flat Single-Surface Model

Since we are limiting our consideration to physical adsorption due to van der Waals interaction [60], we do not need to consider the formation of chemical bonds. We will limit our work to nonspecific intermolecular forces and use the simplest possible potential to describe the adsorbate–adsorbent interactions. The Lennard-Jones (m, n) potential has been used to represent gas–gas as well as gas–solid interactions. While various simple potentials can be used to represent the interaction of a molecule with a solid, we are concerned with obtaining from thermodynamic adsorption data a set of simple parameters that can be used to characterize the gas–solid interaction potential and solid surface and used to predict B_{2s} values. For these purposes, a Lennard-Jones (m, n) potential has proven adequate [52].

For the special case of a flat, uniform surface, the relation between B_{2s} and the gas–solid interaction potential in Eq. (3) can be expressed as:

$$B_{2s} = A \int_z [\exp(-u_{1s}(z)/kT) - 1]\,dz \tag{11}$$

where A is the area of the surface, z is the axis normal to the surface plane, and $u_{1s}(z)$ is the gas–solid interaction potential as the gas molecule approaches the surface [5,6]. For computational purposes, it is more convenient to express this integral as:

$$B_{2s} = Az^* \int_y [\exp(-u_{1s}(y)/kT) - 1]\,dy \tag{12}$$

given in the terms of the reduced distance, y, where $y = (z/z^*)$ and z^* is the equilibrium distance of the center of the adsorbate molecule to the nuclei of the outermost atoms of the flat surface [48].

If a Lennard-Jones (m, n) potential is assumed between the adsorbate and the surface, then the equilibrium separation can be expressed as

$$z^* = (n/m)^{1/(n-m)} z_0 \tag{13}$$

where z_0 is the adsorbate–adsorbent or gas–solid distance of closest approach where u_{1s}, is equal to zero. The gas–solid interaction potential, u_{1s}, is defined by a Lennard-Jones potential as [50]:

$$u_{1s}(y)/k = E^*[n/(n - m)](n/m)^{[m/(n-m)]}$$
$$\times \{y^{-n}(n/m)^{[n/(m-n)]} - y^{-m}(n/m)^{[m/(m-n)]}\} \tag{14}$$

where ε_{1s}^* is the depth of the gas–solid interaction potential at equilibrium separation and E^* is the interaction energy or depth of the gas–solid potential well at equilibrium separation in temperature units ($E^* = \varepsilon_{1s}^*/k$).

To evaluate the integral in Eq. (12) and calculate B_{2s} it is necessary to specify the parameters m, n, E^*, T, A, and z^*. Notice that if B_{2s} can be calculated then the retention time for an adsorbate peak also can be calculated using Eq. (8) and Eq. (9). Alternatively, if retention times are measured using gas–solid chromatography then Eq. (8) and Eq. (9) can be used to determine B_{2s} at various temperatures. Given an appropriate selection for n and m, B_{2s} values can be used to find z^*, E^*, and the surface area A. Hence, gas–solid chromatography combined with a virial approach provides a useful means to study gas–solid interactions and surface properties.

In general, we may represent the combination of Eq. (12) and Eq. (14) as:

$$B_{2s} = Az^*I(E^*, T) \tag{15}$$

where A is the area, z^* is the equilibrium separation between the adsorbate and adsorbent surface, E^* is the interaction energy divided by the Boltzmann constant, T is the temperature and $I(E^*, T)$ is an integral function of T and E^* [3,5,6]. Az^* should not vary with temperature. Using experimental values of B_{2s} at different temperatures one can find the best value of E^* such that Az^* is the most consistent as the temperature is varied because $Az^* = B_{2s}/I(E^*, T)$. To avoid size bias the best-fit interaction energy, E^*, is selected to minimize the value of the standard deviation of $\log(Az^*)$ for a given adsorbate. Examples of this approach will be provided in Section IV of this chapter.

D. Flat Two-Surface Model

In the two-surface approach, the experimental second gas–solid virial coefficient, B_{2s}, can be separated into two additive contributions:

$$B_{2s} = B_{2s}^1 + B_{2s}^2 \tag{16}$$

where B_{2s}^1 and B_{2s}^2 are the second gas–solid virial coefficients for surface one and surface two, respectively. Surface two has an interaction energy, E_2^*, which is greater than the surface one interaction energy, E_1^*. E_r is the ratio of the gas–solid interaction energies ($E_r = E_2^*/E_1^*$) [61]. This relation in combination with Eq. (15) gives:

$$B_{2s}(T) = A_1z^*I_1(E_1^*, T) + A_2z^*I_2(E_2^*, T) \tag{17}$$

where the total surface area, A, is the sum of A_1 the lower energy surface area and A_2 the higher energy surface area [61].

If x is defined as the percentage of the area for surface two, then $A_1 = (1 - x)A$ and $A_2 = xA$, and

$$Az^* = B_{2s}(T)/[(1 - x)I_1(E_1^*, T) + xI_2(E_2^*, T)] \tag{18}$$

Each of the integral functions above is defined by Eq. (12) and Eq. (14). Since Az^* is not dependent on temperature, the best-fit interaction energy, E_1^*, can be found to minimize the value of the standard deviation of $\log(Az^*)$ for a given adsorbate provided the values of x and E_r are known or selected. For the same solid adsorbent with different gases as adsorbates, both the ratio of E_2^* to E_1^*, E_r, and the surface area, A, should be constant because these parameters depend on the surface structure of the adsorbent and not the adsorbate [50].

Since the surface area is independent of the gas used, x and E_r may be determined by finding the minimum value of the standard deviation of the log of surface area, $SD(\log A)$, of all the gas systems combined. Analysis using the two-surface model uses numerical integration and iterative algorithms to scan a range of x and E_r values. Because the value $n = 16$ gave a more consistent fit of monatomic gases on P33 carbon than any lower value of n [52], generally a Lennard-Jones [3,16] potential has been used in our work to represent the gas–solid interaction potential. Although the repulsive potential value is strictly empirical, the value $m = 3$ has a theoretical basis related to the van der Waals interaction of an adatom with a solid [52].

E. Parallel Plate Model

The integral expression for B_{2s} may be adapted to represent specific adsorbent surfaces. For a parallel plate model, B_{2s} can be expressed as:

$$B_{2s} = A \int [e(-u_{1s}(r)/kT) - 1]dr \tag{19}$$

where A is the surface area of one of the parallel plates and r is the internuclear separation between the adsorbate and the center of the pore. For an adsorbent surface composed of slit-like pores, this model represents the surface structure as two flat parallel plates. Using this model, B_{2s} can now be expressed as:

$$B_{2s} = 2aA \int [e(-u_{1s}(y)/kT) - 1]dy \tag{20}$$

where y is a reduced variable that represents the separation between the adsorbate and the center of the pore ($y = r/a$) and $2a$ is the separation between the parallel plates [49]. The limits of integration, originally $-a$ to a in Eq. (19), were changed to -1 to 1 and then to 0 to 1 by taking advantage of the symmetry between the parallel plates and by introducing a 2 in front of the integral as shown in Eq. (20).

The distance from the molecule to the closer plate is expressed as $a - r$, where a is half the distance between the two plates and r is the distance from the molecule to the center between the plates. The distance from the molecule to the farther plate is then expressed as $a + r$. The two interaction potentials can be expressed as:

$$u_{1s}(a - r)/k = E^*[n/(n - m)][n/m]^{[m/(n-m)]}$$
$$\times \{[n/m]^{[n/(m-n)]}[r^*/(a - r)]^n \quad (21)$$
$$- [n/m]^{[m/(m-n)]}[r^*/(a - r)]^m\}$$

$$u_{1s}(a + r)/k = E^*[n/(n - m)][n/m]^{[m/(n-m)]}$$
$$\times \{[n/m]^{[n/(m-n)]}[r^*/(a + r)]^n \quad (22)$$
$$- [n/m]^{[m/(m-n)]}[r^*/(a + r)]^m\}$$

where n is the repulsive power, m is the attractive power, E^* is the potential at the equilibrium distance, and r^* is the equilibrium distance from the molecule to the closer plate [49]. The combined interaction potential, u_{1s}, is the sum of Eq. (21) and Eq. (22).

At equilibrium, the distance to the closer plate is given as $a - r = r^*$ and to the more distant plate as $a + r = 2a - r^*$. Therefore the interaction potentials at equilibrium separation can be expressed as the sum of the closer $w_1(r^*)$ and more distant $w_2(2a - r^*)$ potentials, and are combined to give:

$$[w_1(r^*) + w_2(2a - r^*)]/k = E^*[n/(n - m)][n/m]^{[m/(n-m)]}$$
$$\times \{[n/m]^{[n/(m-n)]}\{1 + [1/(2V^* - 1)]^n\} \quad (23)$$
$$- [n/m]^{[m/(m-n)]}\{1 + [1/(2V^* - 1)]^m\}\}$$

where the reduced variable, $V^* = a/r^*$, is used [49]. Equation (23) is used for u_{1s}/k when the adsorbate is at the equilibrium separation from the closer plate.

Once specific n and m values have been selected, the parallel plate model may be represented as:

$$B_{2s} = 2aAI_p(E^*, V^*, T) \quad (24)$$

where I_p is the parallel plate model integral using Eqs. (21)–(23).

F. Cavity Model

Stroud et al. [62] suggested that the Lennard-Jones and Devonshire (LJD) cell theory of fluids [63,64] is appropriate to model the behavior of an isolated adsorbate molecule inside a cavity (such as a zeolite cavity). Consider an adsorbent that is composed of a collection of identical cavities. In this case, the second gas–solid virial coefficient may be written as:

$$B_{2s} = N_s b_{2s} \tag{25}$$

where b_{2s} is the virial coefficient for a single cavity and N_s is the number of cavities per gram of adsorbent [51]. Using the LJD theory, b_{2s} may be written as:

$$b_{2s} = \exp(-w(0)/kT) \int_0^a \exp[(-w(r) + w(0))/kT] 4\pi r^2 dr \tag{26}$$

where a is the radius of the spherical cavity, $w(0)$ is the average potential at the center of the cavity, r is the distance of the center of the adsorbate molecule from the center of the cavity, and $w(r)$ is the average of the adsorbate potential when the adsorbate is at a distance r from the center of the cavity [65,66].

Note that the previous equation is analogous to Eq. (3) but the -1 term is not included. The strong gas–solid interaction term for cavity interaction where the adsorbent surrounds the adsorbate gas makes it possible to ignore this term and simplify the expression. In the noble gas–13X zeolite analysis reported later in this chapter, the work was duplicated with and without the -1 term and the difference was found to be insignificant. As in previous work [62], we also ignore the -1 term in the cavity model.

Lattice summation calculations of Derrah and Ruthven [67] have shown that the assumption of a spherical cavity is reasonable for gas–zeolite interactions. Since the LJD cell theory assumes that the liquid nearest neighbors are in fixed positions surrounding one moving liquid molecule, the LJD cell theory is actually more appropriate for the gas–solid cavity system than for a liquid. To apply this theory, we need to express the integral in the previous equation in parameters that can represent the structure of the cavity and the energy of the gas–cavity interaction.

If the Lennard-Jones [6,12] potential is used to represent the pairs of gas atom and cavity atom interaction, the potential $w(r)$ can be written as:

$$w(r) = \varepsilon_{1s}^* \int_0^\pi [(r^*/R)^{12} - 2(r^*/R)^6][\sin \theta/2] d\theta \tag{27}$$

where ε_{1s}^* is the adsorbate molecule–cavity interaction energy at equilibrium separation, r^* is the equilibrium separation between the center of the adsorbate molecule and the atoms forming the interior wall of the cavity, θ is the angle used to define the position of the adsorbate relative to a line of length a from the center of the cavity to the nucleus of a cavity wall atom, and R is the adsorbate center–cavity wall atom nucleus separation given by $R^2 = r^2 + a^2 - 2$ arc cos θ [68].

An evaluation of Eq. (26) and Eq. (27) [51,65,66,69] gives:

$$w(0) = \varepsilon_{1s}^*[(V^*)^{-4} - 2(V^*)^{-2}] \tag{28}$$

$$w(r) - w(0) = \varepsilon_{1s}^*[(V^*)^{-4}L(y) - 2(V^*)^{-2}M(y)] \tag{29}$$

where y, the reduced distance, is

$$y = (r/a)^2 \tag{30}$$

V^* reduced volume is

$$V^* = (a/r^*)^3 \tag{31}$$

$$L(y) = (1 + 12y + 25.2y^2 + 12y^3 + y^4)(1 - y)^{-10} - 1 \tag{32}$$

and

$$M(y) = (1 + y)(1 - y)^{-4} - 1 \tag{33}$$

The interaction potential at equilibrium separation, $w(r^*)$, is given by

$$w(r^*) = \varepsilon_{1s}^*\{(1/20)V^{*-11/3}[(1 - V^{*-1/3})^{-10} - (1 + V^{*-1/3})^{-10}]$$
$$- (1/4)V^{*-5/3}[(1 - V^{*-1/3})^{-4} - (1 + V^{*-1/3})^{-4}]\} \tag{34}$$

Changing the variable from r to the reduced variable y in Eq. (26), using the reduced volume, V^*, in Eq. (31), substituting Eq. (28) and Eq. (29) into Eq. (26), and finally substituting into Eq. (25), gives:

$$B_{2s} = cI_c(E^*, V^*, T) \tag{35}$$

$$c = 2\pi a^3 N_s \tag{36}$$

and

$$I_c(E^*, V^*, T) = \exp\{-(E^*/T)[V^{*-4} - 2V^{*-2}]\}$$
$$\int_0^1 \exp\{-(E^*/T)[V^{*-4}L(y) - 2V^{*-2}M(y)]\}y^{1/2}dy \tag{37}$$

where a is the radius of the spherical cavity, N_s is the number of cavities per gram of adsorbent, and $I_c(E^*, V^*, T)$ is an integral dependent on E^*, V^*, and T as shown above.

To evaluate Eq. (35), it is necessary to carry out a numerical integration where the parameters V^* and E^* are specified. If a best choice can be made for these parameters, then a value of the constant c (independent of temperature and adsorbate) is found from $c = B_{2s}/I_c(E^*, V^*, T)$. If a value for r^* can be estimated based on the size of the adsorbate molecule and adsorbent surface atoms, then a can be found from V^* and Eq. (31). Finally N_s can be calculated from values of a, c, and Eq. (36).

The cavity model potential is based on an integral summation of pairwise Lennard-Jones [6,12] potentials between the adatom and cavity. However, the flat model and parallel plate model are based on a Lennard-Jones [3,16] or other large value of the repulsive parameter n. A summation of adatom–solid atom [6,12] potentials yield a [3,9] potential but a larger repulsive term has been shown to be more appropriate [52].

Finally, it is useful to make the connection between the isosteric heat in the limit of low coverage, q_{st}^0, the gas–cavity interaction at equilibrium separation, $w(r^*)$, and the B_{2s} temperature dependence. Derrah and Ruthven [67] used

$$q_{st}^0 = RT - w(r^*) \tag{38}$$

to determine the isosteric heats of monatomic gases with a 5A zeolite. The isosteric heat in the limit of low coverage can be given as [6]:

$$q_{st}^0 = RT + R[d \ln B_{2s}/d(1/T)] \tag{39}$$

Combining these two equations allows us to write $w(r^*)$ as

$$|w(r^*)| = [d \ln B_{2s}/d(1/T)] \tag{40}$$

and using these relationships, a plot of $\ln B_{2s}$ versus $1/T$ can be used to find both q_{st}^0 and $w(r^*)$.

III. EXPERIMENTAL

A. Apparatus

Determination of B_{2s} requires accurate experimental measurement of corrected retention time, flow rate, and adsorbent temperature. In addition, in order to correct the flow rate, measurements must be made of room temperature, column inlet pressure, and column outlet pressure.

Retention times are readily and conveniently determined using an electronic integrator, e.g., a Hewlett-Packard 3392A or later model. Electronic integrators are useful in that they also provide a measure of the adsorbate sample size which, it will be noted later, is required for some systems.

In a situation that is somewhat unusual for chromatographic measurements, determination of flow rate is as important as retention time. Flow rates have been determined using a soap bubble flow meter connected to the exhaust vent of the detector. This method of measurement requires that the detector not affect the flow rate. The detector that has been used in most studies is the thermal conductivity detector. The more sensitive flame ionization detector cannot be used because it introduces hydrogen and air into gases exiting the column and hence changes the flow rate. Use of the soap bubble flow meter is somewhat tedious and for some measurements it could be replaced by a far more convenient electronic digital flow meter. The electronic flow meter has the advantage of providing a continuous reading of the flow rate, although it is not quite as precise as a bubble flow meter.

The room temperature, needed in the correction of flow rates, is the temperature of the soap bubble flow meter. This temperature is conveniently measured by attaching a mercury thermometer or a calibrated thermocouple to the flow meter tube at the midpoint of the monitored volume. Measuring the flow rate using a bubble flow meter saturates the gas from the column with water vapor. A correction is made for this, assuming that the soap solution has a vapor pressure not significantly different from that of pure water.

It should be mentioned that some modern thermal conductivity detectors work by alternating reference and adsorbent column helium flow through the detector. This type of detector does not allow an accurate measurement of the required adsorbent column flow rate.

Correction of the measured room temperature flow rate to a column flow rate requires measurements of the inlet and outlet gas pressures on the chromatographic column. The outlet pressure is assumed to be equal to that in the flow meter, which equals atmospheric pressure less the vapor pressure of water. Barometric atmospheric pressure readings are corrected to standard temperature using available barometer temperature correction tables [70]. Some gas chromatograms have built-in pressure gauges that measure the column head pressure directly or a calibrated electronic pressure gauge can be used. As a simple alternative, a device to measure the pressure can be made by epoxying a syringe needle to the open port of a pressure gauge. The pressure can then be measured by inserting the needle through the septum. This measurement disrupts the gas flow in the system, especially if the pressure gauge has a large internal volume, and it can take some time for the pressure to return to its steady-state value. Because measurement disturbs the steady-state system, it should be carried out only after retention time and flow rate measurements have been completed for a particular adsorbate–temperature combination. The measured pressure, or "septum pressure," depends on the carrier gas pressure, the column packing, and the temperature. The inlet pressure equals the septum pressure plus atmospheric pressure.

Ultrapure helium is used as a carrier gas. It is led through a two-stage gas regulator, a 5A molecular sieve drying tube, and a needle value to the injection port on the gas chromatograph.

B. Experimental Considerations

In early studies, columns were packed with adsorbent in house. However, these columns sometimes gave inconsistent results and occasionally became blocked. Now it is preferable to have the columns packed commercially. Companies such as Supelco will pack the columns and will also give the mass of adsorbent used in the packing to 0.01 g.

Initial outgassing of the adsorbent is carried out under helium flow using the manufacturer's specifications with regard to temperature gradient and maximum temperature. During data collection, it is routine to outgas the column overnight under helium flow.

The easiest method of injecting adsorbate samples is to use a 25-μL gas syringe. Adsorbate gas samples are stored in 250-mL round-bottomed flasks that are sealed with rubber septa. It is important to be able to vary the amount of adsorbate injected and it is often the case that very small sample sizes are needed. These goals are achieved by varying the size of sample injected and by varying the pressure of adsorbate gas in the storage flask. Adsorbate gas pressure in the storage flask is varied by initially filling the flask with adsorbate gas and then adding an inert gas, usually neon, to dilute the adsorbate gas until injections give sample sizes in the right range. An added advantage of this dilution method is that neon is a convenient marker gas, one that does not interact significantly with the adsorbent at the temperatures studied. The difference between the pulse maximum peak of the adsorbate and that of neon gives the corrected retention time for the adsorbate. A more sophisticated gas dosing system, which would allow a more controlled variation of sample size, would consist of a gas injection valve with a vacuum/gas-handling system attached.

In order to conform to the theoretical model used in analysis, the adsorbate chromatographic peaks should be Gaussian in shape. Some systems, particularly those where the adsorbent has a low surface area or where the corrected retention time is long, tend to give non-Gaussian peaks, often characterized by pronounced "tailing" of the peaks. It is believed that this phenomenon arises because of nonequilibrium distribution between the mobile and stationary phases or the effect of surface heterogeneity. These systems have been avoided, either by moving to a higher temperature range or by using another adsorbent.

With low area solids, it is often found that B_{2s} varies significantly with the amount of adsorbate injected. It is possible that this variation is caused by significant contributions from B_{3s} and other higher level virial coefficients, but this has not been investigated. Study of this phenomenon for several columns has indicated that the dependence of B_{2s} on amount of sample (as measured by peak area) is linear if peak areas are 2000 units or less as measured on a Hewlett-Packard 3392A integrator. If a certain system is found to have this dependence, several B_{2s} measurements are made, each using different amounts of adsorbate so as to cover a range of peak areas all below 2000 units. Values of ln B_{2s} are plotted versus peak area and the required value of B_{2s} is determined by extrapolation to zero area. An example of this plot is shown in Fig. 1 for butane adsorbed on Carbopack C. In cases where B_{2s} does not depend on adsorbate sample size, three or four replicate runs are done and the average B_{2s} value is calculated.

The column temperature has been measured using a thermocouple placed at the center of the coiled column. In a calibration before adsorption measurements

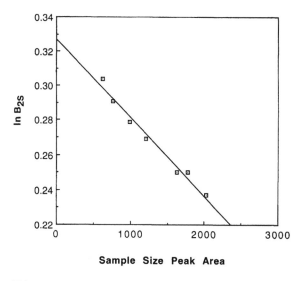

Sample Size Peak Area

FIG. 1 Sample size dependence for butane adsorbed on Carbopack C at 398 K. Ln B_{2s} versus peak area with areas determined from Hewlett-Packard 3392A integrator.

are started, the reading of the thermocouple in the center of the coil is related to the average temperature of the column obtained by averaging the temperature measured at four points around the coil. In this way, allowance is made for possible temperature variations in the thermostatting oven.

One factor that can compromise measurements very quickly is a leaking septum. The performance of the septa is carefully monitored so that, if possible, septa can be replaced before they fail. Leaking septa result in irreproducible flow rates and retention times and non-Gaussian peaks.

C. Data Collection

Prior to a day's measurements, the column is outgassed overnight under helium flow at a temperature at or below the maximum temperature specified for the adsorbent. Most adsorbents have been outgassed at 623 K. The column is detached from the detector during this outgassing.

To start a series of runs, the column is connected to the detector and the helium carrier gas is set to the required flow rate. The column temperature, detector temperature, and interface temperature controls are set to their correct values. To prevent possible condensation, the detector block should be the hottest of these regions. Initial thermal equilibration takes an hour or more, largely due to the slow heating of the detector block. Subsequently, reequilibration after a column temperature change takes no more that about 20 min.

Adsorbate samples are drawn with a gas syringe from the storage flasks and injected into the gas chromatograph while at the same time the integrator is started. Retention times and peak areas are measured for both neon, the marker gas, and the adsorbate. After the integrator has completed measurements of the adsorbate peak area, the flow rate is determined by measuring the time taken for 10 or 20 mL of gas flow through the soap bubble flow meter. Room temperature at the flow meter is also recorded as is the column temperature. If B_{2s} is not dependent on sample size, these measurements are repeated two or three times and an average value computed. If B_{2s} is sample size–dependent then a series of measurements is made at different sample sizes and the value of B_{2s} is determined by extrapolation to zero, as shown in Fig. 1.

After a series of runs on a given system have been completed, a measurement of septum pressure is taken. This measurement takes about 10 min to equilibrate. The helium pressure at the regulator is kept constant, but the inlet pressure for a given column depends on the column temperature.

If a number of adsorbates are being studied, it is most efficient to run all adsorbates at one temperature before setting the gas chromatograph to another temperature. Using mixtures of gases to get two or three B_{2s} values per run has not been efficient in our measurements.

When studying the temperature dependence of B_{2s} it is prudent to randomize selection of temperatures rather than do measurements in order of increasing or decreasing temperatures. Also, it is prudent to check one of the first measurements toward the end of data collection to check for the possibility of gradual surface changes.

The calculation of B_{2s} can be easily programmed into a computer or hand-held calculator so one may follow the reliability of data as they are collected.

D. Conversion to B_{2s}

Calculation of B_{2s} from laboratory data can best be illustrated by example. A typical set of raw laboratory data is given below. Here we have chosen to illustrate the calculation using units encountered in typical laboratory measurements rather than convert to SI units.

Experimental Data

Column temperature	130.5°C
Soap bubble flow time	54.2 s (10.0 mL volume)
Retention time of butane	1.153 min
Retention time of neon	0.471 min
Corrected retention time	0.682 min
Atmospheric pressure	742.3 mm Hg
Vapor pressure water	21.1 mm Hg (at 23°C)

Mass of adsorbent	2.96 g
Room temperature	23.0°C
Septum pressure	21.1 psi

Calculation of B_{2s} can be broken down to 10 steps:

1. Temperature correction of measured barometric pressure: $(742.3 - 2.9) = 739.4$ mm Hg $= (739.4/760) \times 14.696 = 14.298$ psi. The 2.9 value is a correction based on the actual temperature of the barometer and barometric temperature correction tables for thermal expansion of mercury since pressure scale is based on 0°C column of mercury.
2. Calculation of outlet pressure, which equals atmospheric pressure minus the vapor pressure of water, converted to psi: $[(739.4 - 21.1)/760] \times 14.696 = 13.89$ psi.
3. Calculation of inlet pressure, which equals atmospheric pressure (in psi) plus septum pressure (in psi): $14.298 + 21.1 = 35.4$ psi.
4. Column temperature: $130.5 + 273.15 = 403.7$ K; room temperature $23.0 + 273.15 = 296.2$ K.
5. Flow rate at room temperature $= 10.00$ mL/flow time(s) $= 10.00/54.2 = 0.1845$ mL/s.
6. Flow rate corrected to column temperature from Eq. (10): $0.1845 \times (403.7/296.2) = 0.2515$ mL/s
7. Flow rate corrected for pressure drop across column from Eq. (10): $0.2515 \times 1.5 \times \{[(35.4/13.89)^2 - 1]/[(35.4/13.89)^3 - 1]\} = 0.2515 \times 1.5 \times 0.3533 = 0.1333$ mL/s
8. Calculation of B_{2s} for whole sample from Eq. (8): $B_{2s} =$ column flow rate $\times$ corrected retention time $= 0.1333$ mL/s $\times 0.682$ min $\times 60$ s/min $= 5.455$ cm^3 per sample
9. Pressure ratio correction for presence of water vapor in flow meter: $= 5.455 \times [(742.3 - 21.1)]/742.3 = 5.30$ cm^3 per sample
10. Calculation of B_{2s} per gram adsorbent $= 5.30/2.96 = 1.79$ cm^3/g

Plots can be made to check the consistency of the data prior to detailed analysis. The plot of $\ln B_{2s}$ versus $1/T$ should be linear for a given adsorbate–adsorbent system in the temperature ranges that are normally studied. An example of this type of plot is shown in Fig. 2 where $\ln B_{2s}$ values for pentane, butane, propane, and ethane [48] are plotted and linear regressions are produced. The slopes of the lines are 2667 K, 3501 K, 4235 K, and 5146 K for ethane, propane, butane, and pentane, respectively. As shown in Eqs. (39) and (40), these slopes are proportional to the isosteric heats. A comparison of these values to some other property to which they should correlate, such as the number of carbon atoms in a homologous series or possibly the boiling point of the adsorbates, provides a useful check on the consistency of B_{2s} data.

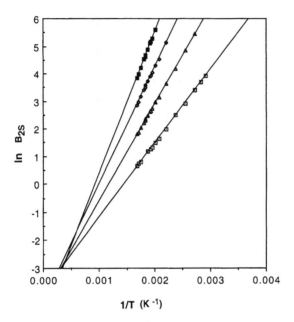

FIG. 2   For 13X zeolite adsorption ln B_{2s} versus the inverse of temperature for, from left to right, pentane, butane, propane, and ethane.

IV.   ANALYSIS AND DISCUSSION

A.   Single-Surface Approach

In physical adsorption, van der Waals forces are responsible for interactions between gas molecules and solid [60,71]. If only the fluctuating dipole–dipole interactions are considered between an adatom and an adsorbent atom, then the attractive portion of the pair potential is inversely related to the sixth power of the atom–atom separation [72]. If the attractive forces are assumed to be pairwise additive, the attraction of a single atom or molecule to solid decays as the third power of the adsorbate–adsorbent separation [73].

The repulsive forces do not lend themselves to the straightforward approach outlined above and the common choice of a Lennard-Jones (6, 12) potential to describe pairwise interaction in physical adsorption is based primarily on mathematical convenience. An adsorbate atom attracted to a solid can be approximated as a summation of pairwise attractions to all of the atoms in the solid and an integration considering a semiinfinite solid would result in the conversion of an LJ (6, 12) potential to an LJ (3, 9) potential of adsorbate–solid attraction. Although an LJ (3, 9) potential can be used in conjunction with Eqs. (12) and (14) to model the B_{2s} values [32], given the arbitrary nature of the repulsive potential there is no reason to expect this to be the best potential [51].

In previous work [52], an LJ (3, 16) or LJ (3, 17) potential was found to give the best fit of Ne, Ar, Kr, and Xe B_{2s} values adsorbed on the graphitized carbon black P33 (2700). The gas–solid interaction energies were also found using a modified Buckingham potential, which uses an exponential decay. The Lennard-Jones potential is used more frequently because it is more convenient to implement. In this work [52], a surface area was found for the P33 (2700) carbon black of 10.8 m²/g. This solid adsorbent consists of truncated polygonal bipyramids made up of stacked graphite lattices. It presents a nearly homogeneous surface to adsorbate molecules and has been used in a variety of studies [2,74,75]. The surface area of this powder was obtained without any assumption of the packing area of the adsorbate molecules, but it is necessary to have a value for z^*. The gas–solid equilibrium separation, z^*, can be based on use of Eq. (13) to determine z^* from z_0 and finding z_0 from $z_0 = (r_g + r_s)/2$ where r_g is the gas radius and r_s is the surface atom radius. The ability to determine surface area independent of a cross-sectional packing area is one advantage of the virial approach.

Kiselev and coworkers included dipole–quadrupole and quadrupole–quadrupole terms in an expression of the gas–solid interaction [76]. From studies of the heats of adsorption of gases on a graphitic surface they found these terms contributed 10% and 1%, respectively, to the overall interaction. A Lennard-Jones (3, n) potential includes only dipole–dipole interactions; however, the constant in front of the attraction term contains implicit contributions from higher multipole terms and so the depth of the potential well and the shape of the potential when substituted in Eq. (12) gives suitable solid surface areas and appropriate estimates of the gas–solid interaction energies.

The single-surface analysis is based on Eqs. (12), (14), and (15) where numerical integration uses an approach such as QUANC8, a quadrature adaptive Newton-Cotes eight-panel method [77], as expressed in Fortran or other suitable software language. If a Lennard-Jones potential in Eq. (14) can be specified by selecting appropriate n and m values such as $n = 16$ and $m = 3$, then $u_{1s}(y)/k$ can be found as the reduced gas–solid separation, y, is varied over a range from 0 to 10. Beyond $y = 10$, there is no significant contribution toward the gas–solid interaction potential.

Since the value of E^* is not known, the program used must cycle through values of E^* until the best-fit value is found. At each trial value of E^* and for a specific experimental temperature T, the integral $I(E^*, T)$ is calculated. $B_{2s}/I(E^*, T)$ is equal to a constant, c, which is Az^*. Since the constant should not vary with temperature, the correct choice of E^* is the one that minimizes the variation of c. To avoid size bias, the minimization of the standard deviation of log c, SD(log c), is used as a measure of the variation of c. Experimental values of B_{2s} at different temperatures are used. A Fortran or other suitable software language program is designed to cycle through values for E^* by successively smaller increments until the minimum SD(log c) is found.

Once the value of the constant c is found for each gas, then the surface area is determined from the z^* value [28]. This process can be repeated for a series of adsorbate molecules on a given solid providing a series of E^* and A values. Variation in the calculated A values provides a further measure of the appropriateness of the selected parameters.

B. Two-Surface Approach

1. Computational Strategy

To calculate the two-surface area of Carbopack C-HT, values of B_{2s}, T, z^*, n, and m must be known or selected. As shown in Eqs. (17) and (18), the remaining variables are x, E_1^*, and E_2^*. E_1^* and E_2^* are linked as the ratio E_r where $E_r = E_2^*/E_1^*$. Searches are made that maximize the agreement given by different adsorbates for the surface area. These searches involved the use of a program that calculates the surface area of the solid for each adsorbate at chosen values of x and E_r. The average and standard deviation of these areas for a series of adsorbates are calculated and reported. This process is repeated to find the combination of x and E_r that gives the lowest value for the standard deviation of log A, SD(log A), and thus the best fit. Repeated evaluations are made by scanning through different ranges of x and E_r values while making incremental changes in E_1^* to minimize SD(log A) [50,61].

2. Basal Plane and Edge Sites of Boron Nitride

As noted previously, P33 graphite consists of truncated polygonal bipyramids made up of stacked graphite lattices and presents a nearly homogeneous surface where physical adsorption can take place [52]. In contrast, hexagonal boron nitride (BN), consists of basal plane and edge surfaces; thus these BN platelets provide two distinct surfaces where physical adsorption occurs [61,78]. In this section we describe how B_{2s} data for multiple adsorbates over a range of temperatures can be used to distinguish and characterize these two types of surfaces. Jaroniec et al. [79] have shown that the effect of heterogeneity on the second gas–solid virial coefficient is significant.

Since Az^* in Eq. (18) is not dependent on the temperature, the best-fit interaction energy, E^*, can be found to minimize the value of the standard deviation of $\log(Az^*)$ for a given adsorbate provided values of x and $E_r(E_r = E_2^*/E_1^*)$ can be selected. For a solid adsorbent with different gases as adsorbates, both the E_r ratio and the surface area, A, should be constant because these parameters depend on the surface structure of the adsorbent and not the absorbate [50].

Since the surface area is independent of the gas used, the best-fit percentage area, x, and the ratio of the interaction potential of two surfaces, E_r, may be determined by finding the minimum value of the standard deviation of the surface area of all gas systems. The typical analysis involves numerical integration and iterative algorithms used to scan a range of x and E_r values. Based on previous studies, an LJ (3, 16) was chosen to represent the potential. A value of $n = 16$ gave a more consistent fit of monatomic gases on P33 carbon than any lower value of n [52].

A pair of x and E_r values is selected and for this pair a series of E_1^* values is used to calculate integrals I_1 and I_2 indicated in Eq. (17). Experimental values of B_{2s} at each available temperature are divided by the sum of the integrals as indicated in Eq. (18) over the range of experimental temperatures. A series of Az^* values are produced for each gas–solid system and the minimum of the standard deviation of the logarithm of Az^*, SD(log Az^*), is found for each gas.

The best value of E_1^* is selected by minimizing SD(log Az^*) at each value of x and E_r. By repeating this for a series of gases adsorbed on the same solid, it is possible to select the best x and E_r values by minimizing SD(log A) for a series of gases. E_r and x should be independent of the gas adsorbate and the surface area, A, should remain consistent as the gas is varied. The best parameters x, E_r, and E_1^* are the ones that give the minimum variation in surface area as the adsorbate is varied. The equilibrium separation must be estimated by some other means.

An LJ (3, 9) potential was used and monatomic gas–boron nitride data were analyzed using a single-surface approach ($x = 0$). This gave the surface areas (m^2/g) equal to 1.87, 1.55, and 0.99 for Ar-BN, Kr-BN, and Xe-BN, respectively. A two-surface approach gave surface areas (m^2/g) equal to 2.419, 2.422, 2.424 for Ar-BN, Kr-BN, and Xe-BN, respectively [80]. Obviously, the two-surface model gives more consistent surface areas and is most appropriate for the boron nitride [50].

For BN with an LJ (3, 16) the best-fit values were: $E_r = 1.71$, $x = 0.030$, and corresponding average surface area of 3.56 (m^2/g). This surface area based on the LJ (3, 16) potential is in closer agreement to the surface area measured by the independent BET method of 5.0 (m^2/g) than the previous graphical BN analysis [61] based on the LJ (3, 9) potential, which gave 2.62 (m^2/g). For P33 with LJ (3, 16), and using a two-surface approach, the best-fit values were found to be $E_r = 1.0$ and $x = 0$, and the corresponding average surface area of 10.9 (m^2/g). By definition, either $x = 0.0$ or $E_r = 1.0$ is equivalent to a single-surface solid. Values of x from 0.00 to 0.10 and E_r from 1.0 to 2.5 were used in this analysis because all the best-fit parameters were found to fall within this range. These results demonstrate that the P33 graphite presents a single uniform surface,

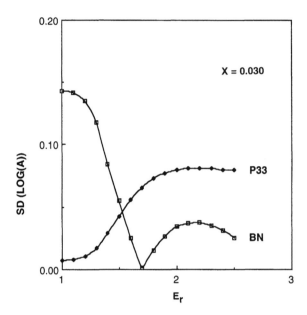

FIG. 3 Standard deviation of log A for argon, krypton, and xenon with single-surface P33 graphite and two-surface boron nitride versus the high- to low-energy surface interaction energy ratio, E_r.

whereas the boron nitride presents two different surfaces, each with their own unique gas–solid interaction energy.

Figure 3 shows a comparison of the behavior of the P33 graphite and boron nitride for values of $x = 0.030$ as E_r is varied [80]. The SD(log A) values provide a comparison of the variation or consistency of the areas obtained for argon, krypton, and xenon adsorbates. Clearly, the areas for the P33 graphite surface become more consistent among the gases (argon, krypton, xenon) as E_r approaches 1.0. For BN the variation of surface area for various adsorbates, as measured by SD(log A), decreases as E_r is increased, reaches a minimum at 1.71, and then increases before declining slightly.

The interaction potential energies of rare gases adsorbed on the basal plane of boron nitride, E_1^*, 1026 K for Ar, 1302 K for Kr, and 1660 K for Xe, compared well with the values from potentials calculated by Karimi and Vidali [81] for energies of rare gases adsorbed on the basal plane of boron nitride: 1082 K for Ar, 1306 K for Kr, and 1808 K for Xe. On P33 graphite the E_1^* values found in the virial analysis (1114 K for Ar, 1468 K for Kr, and 1929 K for Xe) are similar to Karimi and Vidali [81] calculated values (1218 K for Ar, 1614 K for Kr, and 1913 K for Xe). In addition, Vidali et al. [82] used a universal reduced adsorption potential to calculate 1103 K, 1428 K, and 1912 K for Ar, Kr, and Xe with graphite.

The differences between the graphite and boron nitride have been previously characterized [83,84] and the distinctive differences in behavior have been characterized because of the known heterogeneity of the BN [78,84–86]. These two adsorbents are useful model systems to distinguish differences in surface structure and test models that do not make prior assumptions of surface differences. No prior knowledge of the nature of the surface structure is assumed in the analysis and yet it is possible to clearly identify the P33 graphite as a one-surface solid and the boron nitride as a two-surface solid. The results given above indicate that a virial analysis of B_{2s} data for several adsorbates over a range of temperature can be used to determine gas–solid interaction energy parameters and solid structural parameters, as well as to determine something of the surface heterogeneity.

3. Chlorofluorocarbons and a Microporous Carbon

To further test this two-surface model, gas–solid virial coefficients obtained from gas–solid chromatography for ethane (C_2H_6), propane (C_3H_8), fluoromethane (CH_3F), chloromethane (CH_3Cl), Freon-22 ($CHClF_2$), Freon- 12 (CCl_2F_2), and dichloromethane (CH_2Cl_2) with a microporous carbon, Super Sorb, were analyzed [37]. Previous studies of chlorofluorocarbon adsorption have been of both theoretical and applied interest, including several patents directed at removing low concentrations of these molecules from the air [37]. Using the same approach outlined previously, the best two-surface approach yielded $E_r = 2.5$ (range 1.0–2.9 examined) and $x = 0.03$.

For every E_r it is possible to obtain the fraction of high-energy surface area, x, that gives a best fit of the data as judged by minimizing SD(log A). As E_r is increased, the corresponding x value from the best-fit tends to decrease and the average surface area found for the seven gas–solid systems increases. The z^* values ranging from 0.409 nm for CH_3F to 0.478 nm for CCl_2F_2 were calculated based on the additivity of carbon and gas radii. The gas radii were found from molecular weight, liquid bulk density, and an empty space correction factor [37]. The radii for the adsorbate molecules determined by this approach agreed within 10% of values obtained from an energy minimization molecular modeling program.

The two-surface model gives a better representation of B_{2s} data as judged by the value of SD(log A), although the difference is not nearly as significant as for the monatomic gas and boron nitride systems. The microporous carbon is a complicated structure; however, assumptions of flat uniform surface have been useful in providing relative surface areas. While maintaining a simple model, the addition of a second, high-energy surface yields a significant improvement in surface areas determined.

The one-surface model gives an area (m^2/g) and standard deviation of 127 $\pm$ 83 and the two-surface model gives 1513 $\pm$ 502 for the seven gas–solid systems. The two-surface model gives a more consistent area and a value much closer to the traditional BET area of 3169 m^2/g for the Super Sorb. The one-surface model gives an area that is 4.0% of the value of the BET area. The two-surface model gives an area that is 48% of the area of the BET area. As previously observed, the virial surface area is expected to be less than the BET area because the BET area includes a contribution from condensation in micropores and the virial area is more representative of the geometric area [25,87]. For a high-surface-area microporous carbon, the virial area might be expected to be in the range of one-fourth to one-half of the BET area.

The application of this two-surface model to the halogenated hydrocarbon–microporous B_{2s} data shows that the two-surface model can be applied to more complicated adsorbate–adsorbent systems and provides a method to explore the heterogeneity of solid surfaces. In the next section we will examine one more application of the two-surface model to study surface heterogeneity.

4. Lower and Higher Energy Sites on Carbopack C and Carbopack C-HT

Gas–solid chromatography was used to determine the second gas–solid virial coefficients for normal propane, butane, pentane, and hexane with Carbopack C (Supelco, 10 m^2/g) and Carbopack C-HT (Supelco, 10 m^2/g), and one-surface and two-surface virial analyses were compared [53]. Carbopack C is a graphitized carbon black powder and Carbopack C-HT is produced by flowing hydrogen over Carbopack C heated to temperatures above 1273 K. Previous work using these solids as adsorbents showed that there are high-energy ''active sites'' located on the surface of both Carbopack C [88–93] and Carbopack C-HT [91,93], which implies that there are two types of surface. Since the treatment of Carbopack C to produce Carbopack C-HT removes some of the high-energy sites from the surface, these two surfaces provide a useful comparison of single-surface and two-surface models. Lin and Parcher [88–90] found that modifying the surface of Carbopack C removed many of the active sites and that these are due to both chemical and physical imperfections in the surface.

The alkanes used in this study were selected to provide nonspecific adsorbates. As expected, due to the van der Waals interactions in physical adsorption, retention time increased with increasing adsorbate size Also as is typical, retention times decreased with increasing temperature. Plots of $\ln B_{2s}$ versus $1/T$ were linear and the slopes increased in order from propane, butane, pentane, and hexane.

Using the single-surface equations and analysis procedure discussed in the previous section, the average single-surface area and standard deviation for the four gases were found to be A = 0.69 $\pm$ 0.81 m^2/g for Carbopack C and A = 1.6 $\pm$ 1.5 m^2/g for Carbopack C-HT [53]. This single-surface model assumes a homogeneous surface. On the other hand, the two-surface model assumes the majority of the solid surface has a homogeneous energy with isolated regions of high-energy sites, or "hot spots." Higher energy sites could be due to the presence of oxygen and sulfur on the surface or due to surface features such as steps, crevices, or edge sites.

Results of the two-surface method discussed above gave the best agreement of the adsorbate areas for the four gases when $x = 0.015$ (1.5% of the surface area due to higher energy sites) and $E_r = 1.80$ for Carbopack C-HT. The average surface area was found to be 13.9 ± 2.0 m^2/g.

For Carbopack C-HT and C, values of E_1^* and E_2^* (and thus E_r) should be the same because hydrogen treatment only removes some of the active sites but does not alter the energy of the rest of the surface. Using the same values of E_r, E_1^*, E_2^*, and z^* as in the Carbopack C-HT analysis, only the value of x was varied. For each value of x, the standard deviation of the areas for the four adsorbates was divided by the corresponding average area. An $x = 0.026$ gave the minimum value of this ratio. For this x the surface area was 13.2 ± 1.7 m^2/g for Carbopack C. The best E_1^* values for this two-surface analysis were 2030 K for propane, 2488 K for butane, 2839 K for pentane, and 3140 K for hexane [53].

Earlier literature BET values for the surface areas of Carbopack C of 13.7 m^2/g [94] and 12.5 m^2/g [95] agree with this two-surface area of 13.2 m^2/g. For Carbopack C-HT the reported BET value of 13.6 m^2/g by Vidal-Madjur et al. [96] is in good agreement with our value of 13.9 m^2/g found using the two-surface method. Confirmation of the presence of high-energy active sites on both the Carbopack C and the Carbopack C-HT causes dramatic improvement in the surface area calculations by changing from a single-surface to a two-surface model—1.6 to 13.9 m^2/g for Carbopack C-HT and 0.7 to 13.2 m^2/g for Carbopack C.

The high-temperature hydrogen treatment removes some of the higher energy sites, or hot spots, from the surface of the Carbopack C ($x = 2.6\%$) when it is changed into Carbopack C-HT ($x = 1.5\%$). The higher energy sites were found to have an average energy 1.8 times greater ($E_r = 1.8$) than the more common low-energy sites.

Previous work [88–93] suggested that these graphitized carbon black powders have surface impurities that create high-energy sites. An ESCA (electron spectroscopy for chemical analysis) analysis of these two surfaces showed only a carbon peak and no oxygen or sulfur [53]. However, they could be present in amounts, as reported by Kraus [94] and DiCorcia et al. [92], which are below the ESCA detection limits (0.2% sulfur atoms and 0.1% oxygen atoms) of the analysis on these samples. The fraction of the surface that could be covered with sulfur or oxygen atoms is too small to account for the number of high-energy sites. So an additional factor could be due to surface imperfections. It is possible that the hot spots are a mixture of surface imperfections and surface impurities.

The isosteric heats of adsorption at zero coverage on Carbopack C have been reported as 27.4, 34.4, 41.7, and 49.6 kJ/mol for propane, butane, pentane, and hexane, respectively [97]. The E^* (depth of gas–solid interaction) values from the one-surface analysis are 26.6, 35.9, 45.3, and 51.7 kJ/mol, respectively.

C. Cavity Model

1. Zeolites

Zeolite molecular sieves are of great interest in examining virial theories of adsorption for microporous solids because they possess regular crystalline structure [98–101]. The low coverage, Henry's law, physical adsorption of argon, krypton, and xenon with the 5A and 13X zeolites has been represented with a virial equation [51]. In this section we show how B_{2s} values such as those obtained by gas adsorption or gas–solid chromatography can be used to obtain the energetic and structural parameters that characterize Henry's law behavior in zeolite systems.

Henry's law B_{2s} data for zeolites have the advantage that the adsorbate occupation of cavities averages less than one gas molecule per cavity and thus adsorbate–adsorbate interactions can be ignored. If we consider a zeolite to consist of a series of interconnected spherical cavities, the structure can be characterized by only two parameters—the cavity radius, a, and the number of cavities per gram of adsorbent, N_s [69].

X-ray crystallographic data for the 5A and 13X zeolites show that their structures are a collection of identical approximately spherical cavities [102–104]. Each cavity in the 5A structure is formed by eight interconnected sodalite units with access to each central cavity via six openings or windows of mean diameter less than 0.5 nm. Each cavity in the 13X structure is formed from 10 interconnected sodalite units with access to the central cavity via four windows of mean diameter about 1.0 nm [105].

Many previous studies modeling adsorbate–zeolite interactions were based on lattice summation calculations [62,67,98,106–112]. These calculations are based on atom–atom and atom–ion interactions and include attractive and repulsive dispersion as well as electrostatic induction terms. Such calculations require exact structural information, i.e., the location of all cavity atoms and ions that are considered to contribute significantly to the adsorbate–cavity interaction. These lattice summation calculations have been used to generate potential profiles along different symmetry axes as well as to calculate a variety of thermodynamic properties. Kiselev and Du [109] have compared experimental and calculated Henry constants for Ar, Kr, and Xe in type X zeolites and found that the experimental Henry constants average 0.93 of the calculated lattice summation values. Including this correction term, they were able to correctly calculate thermodynamic values from lattice summation calculations.

2. Application of the Lennard-Jones and Devonshire Cell Model

Stroud [62] used an integral procedure along with the Lennard-Jones and Devonshire (LJD) cell model [63,64] to generate a potential to represent the sorbate-cavity interaction and included multiple cavity occupancy so that adsorbate-adsorbate interactions were also included. Soto [105,111] used the LJD cell model to replace the dispersion summation with an integral and used lattice summation to represent only the adsorbate-cation electrostatic inductive interaction. Yang and Pierotti [69] began with thermodynamic data in the form of second gas-solid virial coefficients and used the LJD model to extract from these data the molecular and structural parameters for the Ar, Kr, and Xe zeolite 5A systems.

Soto [105] observed that ignoring electrostatic interactions and using only the LJD cell model integrand to represent the adsorbate–cavity interaction led to values of the cavity diameter and the gas–cavity interaction energy that were too small. However, this result was based on fixing the strength of the adsorbate cavity atom interactions based on dispersion forces.

As we show below, a modification of the approach developed by Yang and Pierotti [69] has been used to extract reasonable molecular and structural parameters from the temperature dependence of virial coefficient thermodynamic data [51]. In this work, it is essential to allow the depth of the potential to be determined by fitting the data and not by making prior assumptions about the nature of the atom–atom interactions.

Equations (35)–(37) can be used to express the Henry's law isotherm, Eq. (1), in terms of four parameters: a, N_s, r^*, and E^*. To evaluate Eq. (37), one must carry out a numerical integration where the parameters V^* and E^* are specified. With a unique selection of values for these parameters and a value of r^* based on atomic radii, it is possible to determine the cavity radius, a, from Eq. (31) and finally the value of N_s from the constant, c, from Eq. (36). Equation (34) provides a connection between E^* and V^* values that can be utilized if the interaction at equilibrium separation, $w(r^*)$ is known. Any generated pair of E^* and V^* values must give the expected value of $w(r^*)$ as shown in Eq. (40).

If a value of V^* is chosen, then a unique selection for E^* can be made from the B_{2s}–temperature data for a given gas–solid system. Since c in Eq. (36) involves only the structural parameters a and N_s, then c should not vary with temperature. An iterative procedure is employed to find the value of E^* that gives the minimum value of the standard deviation of $\log(c)$ for selected values of V^* for each of the six gas–zeolite systems. Evaluation of the integral, Eq. (37), is carried out by numerical integration [77].

For V^* values (in a range of 4–12) for each gas–zeolite system the best value of the interaction energy, E^* was found, within the nearest 1 K, by finding the standard deviation of $\log(c)$ for each trial of E^*. Values of the constant c are found from Eq. (35) by dividing experimental values of B_{2s} by the calculated functions I_c at each temperature for which B_{2s} values are available. The minimum of the standard deviation of $\log(c)$ is associated with the best fit of E^* because this indicates that the structural parameters a and N_s are—for this value of E^*—the least affected by changes in the temperature.

For any selected value of V^* there is an unambiguous choice of the best-fit E^* value. However, it is not possible to select the best-fit pair of V^* and E^* values in this same manner because a series of such pairs are not much different in the quality of their fit. In other words, for any V^* an E^* can be found that gives a good fit of the data, although some values of V^* and E^* may be unreasonable. As V^* becomes smaller then the selected E^* will also become smaller. It is necessary to find a unique pair of V and E^*.

Equation (34) is used to calculate $w(r^*)$ for each V^* and E^* ($E^* = \varepsilon_{1s}^*/k$) pair and find the pair that gives the appropriate $w(r^*)$ value based on the B_{2s} data and Eq. (40). For any gas–zeolite system $w(r^*)$ can be determined as a function of V^*. Table 1 shows the results of this approach with values of $w(r^*)/k$, E^*, V^*, and SD($\log c$). For the 5A and 13X zeolites the interaction energies follow the expected trend of increasing as the size of the monatomic adsorbate is increased [51].

The mean values of c along with the corresponding structural parameters—the cavity radius, a, and the number of cavities, N_s—are shown in Table 2. The best-fit V^* and E^* values from Table 1 for each gas–zeolite system and the corresponding experimental values of $B_{2s}(T)$ are used with Eq. (35) to generate a series of c values from which the average c is determined. Since $V^* = (a/r^*)^3$, Eq. (31), one can calculate the cavity radius from r^* and V^* values for each gas–zeolite system. For a Lennard-Jones [6,12] potential the equilibrium separation, r^*, is related to the distance of closest approach, r_0, as $r^* = (2)^{1/6} r_0$.

It is commonly assumed that the Lorentz-Berthelot combining rules are valid and that gas–solid hard sphere separation is the arithmetic mean of the gas–gas

TABLE 1 Cavity Model Parameters from Virial Analysis of Zeolites

Gas	$w(r^*)/k$	V^*	$E^*(K)$	SD($\log c$)
5A Zeolite				
Argon	1213	8.18	12901	0.07275
Krypton	1859	7.90	18404	0.07447
Xenon	2694	7.72	25555	0.14040
13X Zeolite				
Argon	1267	8.04	13006	0.03508
Krypton	1780	7.79	17517	0.08173
Xenon	2267	7.77	21754	0.08089

TABLE 2 Cavity Model Zeolite Structure Parameters from Virial Analysis

Gas	$c(cm^3/g)$	$r*(nm)$	$a(nm)$	$N_s \times 10^{-20}$ (cavities/g)
5A Zeolite				
Argon	0.619	0.343	0.691	2.99
Krypton	0.304	0.355	0.708	1.37
Xenon	0.154	0.374	0.739	0.608
13X Zeolite				
Argon	0.298	0.343	0.687	1.47
Krypton	0.304	0.355	0.704	1.38
Xenon	0.460	0.374	0.740	1.80

and solid–solid diameters [76]. Soto [105] determined gas–solid r_0 values using the oxygen–oxygen surface diameter as 0.276 nm and estimating gas diameters of Ar, Kr, and Xe as 0.335, 0.357, and 0.390 nm, respectively. These values are used to find the $r*$ values shown in Table 2. N_s values are found from cavity radii and c values using Eq. (36). The average values for the 5A zeolite are $a = 0.712$ nm and $N_s = 1.66 \times 10^{20}$ cavities/g and for the 13X zeolite are $a = 0.711$ nm and $N_s = 1.55 \times 10^{20}$ cavities/g.

The structural parameters obtained from this approach can be compared to the values obtained from x-ray crystallographic studies. The best-fit average values of 0.712 nm for the 5A and 0.711 nm for the 13X compare well to experimental cavity radii for the closest oxygen atoms 0.704 nm and 0.709 nm for the interior of the 5A and 13X, respectively [102,103,105].

The values of N_s, the number of cavities per gram, are not as accurate as the cavity radius, a, values. Values for $N_s \times 10^{-20}$ (cavities/g) were 1.66 ± 1.22 for the 5A and 1.55 ± 0.22 for the 13X. The corresponding exact values from x-ray crystallographic measurements are 3.59 and 3.64 for the 5A and 13X, respectively. The calculated values represent only 46% and 43% of the expected values for the 5A and 13X. However, the interior cavity wall surface is effectively reduced due to the six 5A windows and four 13X windows by approximately 40–60%. If this factor were included in Eq. (36) for the constant c then the values of N_s would be approximately doubled and agree closely with the known values.

Lattice summation calculations for the 5A system by Derrah and Ruthven [67] showed a significant variation in calculated values of $w(r*)$ depending on the formula used to find the dispersion interaction energies. The Slater-Kirkwood formula generated values of $w(r*) = 1700$, 2150, and 2500 K, for Ar, Kr, and Xe, respectively [69]. This compares to our values of 1213, 1859, 2694 in Table 1. A much better comparison is obtained for Ar values by using their London value of 1100.

Work by Soto and others [105,111] using atomic parameters to calculate Henry's law constants led them to conclude that the LJD cell model could be used to represent dispersion forces, but that electrostatic forces must be treated by a lattice summation and could not be replaced by a sphericalization procedure. While the propriety of including the electrostatic induction in the potential is clear in a procedure beginning with atomic parameters and attempting to calculate thermodynamic data, it is interesting to note that in the reverse procedure (thermodynamic to atomic) one can ignore the detailed origins of the interaction energy and merely allow the fitting procedure to pick out a potential well of sufficient depth to include whatever forces that may contribute to the adsorbate–cavity interaction. For the Kr-13X system Soto [111] found the interaction energy at the maximum potential well depth, $E^* = 22600$ K, to be due to 8800 from dispersion energy and 13800 from electrostatic energy. Using our fitting procedure we found a value of $E^* = 17157$ K. While not identical, our E^* value is including an implicit contribution from electrostatic interactions. This approach that was found to be useful for the zeolites can be modified to be used with other microporous solids where the structure cannot be characterized as well.

D. Comparison of Flat, Parallel Plate, and Cavity Models

In previous research [49] a comparison was made of three different virial models (flat, parallel plate, and cavity) applied to argon B_{2s} values for seven different solids, including graphitized carbon black P33 (P33), Mexican graphite (MG), 5A-zeolite (5A), 13X-zeolite (13X), Nuchar-SC (SC), Nuchar-SA (SA), and Super Sorb (SS). The seven powders represented three surface types: low-surface-area carbons with flat surfaces (MG, P33), microporous carbons (SC, SA, SS), and zeolites containing interconnected spherical cavities (5A, 13X).

For each model, B_{2s} is equal to some constant, c, times the value of an integral dependent on the selected n and m values, T, and E^*. For the parallel plate and cavity models there is also the structural parameter, a, which represents the distance from the wall to the center of the parallel plates or the radius of cavity. The analysis involves an iterative computer process where the E^* value is varied, the integral is computed at each temperature using numeric integration [77], and the B_{2s} value at each temperature is divided by the integral to find the constant, c. The constant c is Az^*, $2aA$, or $2\pi a^3 N_s$ for the flat, parallel plate, and cavity models, respectively.

E^* is cycled through a range of values to give the parameters for the integrals that produce c values that give the smallest standard deviation of the $\log(c)$. Since c depends on surface structure and area, it should not depend on temperature. Therefore, the best parameters give the most consistent c values for B_{2s}–temperature data over a range of temperatures.

For the parallel plate model, the attractive and repulsive parameters were 3 and 16. For this research, the interaction energy is determined from the overlapping of the two interaction potentials based on the selected V^* and Eq. (23). For a selected V^* value, E^* values are cycled until the minimum SD(log c) is found. Sets of V^*, E^*, and $w1 + w2$ values are generated for various V^* values. The values of V^* are cycled until the resultant value gave an E^* consistent with the expected interaction for a single surface from the prior flat model analysis. As the plates are brought closer together, the two interaction potentials overlap. This overlapping creates a combined interaction potential that is deeper than that for a single flat surface and that is dependent on the distance between the two parallel plates. The area, A, is found from $c = 2aA$ where a is determined from $V^* = a/r^*$. The volume between the parallel plates is found from the total area of a plate times the separation as given by $V = 2aA$.

For the cavity model, the analysis evaluates a range of E^* values for each V^* value. By choosing the E^* that produces the smallest standard deviation for a particular V^* value, sets of V^* and E^* pairs are created. A value is then calculated for the average interaction potential, $w(r^*)$, for each V^* and E^* pair. An experimental value of $w(r^*)$ for each adsorbate–adsorbent system is obtained from the slope of a graph of $\ln(B_{2s})$ versus $1/T$, Eq. (40), so the best V^* and E^* pair is chosen such that the calculated $w(r^*)$ value best agrees with the experimental $w(r^*)$ value [51].

Since the structural constant c is expressed as $c = 2\pi a^3 N_s$, the radius of the sphere, a, is found from V^* values and $V^* = (a/r^*)^3$. The number of cavities per gram of adsorbent, N_s, is calculated by taking the average c for each argon–solid system and dividing it by $2\pi a^3$. The surface area, A, is given by

$$A = 4\pi(a - r_s)^2 N_s \tag{41}$$

and the volume of the cavity as

$$V = (4/3)\pi(a - r_s)^3 N_s \tag{42}$$

where r_s is the hard sphere radii of the atoms on the surface of the adsorbent. The r_s value used in the calculations for the 5A and 13X zeolites was 0.138 nm [8], and the r_s value used in the calculations for the other five powders was 0.170 nm [25].

As shown in Table 3, the interaction energies are consistent with the virial models. The flat model gave the lowest values for the depth of the interaction potential, E^*, while the cavity model in which the molecule is surrounded by surface atoms gave the highest values. The parallel plate model where the molecule is between two surfaces gave intermediate values of interaction energy.

Judging the standard deviation of (log c), each model was able to choose parameters that fit the experimental data equally well. The parameters (E^* for the flat, E^* and V^* for the cavity and plate models) are able to be adjusted to

TABLE 3 Comparison of Gas–Solid Interaction Energies from
Flat, Parallel Plate, and Cavity Models

Solid	Flat	Parallel	Cavity
Low-surface carbon			
P33	1113	1374	10266
MG	942	1163	9969
Microporous carbon			
SC	1861	2296	17121
SA	1477	1812	13863
SS	1482	1819	13922
Zeolite			
5A	1563	1928	14458
13X	1417	1753	13007

generate consistent B_{2s} values over a range of temperatures. More complicated
models such as a two-surface model have been shown to bring about improvement in quality of fit where appropriate [50]. However, for a single-surface model
the quality of fit, as judged by the consistency of the structural constant c, cannot
be used as a criterion to match a specific virial model to a specific surface structure.

Surface areas calculated from each model are summarized in Table 4. BET
surface areas, based on nitrogen adsorption at liquid nitrogen temperatures, are
11, 26, 903, 1661, and 3169 for P33, MG, SC, SA, and SS, respectively [32,52].
The surface areas calculated from the flat model agreed with the BET areas of

TABLE 4 Comparison of Surface Areas Obtained from Flat,
Parallel Plate, and Cavity Models

Solid	Flat	Parallel plate	Cavity
Low-surface carbon			
P33	11	3.5	22
MG	21	7.0	31
Microporous carbon			
SC	194	62	403
SA	441	154	894
SS	817	282	1656
Zeolite			
5A	233	76	546
13X	237	74	555

the flat surfaces, P33 and MG, while the surface areas calculated from the cavity model were approximately half of the BET areas of the surfaces composed of microporous carbons. This result is consistent with the observation that the virial model represents geometric area whereas the BET area includes a significant contribution from condensation in pores, and so the BET areas are greater than the virial areas.

The surface areas from the parallel plate model underestimated the areas relative to the flat and cavity models. However, the surface areas for all three models showed a clear distinction between the flat surfaces (P33 and MG) relative to the microporous surfaces. In other words, even without any prior knowledge of surface structure, a virial analysis can be expected to distinguish between a low- and high-surface-area solid. A low-surface-area solid can be more appropriately analyzed by the flat model, whereas a high-surface-area solid can be more appropriately analyzed by the cavity model.

E. Correlations with Other Physical Properties

Given the ability to obtain B_{2s} values through gas–solid chromatography and to use the temperature and adsorbate variation of these values to determine something of surface area, adsorbent structure, and the nature of the gas–solid interaction, it would also be advantageous to be able to predict B_{2s} values from adsorbate properties. By predicting B_{2s} we could predict the extent of adsorption, n_a, at any temperature using Eq. (2) and also the adsorbate retention time, t, and the ability to separate a mixture of gases using Eqs. (8)–(10) where $t = mB_{2s}^3/F_1$.

Gas–solid chromatography was used to obtain virial coefficients for seven alkanes, seven halogenated hydrocarbons, two ethers, and sulfur hexafluoride [28]. B_{2s} values in the temperature range 314–615 K were found with the adsorbent Carbopack B (Supelco, 100 m^2/g). In the analysis of these data, an alternative to the numerical integration is provided by an equation derived by Hansen [113–115]. This approximation is most applicable as E^*/T takes on larger values. Hansen has shown that the integral in Eq. (12) that has been expressed as a power series of gamma functions [6] can be approximated by the simple analytical expression and it has been shown [28] that combining his expression with Eq. (13) gives

$$B_{2s} = Az^*(2\pi/mn)^{1/2}(T/E^*)^{1/2}\exp(E^*/T) \tag{43}$$

This equation can be used as an approximation to Eq. (15) which represents an exact integral.

Combining Eq. (43) with Eq. (18) gives an approximation of the two-surface model where

$$B_{2s} = Az^*(2\pi/mn)^{1/2}[(1 - x)(T/E_1^*)^{1/2}\exp(E_1^*/T)$$
$$+ x(T/E_2^*)^{1/2}\exp(E_2^*/T)] \tag{44}$$

The use of Hansen's analytical solution for Eq. (18) in place of numerical integration greatly simplifies the computation required to find the best-fit x, E_1^*, and E_2^* parameters. Although Eq. (44) is an approximation, it may provide a useful means to correlate and predict B_{2s} values for a set of molecules.

As shown before, the gas hard sphere size can be calculated by combining the gas and solid radii $r_{gs} = r_g + r_s$, where $r_s = 0.170$ nm is the radius of carbon based on graphite layer size [25]. The gas radius, r_g, may be estimated by

$$r_g = [(3/4\pi)(CW/DL)]^{1/3} \tag{45}$$

where W is the molecular weight of adsorbate molecule, D is the liquid bulk density, L is Avogadro's constant, and C is an empty space packing correction factor [28]. In one study [37], the best correction factor, C, was found to be 0.4590 by comparing calculated r_g values for methane, ethane, propane, dichlorodifluoromethane, argon, krypton, and xenon with available values [28]. Then z^* can be calculated from $z_0(z_0 = r_{gs})$ and Eq. (13).

Since the surface area, A, the percentage of the area for surface two, x, and the ratio of E_2^* to E_1^*, E_r, are structural features of the solid, they should be constant for the same solid with different gases [50]. The E_r and x values that best described the Carbopack B solid were determined using the previously described two-surface method. The best pair of E_r and x, $E_r = 2.12$ and $x = 0.008$, gave an area of 101 m^2/g and a standard deviation of 43. However, with the methylether outlier removed, the area and SD(A) became 92 $\pm$ 24 m^2/g. The BET area is 100 m^2/g. The two-surface model was a significant improvement over the area determined by the single-surface approach.

In previous work, Berezin [116] has shown that there should be a correlation between gas–solid interaction energy and the ratio of critical temperature divided by the square root of critical pressure, $T_c/P_c^{1/2}$. As shown in Fig. 4, E_1^* adsorption energies can be correlated well by $T_c/P_c^{1/2}$ for 17 gases—seven alkanes, seven halogenated hydrocarbons, two ethers, and sulfur hexafluoride on adsorbent Carbopack B—(correlation coefficient, $r = 0.985$). A microporous solid like the Carbopack B probably allows for a diverse set of molecules to be grouped together in a way that may not be possible for a low-surface-area solid. However, since B_{2s} is very sensitive to interaction energy values, it is difficult to group all of the gases together to get E_1^* for different groups precisely enough to predict exact B_{2s} values. Only for a structurally similar series is it possible to have a general equation to get B_{2s} values.

In this set of 17 molecules, the structure of the molecules are too different to be combined together. E_1^* can be approximated from the critical constant ratio and z^* approximated from molecular modeling. With a knowledge of A, x, E_r, n, and m the two-surface Hansen equation can only be used to accurately calculate B_{2s} for a similar series of adsorbates. While the expression of B_{2s} based on Hansen's equation is most valid at higher E^*/T ratios, this type of analytical expres-

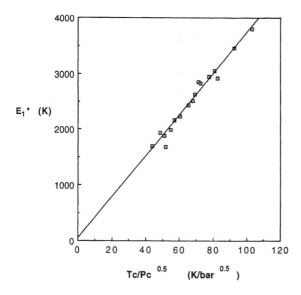

FIG. 4 Correlation of gas–solid interaction energy for lower energy surface, E_1^*, with ratio of critical constants $T_c/P_c^{0.5}$ for 17 adsorbate molecules with a microporous carbon, Carbopack B.

sion provides a simplicity lacking in the numerical expression and provides a rational basis for a simple graphical approach of plotting Ln B_{2s} versus $1/T$.

In previous work [37] the E^* values based on B_{2s} values determined from gas–solid chromatography for seven different hydrocarbons and chlorofluoro-carbons were correlated with the adsorbate molecular structures. The structure was represented by

$$E_{cal}^* = n_C E_C + n_H E_H + n_F E_F + n_{Cl} E_{Cl} + (DM) E_{DM} \qquad (46)$$

where n_C is the number of carbon atoms, n_H is the number of hydrogens, n_F is the number of fluorines, n_{Cl} is the number of chlorines, DM is the dipole moment in Debye, E_C is the energy contribution of each carbon atom, E_H is the energy per hydrogen, E_F is the energy per fluorine, E_{Cl} is the energy per chlorine, and E_{DM} is the energy per Debye of dipole moment. This equation assumes that each atom makes a unique and additive contribution to the overall adsorbate–solid interaction energy [28].

Experimental values of E^* are used in conjunction with known values of n_C, n_H, n_F, n_{Cl}, and DM to solve a series of simultaneous equations to find the best values of E_C, E_H, E_F, E_{Cl}, and E_{DM}. Once the energy factors are determined, E_{cal}^* is calculated from Eq. (46) and then the experimental E^* is plotted versus

E_{cal}^* to find out how well the equation works to correlate E^* values. A plot of E^* versus E_{cal}^* gives a slope of 1.000 and a correlation coefficient of 0.976. When the dipole moment factor is excluded the fit is much poorer.

A plot of E^* values for the same seven adsorbates versus the boiling point of the molecules gives $r = 0.932$ [37]. Both adsorption and boiling point are related to the strength of van der Waals attraction, so it is reasonable that they correlate with each other. Critical constant, boiling point, and structural correlations have been compared for other gas–solid systems and are all reasonably effective. However, on a silica gel surface where specific interactions with OH bonds are important, a comparison of aromatics, aliphatic hydrocarbons, and alcohols reveals that only the structural approach is useful in correctly correlating B_{2s} values [117].

In future work, it may be possible to develop a more general means of predicting B_{2s} values from a limited set of parameters that are dependent on molecular structure and physical properties. However, more success is likely in dealing with a set of structurally similar molecules and developing an equation to calculate B_{2s} that is specific for a limited set of similar adsorbate molecules.

Another useful approach to correlating gas–solid chromatographic data is to consider relative retention times and correlate with some appropriate property or combinations of properties. For a given flow rate and column temperature, the retention times are converted to a series of relative values by, for example, scaling the longest retention time to a value of 500 and then scaling the other times proportionally. The resulting values are sometimes referred to as retention index (RI) values or relative retention times (RRT) Previous studies have used a variety of physical properties or molecular descriptors to correlate or predict RRT or RI. These properties have included vaporization enthalpy [118]; connectivity index and bending energy of molecules [119]; connectivity index, ionization potential, molecular size, and quadrupole moment [120]; molecular polarizability [121]; and molecular mass and selected structural fragments [122]. These studies are examples of a specific type of quantitative structure–activity relationship (QSAR), sometimes referred to as quantitative structure–retention relationship (QSRR). The prediction of relative chromatographic retention times and separation effectiveness is a key objective of QSRR.

Future QSRR studies may benefit from a virial approach that takes advantage of the theoretical basis of B_{2s} as it relates to surface structure and gas–solid interactions. The combination of the B_{2s} virial approach with QSRR may lead to better selections of descriptors for predicting chromatographic separation parameters and also help us better understand and utilize gas–solid chromatography.

ACKNOWLEDGMENT

We would like to express deep appreciation for the inspiration provided by our respective research advisors, Robert A. Pierotti and Douglas H. Everett, and recognize the valuable contributions of the many undergraduate students who have worked with us on related research projects over the years.

REFERENCES

1. W. A. Steele and G. D. Halsey, Jr. J. Chem. Phys. *22*:979 (1954).
2. J. R. Sams, Jr., G. Constabaris, and G. D. Halsey, Jr. J. Phys. Chem. *64*:1689 (1960).
3. J. R. Sams, Jr., G. Constabaris, and G. D. Halsey, Jr. J. Chem. Phys. *36*:1334 (1962).
4. J. A. Barker and D. H. Everett, Trans. Faraday Soc. *58*:1608 (1962).
5. W. A. Steele, in *The Solid-Gas Interface Volume 1* (E. A. Flood, ed.), Dekker, New York, 1967, pp. 199–220.
6. R. A. Pierotti and H. E. Thomas in *Surface and Colloid Science Vol. 4.* (E. Matijevic, ed.), Wiley-Interscience, New York, 1971, pp 93–259.
7. V. A. Bakaev and O. V. Chelnokova, Surf. Sci. *215*:521 (1989).
8. W. Rudzinski, J. Jagiello, J. Michalek, S. Milonjic, and M. Kopecni. J. Colloid Interface Sci. *104*:297 (1985).
9. E. A. Boucher and D. H. Everett. Trans. Faraday Soc. *67*:2720 (1971).
10. W. A. Rudzinski, A. Waksmundzki, Z. Suprynowicz, and J. Rayss. J. Chromatogr. *72*:221 (1972).
11. S. Ross, J. K. Saelens, and J. P. Oliver. J. Phys. Chem. *66*:696 (1962).
12. S. J. Gregg. Colloids Surf. *21*:109 (1986).
13. M. Jaroniec, R. Madey, J. Choma, B. McEnaney, and T. J. Mays. Carbon. *27*:77 (1989).
14. F. Rodriguez-Reinoso and A. Linares-Solano. Chem. Phys. Carbon. *21*:1 (1989).
15. W. van Megan and I. K. Snook. Mol. Phys. *54*:741 (1985).
16. N. Avgul, A. G. Bezus, E. S. Dobrova, and A. V. Kiselev. J. Colloid Interface Sci. *42*:486 (1973).
17. V. Biba, Z. Spitzer, and O. Kadlec. J. Colloid Interface Sci. *69*:9 (1979).
18. M. Jaroniec and J. Choma. Mater Chem. Phys. *19*:267 (1988).
19. R. W. Innes, J. R. Fryer, and H. F. Stoeckli. Carbon. *27*:71 (1989).
20. J. Fernandez-Colinas, R. Denoyel, Y. Grillet, F. Rouquerol, and J. Rouquerol. Langmuir. *5*:1205 (1989).
21. B. McEnaney and K. J. Masters. Thermochim Acta. *82*:81 (1984).
22. S. Ozeki. Langmiur. *5*:186 (1989).
23. P. B. Balbuena and D. A. McQuarrie. J. Phys Chem. *92*:4165 (1988).
24. J. H. Cole, D. H. Everett, C. T. Marshall, A. R. Paniego, J. C. Powl, and F. Rodriguez-Reinoso. J. Chem. Soc. Faraday Trans. I. *70*:2154 (1974).
25. D. H. Everett and J. C. Powl. J. Chem. Soc. Faraday Trans I. *72*:619 (1976).
26. L. F. Smirnova, V. A. Bakaev, and M. M. Dubinin. Carbon. *25*:599 (1987).
27. A. Patrykiejew, S. Sokolowski, J. Stawinski, and Z. Sokolowski. J. Colloid Interface Sci. *124*:371 (1988).
28. T. R. Rybolt, M. T. Epperson, H. W. Weaver, H. E. Thomas, S. E. Clare, B. M. Manning, J. T. McClung. J. Colloid Interface Sci. *173*:202 (1995).
29. T. R. Rybolt, M. D. Wall, H. E. Thomas, J. W. Bramblett, M. Phillips. J. Colloid Interface Sci. *138*:113 (1990).
30. C. Vidal-Madjar and A. Jaulmes. Pure Appl. Chem. *61*: 2005 (1989).

31.   D. Poskus. Zh. Fiz. Khim. 61:3303 (1987).
32.   T. R. Rybolt and R. A. Pierotti AIChE J. (1984), 30(3), 510–13
33.   R. Leboda. Chem. Anal. (Warsaw) 21:1001 (1976).
34.   S. Sokolowski, R. Leboda, A. Waksmundzki. Rocz. Chem. 50:1565 (1976).
35.   Z. Suprynowicz, M. Jaroniec, and J. Gawdzik, Chromatographia 9:161 (1976).
36.   W. Rudzinski, Z. Suprynowicz, and J. Rayss. J. Chromatogr. 66:1 (1972).
37.   T. R. Rybolt, X. Zhang, M. D. Wall, H. E. Thomas, L. E. Mullinax and J. R. Lee, J. Colloid Interface Sci., 149:359 (1992).
38.   T. Takaishi, T. Okada, and K. Nonaka. J. Chem. Soc., Faraday Trans. 93:1251 (1997).
39.   M. S. Sun, O. Talu, D. B. Shah. J. Phys. Chem. 100:17276 (1996).
40.   T. Ohgushi, H. Yokoyama. J. Chem. Soc., Faraday Trans. 88:3095 (1992).
41.   H. B. Abdul-Rehman, M. A. Hasanain, and K. F. Loughlin, Ind. Eng. Chem. Res. 29:1525 (1990).
42.   O. M. Dzhigit, A V. Kiselev, and T. A. Rakhmanova. Zeolites 4:389 (1984).
43.   R. Harlfinger, D. Hoppach, U. Quaschik, and K. Quitzsch. Zeolites 3:123 (1983).
44.   R. Harlfinger, D. Hoppach, H. P. Hofmann, R. Schoellner, and K. Quitzsch. Z. Phys. Chem. (Leipzig) 261:65 (1980).
45.   J. R. Conder. Advan. Anal. Chem. Instrum. 6:209 (1967).
46.   S. Ross, I. D. Morrison, H. B. Hollinger. Adv. Colloid Interface Sci. 5:175 (1976).
47.   R. A. Pierotti. Chem. Phys. Lett. 2:420 (1968).
48.   T. R. Rybolt and D. R. Olson. J. Colloid Interface Sci. 163:303 (1994).
49.   D. R. Olson, T. R. Rybolt, D. S. Bodine, and M. D. Wall. J. Colloid Interface Sci. 159:205 (1993).
50.   X. Zhang and T. R. Rybolt. J. Colloid Interface Sci. 150:575 (1992).
51.   T. R. Rybolt, R. L. Mitchell, and C. M. Waters. Langmuir. 3:326 (1987).
52.   T. R. Rybolt and R. A. Pierotti. J. Chem. Phys. 70:4413 (1979).
53.   C. D. Bruce, T. R. Rybolt, H. E. Thomas, T. E. Agnew, and B. S. Davis. J. Colloid Interface Sci. 194:448 (1997).
54.   S. A. Greene in Principles and Practice of Gas Chromatography (R. L. Pecsok, ed.), John Wiley & Sons, New York, 1959. pp. 28–47.
55.   A. J. P. Martin and R. L. M. Synge. Biochemical J. 35:1358 (1941).
56.   P. E. Eberly and E. H. Spencer. Faraday Soc. London Trans. 57:289 (1961).
57.   D. Atkinson and G. Curthoys. J. Chem. Educ. 55:564 (1978).
58.   A. T. James and A. J. P. Martin. Biochemical J. 50:679 (1951).
59.   A. B. Littlewood, C. S. G. Phillips, and D. T. Price. J. Chem. Soc. London. 55: 1480 (1955).
60.   J. N. Israelachvili, Intermolecular and Surface Forces, Academic Press, London, 1985, pp. 65–85.
61.   A. C. Levy, T. R. Rybolt, and R. A. Pierotti. J. Colloid Interface Sci. 70:74 (1979).
62.   H. J. F. Stroud, E. Richards, P. Limcharoen, and N. G. Parsonage. J. Chem. Soc. Faraday Trans. I 72:942 (1976).
63.   J. E. Lennard-Jones and A. F. Devonshire, Proc. Roy. Soc. London A. 163:53 (1937).
64.   J. E. Lennard-Jones and A. F. Devonshire, Proc. Roy. Soc. London A. 165:1 (1938).

65. J. O. Hirschfelder, C. F. Curtiss, and R. B. Bird, *Molecular Theory of Gases and Liquids*; John Wiley & Sons, New York, 1954, pp. 293–298.
66. R. Fowler and E. A. Gugenheim, *Statistical Thermodynamics*, Cambridge University Press, London, 1949, pp. 336–342.
67. R. I. Derrah and D. M. Ruthven. Can. J. of Chem. *53*:996 (1975).
68. T. L. Hill, *An Introduction to Statistical Thermodynamics*; Addison-Wesley, Reading, Massachusetts, 1960, p. 293.
69. C. C. Yang, *Ph. D. Thesis; Georgia Institute of Technology*, Atlanta, Georgia, 1979.
70. David R. Lide, ed. *CRC Handbook of Chemistry and Physics 72nd edition*, CRC Press, Boca Raton, Florida, 1991.
71. A. D. Crowell, in *The Solid -Gas Interface Volume 1* (E. A. Flood, ed.), Dekker, New York, 1967, pp. 175–202.
72. F. London. Z. Phys. *63*:245 (1930).
73. F. London. Z. Phys. Chem. Abt. B. *11*:222 (1931).
74. R. N. Ramsey, *Ph.D. Thesis; Georgia Institute of Technology*, Atlanta, Georgia, 1970.
75. A. C. Levy, *Ph.D. Thesis; Georgia Institute of Technology*, Atlanta, Georgia, 1976.
76. D. M. Young and A. D. Crowell, *Physical Adsorption of Gases*, Butterworth, London, 1962.
77. G. E. Forsythe, M. A. Malcolm, and C. B. Moler, *Computer Methods for Mathematical Computations*; Prentice-Hall, Englewood Cliffs, New Jersey, 1997; pp. 84–105.
78. R. N. Ramsey, H. E. Thomas, and R. A. Pierotti. J. Phys. Chem. *76*:3171 (1972).
79. M. Jaroniec, X. Lu, and R. Madey. J. Colloid Interface Sci. *146*:580 (1991).
80. T. R. Rybolt and X. Zhang in *Fundamentals of Adsorption: Proceedings of Fourth International Conference on Fundamentals of Adsorption* (M. Suzuki, ed.), Kodansha, Tokyo, 1993, pp. 553–558.
81. M. Karimi and G. Vidali, Phys. Rev. B. *36*:7576 (1987) Erratum: M. Karimi and G. Vidali, Phys. Rev. B. *42*:1462 (1990).
82. G. Vidali, M. W. Cole, and J. R. Klein. Phys. Rev. B. *28*:3064 (1983).
83. D. Graham, and W. S. Kay, J. Coll. Sci. *16*:182 (1961).
84. N. Dupont-Pavlovsky, C. Bockel, and A. Thomy Surf. Sci. *160*:12 (1985).
85. M. Karimi and G. Vidali. Phys. Rev. B, Condensed Matter. *34*:2794 (1986).
86. H. E. Thomas, R. N. Ramsey, and R. A. Pierotti. J. Chem. Phys. *59*:6163 (1973).
87. K. S. Sing, D. H. Everett, R. A. W. Haul, L. Moscou, R. A. Pierotti, J. Rouquerol and T. Siemieniewska. Pure Appl. Chem. *57*:603 (1985).
88. P. J. Lin and J. F. Parcher. J. Colloid Interface Sci. *91*:76 (1983).
89. P. J. Lin and J. F. Parcher. Anal Chem. *57*:2085 (1985).
90. J. F. Parcher and P. J. Lin. J. Chromatography *250*:21 (1982).
91. F. Bruner, G. Bertoni, and P. Ciccioli. J. Chromatography *120*:307 (1976).
92. A. D. DiCorcia, R. Samperi, E. Sebastiana, and C. Severini. Anal. Chem. *52*:1345 (1980).
93. K. J. Hyver, J. F. Parcher, D. M. Johnson, and P. J. Lin. J. Chromatography *328*: 63 (1985).
94. G. Kraus. J. Phys. Chem. *59*:343 (1955).
95. M. H. Polley, W. D. Schaeffer, and W. R. Smith. J. Phys. Chem. *57*:469 (1953).

96. C. Vidal-Madjur, M. Gonnard, F. Benchah, and G. Guichon, J. Chromatogr. Sci. *16*:190 (1978).
97. M. Lal and D. Spencer. J. Chem. Soc. Faraday Trans. II *70*:910 (1974).
98. A. V. Kiselev, in *Molecular Sieve Zeolites–II*, Adv. Chem. Ser. 102; Am. Chem. Soc., Washington, D. C., 1971, pp. 37–68.
99. W. M. Meier, in *Molecular Sieves*; Soc. of Chem. Ind., London, 1968, pp. 10–27.
100. D. A. Whan, Chem. in Britian. *17*:532 (1981).
101. J. M. Newsam. Science. *231*:1093 (1986).
102. L. Brousard and D. P. Shoemaker, J. Am. Chem. Soc. *82*:1041 (1960).
103. K. Seff and D. P. Shoemaker. Acta Cryst. *22*:162 (1967).
104. K. Seff. Acc. Chem. Res. *9*:121 (1976).
105. J. L. Soto, *Ph. D. Thesis; University of Pennsylvania*, 1979.
106. P. Broier, A. V. Kiselev, E. A. Lesnik, and A. A. Lopatkin. Russ. J. Phys. Chem. (English Transl) *42*:1350 (1968).
107. P. Broier, A. V. Kiselev, E. A. Lesnik, and A. A. Lopatkin, Russ. J. Phys. Chem. (English Transl) *43*:844 (1969).
108. A. G. Bezus, A. V. Kiselev, A. A. Lopatkin, and P. Q. Du. J. Chem. Soc. Faraday Trans. 2. *74*:367 (1978).
109. A. V. Kiselev and P. Q. Du. J. Chem. Soc. Faraday Trans. 2. *77*:1 (1981).
110. A. V. Kiselev and P. Q. Du. J. Chem. Soc. Faraday Trans. 2. *77*:17 (1981).
111. J. L. Soto, P. W. Fisher, A. J. Glessner, and A. L. Myers. J. Chem. Soc. Faraday Trans. 1. *77*:157 (1981).
112. A. V. Kiselev and A. A. Lopatkin, in *Molecular Sieves*; Soc. of Chem. Ind., London, 1968, pp. 252–266.
113. R. S. Hansen, J. A. Murphy, and T. C. McGee. Trans. Faraday Soc. *60*:579 (1964).
114. R. S. Hansen. J. Phys. Chem. *63*:743 (1959).
115. R. S. Hansen and J. A. Murphy. J. Chem. Phys. *39*:1642 (1963).
116. G. I. Berezin. Russian J. Phys. Chem. *57*:233 (1983).
117. O. R. Meeks and T. R. Rybolt. J. Colloid Interface Sci. *196*:103 (1997).
118. M. P. Elizalde-Gonzalez, M. Hutfliess, and K. Hedden. J. High Resolut. Chromatogr. *19*:345 (1996).
119. R. Corbella, M. A. Rodriguez, M. J. Sanchez, F. Garcia Montelongo. Chromatographia *40*:532 (1995).
120. A. Bemgard, A. Colmsjoe, and K. Wrangskog. Anal. Chem. *66*:4288 (1994).
121. H. Lamparczyk and A. Radecki. J. High Resolut. Chromatogr. Chromatogr. Commun. *6*:390 (1983).
122. N. Dimov and M. Moskovkina. J. Chromatogr. *552*:59 (1991).

2

Inverse Gas Chromatography as a Tool to Characterize Dispersive and Acid-Base Properties of the Surface of Fibers and Powders

MOHAMED NACEUR BELGACEM Department of Paper Science and Technology, University of Beira Interior, Covilhã, Portugal

ALESSANDRO GANDINI Polymeric Materials, Ecole Française de Papeterie et des Industries Graphiques, St. Martin d'Hères, France

I.	Introduction	42
II.	History and Principles of IGC	44
III.	Advantages and Limitations of IGC	45
IV.	Column Preparation	46
V.	Instrumentation	46
VI.	Methodology and Technical Precautions	47
VII.	Probes and Their Relevant Characteristics	48
	A. Dispersive properties	48
	B. Surface area	48
	C. Acid–base properties	52
VIII.	Calculation Procedures	54
	A. Determination of retention time	54
	B. Determination of retention volume	55
	C. Determination of dispersive properties	56
	D. Determination of thermodynamic parameters	59
	E. Determination of acid–base properties	61
	F. Determination of the fiber–matrix interaction parameter	67
IX.	Dispersive Properties	68
	A. Characterization of fibers	68
	B. Characterization of organic fillers and powders	85

41

    C. Characterization of inorganic fillers and powders   95

 X. Acid–Base Properties   108
    A. Fibers   108
    B. Fillers and powders   112

 XI. Interaction Parameters in Composites and Blends   113

XII. Conclusions   113

    Symbols and Abbreviations   116

    References   118

I. INTRODUCTION

After the end of World War II, science and technology related to polymers underwent a spectacular development that resulted in the production of a wide variety of materials covering a vast domain of applications. These thermoplastic and thermoset materials present many advantages compared with metallic, ceramic, and glassy counterparts, e.g., among others, a much lower density, and the ease of processing by simple procedures. During the last three decades, attention has been focused on the combination of these polymers with natural or synthetic fibers and fillers in order to obtain high-added-value materials.

Whereas fiber-based composites ensure a considerable improvement of the mechanical properties, which can reach those of metallic structures and provide applications, e.g., in transport and space technology, the addition of powders to polymers is particularly useful in the formulation of paints, inks, elastomers, coatings, blends, etc.

It is well known that the performance of both types of composite, particularly their mechanical properties, depends on the properties of the individual components (fiber and matrix) as well as on their surface compatibility. The latter point is of great relevance because the physical and chemical nature of the interactions between the continuous phase (matrix) and the reinforcing fibers or fillers determines the quality of the interface. In fact, the work of adhesion between the matrix and the fibers (or fillers) is directly related to the surface energy (including both the dispersive and the nondispersive contributions) of each component. Therefore, in order to optimize the properties of a composite, it is essential to determine accurately the surface properties of its matrix and reinforcing agent, i.e., the various factors making up the free energy of their surfaces.

In summary, the surface properties of fibers and fillers, used in the elaboration of polymer composites, emulsions, suspensions or blends, must be characterized by three types of analyses, namely (1) their chemical composition, (2) their specific area, and (3) their surface energy.

The chemical characteristics of the surface of fibers and fillers owe their relevance to the possibility of establishing specific interactions with the matrix like acid–base exchanges, hydrogen bonding, or, indeed, covalent linkages. The stronger these interactions, the more adhesive will be the interface and, consequently, the better the mechanical properties of the composite, which will tend to show only cohesive failure.

There are several techniques [1–4] that provide precise information about the surface composition of solids in the form of fibers, powders and films. The most common among them are Fourier transform infrared (FTIR) and Raman spectroscopy, x-ray photoelectron spectroscopy (XPS) and secondary ion mass spectrometry (SIMS). As shown in Table 1, each technique is characterized by a specific depth of penetration from the surface. Thus, for example, the use of FTIR spectroscopy working in the multiple reflection mode gives very poor information about surface chemistry because it analyzes down to about 1 μm, which corresponds to over 1000 molecular layers. Therefore, the inspection of surfaces that have been subjected to chemical modification (grafting, oxidation, etc.) by this technique will not clearly reveal these changes unless the yields are particularly high. Conversely, XPS gives more precise details about the chemical composition of surfaces thanks to the low depth associated with the penetration of the scanning electrons, i.e., approximately 3 nm, which corresponds to a few molecular layers. This advantage is often overshadowed by the considerable difference in cost and in manipulation complexity between the two techniques favoring FTIR spectroscopy.

TABLE 1 Comparison of Different Techniques Used to Characterize Solid Surfaces

Technique	Depth of analysis (nm)	Probe/response	Information
FTIR	1000	hv/hv	Chemical structure (functional groups)
Raman	50	hv/hv	Chemical bonds and structure
XPS	3	x-ray/photoelectrons	Chemical bonds and structure
SIMS	1	Ion/ion	Chemical structure
Wettability	—	Molecule/molecule	Surface energy and acid–base properties
IGC (gas adsorption)	—	Molecule/molecule	Surface energy, acid–base properties and surface area
BET	—	Molecule/molecule	Surface area

FT, Fourier transform infrared; XPS, x-ray photoelectron spectroscopy; SIMS, secondary ion mass spectrometry; IGC, inverse gas chromatography; BET, Brunauer–Emmet–Teller.

The classical method used to characterize the surface area of powders and fibers is the BET (Brunauer–Emmet–Teller) technique, based on the determination of the adsorption isotherms of different gases (nitrogen, argon, krypton, etc.) on the solid surface [5].

The surface free energy of solids is usually evaluated using static and dynamic contact angle (CA) measurements [6–8]. However, this technique is not adapted to porous or rough solid surfaces or to materials such as fibers and powders that cannot be converted to continuous flat surfaces. The macroscopic character of the CA technique can, moreover, be the source of erroneous evaluations of surface properties. Examples in this context are (1) the partial degradation of polymers during the processing of films or plates that will influence CA values more than the actual surface composition; (2) the atmospheric contamination of high-energy surfaces like metals, oxides etc., which again gives rise to abnormally large changes in CA values; and (3) the partial migration to the surface of additives present in polymeric materials.

An additional problem associated with the CA technique is the fact that it cannot intrinsically provide the actual value of the total surface energy, γ. Different authors (Fowkes [8], Van Oss [9], Zisman [10], Owens-Wendt [11]) have proposed alternative approaches aimed at obtaining reliable values of the dispersive and nondispersive contributions to γ, using different assumptions. The results obtained by applying these treatments often differ considerably among each other and it is difficult therefore to assess their respective merits and drawbacks in general terms.

The use of a more recent technique, namely, *inverse gas chromatography* (IGC) [12–15], to characterize surface properties has opened a new stimulating field of research because of the advantages it offers, coupled with a relatively simple experimental setup.

This chapter deals with the use of IGC to determine surface properties of fibrous materials, such as cellulose, glass, carbon, and aramid fibers, and powderous materials, such as synthetic polymers, lignins, and organic and inorganic pigments and fillers. The properties discussed here are on the one hand the dispersive component of the free energy of the surface and on the other hand its acid–base character, as obtained from measurements at infinite dilution conditions. Before dealing with these issues, we will discuss the historical development of the technique, its principle, its advantages and limitations, the experimental methodology it requires, the instrumentation used and the calculation procedures adopted to exploit its data.

II. HISTORY AND PRINCIPLES OF IGC

Conventional gas–liquid chromatography (GLC) was introduced in 1952 by James and Martin and was not applied to the study of polymers because of their low volatility. Fifteen years later, Kiselev [16] invented the IGC technique, which remained unexploitable up to 1969, when Smidsrød and Guillet [17], developed its theory and methodology. The word *inverse* indicates that the material of interest constitutes the stationary phase rather than injected volatile probes. The column containing the material under investigation is placed between the injector and the detector of any conventional GLC equipment, and measurements consist of injecting volatile probes possessing known properties through the column via an inert carrier gas. The retention time of each probe is related to the properties of the surface of the stationary phase.

The IGC technique can be carried out under two different conditions: (1) at infinite dilution (or zero coverage), for which Henry's law is rigorously applicable, and (2) at finite concentration. The infinite dilution conditions are based on the injection of a very small quantity of the vapor of a given probe (10^{-6}–10^{-7} mol) into the column which contains material with a large surface area. In these conditions, the adsorbate–adsorbate interactions are neglected and the adsorption-desorption phenomena occur only within a monolayer. These conditions enable experiments to be carried out at low vapor concentrations with a surface coverage approaching zero. Measurements at finite concentrations are not relevant to the present context.

In addition to the surface properties of solid materials, IGC has been extensively used to characterize bulk properties of polymers, such as transition temperatures, their degree of crystallinity, as well as the interaction parameters of some polymer solvents [18,19] and/or polymer–polymer systems [20,21].

III. ADVANTAGES AND LIMITATIONS OF IGC

IGC was found to be a good complementary method for surface properties characterization that presents numerous advantages, namely:

1. Simplicity
2. Lack of sensitivity to surface rugosity compared to CA measurements
3. Applicability to materials that cannot be cast in the form of films
4. No surface contamination of samples under investigations from the surrounding atmosphere
5. Relative rapidity of data collection
6. Accuracy of results
7. Data collection over an extended temperature range
8. A large variety of probes, with different acid–base or neutral character
9. Ease of temperature control, and
10. Low capital investment for the basic tool, i.e., GLC equipment

However, IGC has also some limitations because it requires a well-defined particle size range and cannot be applied to high-surface-energy materials because of chemisorption. Materials that cannot be easily ground to a fine powder are also not suitable for IGC.

IV. COLUMN PREPARATION

There are three types of IGC columns: packed, capillary and fiber [22]. The first type requires the preparation of a concentrated solution of the substrate and its adsorption onto the solid support or the capillary column walls. This falls outside the scope of this book and will not be treated here.

To carry out the IGC characterization of fibers and powders, the most commonly used columns are made of stainless steel or Pyrex and have an internal diameter of 3–7 mm. Their length can vary from 20 to 400 cm. Packing the column will vary with the size of the material under study. Fibers and particles of 200–400 μm prevent the plugging of the column. Long fibers are cut and sieved before packing. Some synthetic polymers have also been studied in the form of particles with a 200- to 500-μm size. Particles were obtained by crushing and sieving the initial coarse polymer grains [23].

In the case of film-like materials (paper, cellophane, etc.) that are 20–40 μm thick, small "confetti" discs are prepared using a cleaned paper tape punch machine. This confetti (1–1.5 mm) is then collected and introduced into the column [24].

When the average particle size is too small (a few micrometers), two methods of column loading have been reported to avoid plugging: (1) the fine powder is mixed with standard glass beads (200–300 μm) at a known ratio and the columns loaded with this mixture [25]. A dummy column filled with the same glass beads is then used for calibrations [25]. (2) Alternatively, powders can be compressed into pellets, which are then crushed and sieved to select the fraction of agglomerates with diameters between 200 and 400 μm [26].

A small glass fiber stopper is usually placed at each of the two ends of the column. Columns are obviously weighed before and after packing to determine the quantity of packed substrate. The reader should keep in mind that in all cases the goal is to provide the maximum possible fiber or powder surface areas without plug formation.

V. INSTRUMENTATION

A standard GLC apparatus equipped with a flame ionization detector (FID) is commonly used to carry out IGC data collection. Thermal conductivity detectors (TCDs) have been also used for this purpose [27,28]. Inert carrier gases are typically helium or nitrogen, but hydrogen has also been employed. To measure the flow rate of the carrier gas, usually a soap bubble flow meter is used. The pressure at the inlet of the column is given by the internal manometer of the GLC equipment, whereas the outlet pressure of the column is taken as the atmospheric pressure and is usually measured by a mercury barometer. The oven temperature is conventionally measured with a Pt resistance thermometer. Oven, injector, and detector are always thermostated. Recently, automatic systems for sample injection and data storage have become commercially available.

VI. METHODOLOGY AND TECHNICAL PRECAUTIONS

The filled columns are usually conditioned overnight at 105–110°C, but in certain cases the conditioning temperature can be lower or higher. To measure the retention time for a given probe, a small quantity of its own vapor and the marker (1–5 µl) are injected into the column using a 10-µl microsyringe. Experiments are usually repeated at least in triplicate for each probe. Then the retention time is taken from the chromatograms, which are usually recorded with an integrator or a microcomputer. The dead volume is usually determined by the injection of a nonadsorbing marker. The retention volume for the marker should be temperature-independent so that any adsorption phenomenon can be considered negligible.

There are some technical requirements for the success of the IGC experiments and the obtention of exploitable data. Before starting data collection, the validity of the IGC method for a given material should be verified from specific experimental observations, namely:

1. Chromatographic peaks for polar as well as nonpolar probes and for the noninteracting marker should be reproducible, so that permanent sorbtion of the probes onto the solid under investigation can be neglected.
2. Peaks for all probes, including the marker, should be sharp and symmetrical to testify the absence of their diffusion inside the investigated materials.
3. The retention volume of a given probe should remain constant when the flow rate of the carrier gas is increased or decreased, confirming once again the absence of diffusion into the bulk of the materials under scrutiny.
4. Permanent surface contamination of the substrate should be avoided. This can be checked by injecting again a neutral probe at the end of the series of experiments with different neutral, acidic, basic, and amphoteric probes. Its retention time should remain unchanged.
5. The volume of the injected vapor should not affect the retention time, which should be constant within the range of 1–10 µl, in order to confirm that the zero coverage conditions are respected and that there are no adsorbate-adsorbate interactions.

These observations ought to be valid for experiments conducted at all temperatures and under different carrier gas flow rates.

VII. PROBES AND THEIR RELEVANT CHARACTERISTICS

Neutral, acidic, basic, and amphoteric volatile probes are used to investigate the dispersive and acid–base properties of surfaces by IGC. The n-alkane series (neutral probes) used to determine the dispersive energy usually consists of the normal isomers of hexane, heptane, octane, and nonane, but higher n-alkanes, such as decane, undecane, and dodecane, have also been used. Lewis acid probes are usually chloroform, methylene chloride, and carbon tetrachloride. The family of Lewis base probes is formed essentially by tetrahydrofuran and diethyl oxide (ether). Finally, acetone and ethyl acetate are used as amphoteric probes. The most commonly used marker is methane, but propane has also been reported [29–32]. The relevant characteristics of these probes, which are used in the calculation of dispersive and acid–base properties, are discussed below.

A. Dispersive Properties

The dispersive contribution to the surface tension of the probes in the liquid state is determined by CA and Wilhelmy plate measurements. The dispersive energy values of the probes are summarized in Table 2, which also includes their boiling point and vapor pressure at 25°C, because these parameters are also used in IGC calculations.

B. Surface Area

The surface area, a, (Å^2) that a molecule of adsorbate occupies on the solid substrate was first calculated by Emmet and Brunauer [33] from the density of the corresponding liquids according to the following empirical equation:

$$a(\text{Å}^2) = 1.091\,[M/\rho N]^{2/3}\,10^{16} \tag{1}$$

where M is the molecular weight of the sorbate (g/mol), ρ its density in the liquid state (g/m³), and N Avogadro's number. The effect of the density term is to make a temperature-dependent although in practice a appeared to be practically insensitive to temperature.

Gray et al. [24,34,35] determined the specific area of cellulosic materials from adsorbtion isotherms of organic molecules onto the fibers' surface. They used Eq. (1) to estimate the surface area of the adsorbed probes and found that the values obtained from nitrogen adsorbtion were systematically about 20% larger than those obtained from the isotherms of the organic vapors. They concluded that the values obtained from Eq. (1) were underestimated and introduced a correction factor of 22%. Thus:

$$a_c = 1.22a \tag{2}$$

TABLE 2 Values of Boiling Temperatures at 1 Atm, Vapor Pressure at 25°C and the Dispersive Component of Surface Energy of Compounds Commonly Used as Probes in IGC

Molecule	bp (°C)	P_0 (kPa)	γ_L^D (mJ/m^2)	Acid–base properties
n-Pentane	36	91.7	15.5	Neutral
n-Hexane	68.7	20.2	17.9	Neutral
n-Heptane	98.5	6.09	20.3	Neutral
n-Octane	125.6	1.86	21.3	Neutral
n-Nonane	150.8	0.57	22.7	Neutral
n-Decane	174.1	0.17	23.4	Neutral
n-Undecane	195.9	—	24.6	Neutral
n-Dodecane	259	—	—	Neutral
Cyclohexane	80.7	13.0	24.7	Neutral
Methylene chloride	40.0	58.2	27.2	Acidic
Cloroform	61.1	26.2	26.7	Acidic
Carbon tetrachloride	76.8	15.2	26.4	Acidic
Benzene	80.0	12.7	28.2	Acidic
Toluene	110.6	3.79	27.9	Acidic
p-Xylene	138.3	1.19	—	Acidic
Chlorobenzene	131.7	1.60	—	Acidic
Nitromethane	101.1	4.79	—	Acidic
Methanol	64.6	16.9	22.1	Acidic
Ethanol	78.2	7.87	22.0	Acidic
Propanol	97.2	2.76	23.3	Acidic
n-Butanol	117.7	0.86	24.9	Acidic
Tetrahydrofuran	65.0	21.6	22.5	Basic
Diethyl oxide	34.5	71.7	16.7	Basic
Triethylamine	89.0	7.70	20.2	Basic
Pyridine	115.2	2.76	36.6	Basic
n-Butylamine	77.0	12.2	23.4	Basic
1,4-Dioxane	101.5	4.95	—	Basic
Acetonitrile	81.6	11.9	—	Basic
Acetone	56.0	30.8	20.7	Amphoteric
Ethyl acetate	77.1	12.6	20.5	Amphoteric
2-Butanone	79.5	12.6	—	Amphoteric
Water	100	3.17	21.1	Amphoteric

Table 3 reports the values of the surface area of molecules commonly used as probe, as obtained from different approaches.

Hill [36] calculated the surface area of some probes on the basis of the corresponding atomic van der Waals radii (Table 3). This method, which is based on

TABLE 3   Values of the Molecular Surface Area of Compounds Commonly Used as Probes in IGC According to Different Approaches

	Surface area (Å^2)					
	As calculated according to				As measured by	
Probe	Emmet [33]	Hill [36]	Gray [24]	Snyder [39]	BET [38]	Schultz [42]
n-Pentane	36.2	37.2	44.2	50.2	49.2	—
n-Hexane	39.6	42.2	48.3	57.8	56.2	51.5
n-Heptane	42.5	46.6	51.9	65.5	63.1	57.0
n-Octane	45.7	51.6	55.8	73.1	64.6	62.8
n-Nonane	48.5	56.3	59.2	80.8	84.4	68.9
n-Decane	51.4	61.0	62.7	87.6	86.0	—
n-Undecane	54.1	65.5	66.2	95.2	—	—
n-Dodecane	57.0	69.9	69.6	102.9	—	—
Cyclohexane	34.7	36.5	42.4	51.0	41.7	—
CH_2Cl_2	24.5	26.4	29.9	34.9	—	—
Chloroform	28.5	29.4	34.8	42.5	27.9	44.0
CCl_4	32.2	34.2	39.3	42.5	39.2	46.0
Benzene	30.5	32.6	37.2	51.0	43.6	46.0
Toluene	34.3	38.0	41.9	57.8	55.2	—
p-Xylene	37.9	43.3	46.2	64.6	—	—
Chlorobenzene	33.3	37.2	40.7	57.8	—	—
Nitromethane	21.8	29.8	26.6	32.3	—	—
Methanol	18.0	21.9	22.0	26.4	21.9	—
Ethanol	23.0	26.4	28.1	34.0	28.3	—
n-Propanol	27.2	30.5	33.1	41.7	32.8	—
n-Butanol	31.1	34.9	37.9	49.3	35.4	—
THF	28.7	30.6	35.0	42.5	—	45.0
Diethyl oxide	33.8	35.1	41.2	38.3	42.0	47.0
TEA	41.1	43.5	50.1	—	43.0	—
Pyridine	28.5	31.6	34.8	49.3	39.0	—
Butylamine	32.7	34.6	39.8	49.3	—	—
1,4-Dioxane	29.6	32.2	36.1	51.0	—	—
Acetonitrile	21.4	32.2	26.1	85.0	—	—
Acetone	27.1	31.3	33.0	35.7	—	42.5
Ethyl acetate	31.5	36.3	38.5	48.5	—	48.0
2-Butanone	30.6	35.0	37.4	39.1	—	—
Water	10.5	13.1	12.8	—	12.5	—

BET: Brunauer–Emmet–Teller technique.

the size of atoms in the absence of any adsorbate, excludes the interactions of the probe with the substrate. This work showed that the surface areas related to the three-dimensional van der Waals constant, b, through the following equation:

$$a = \frac{\Pi}{2}\left(\frac{3b}{2\Pi}\right)^{2/3} \qquad (3)$$

where b can be expressed as a function of the critical temperature, T_c (°K) and pressure, P_c (atm), to give Eq. (4):

$$a = 6.354\left(\frac{T_c}{P_c}\right)^{2/3} \qquad (4)$$

The values thus obtained are also reported in Table 3.

Avgul et al. [37] have also determined the surface area of the n-alkane probes. They gave a value of 45 Å^2 for n-pentane and an increment of 6 Å^2 for each additional methylene group in the series. McClellan and Harnsberger [38] determined this parameter by projecting the molecular model of the given probe and measuring the surface area of the projected molecular using a planimeter. The values thus obtained were much lower than those obtained from liquid density [33] [eq. (1)] and van der Waals radii [36] Eq. (3)] and depended strongly on the spatial configuration of the probe. McClellan and Harnsberger [38] summarized, for comparison, the average values of the surface area of different probe molecules as obtained from BET measurements as shown in Table 3.

Snyder [39] has also reported the value of the surface area of some probes making the assumption that there is a total adsorption of the probe on a plane surface. He did not take into account the specific conformation that a given probe can acquire when it is in contact with the surface. In this work the author determined the surface area of different groups (methyl, methylene, nitro, mercapto, etc.) and expressed the surface area of the adsorbed molecules as the sum of the contributions from each group (a_i) following Eq. (5):

$$a = \sum^{i} a_i \qquad (5)$$

Finally, Schultz et al. [40–42] have undertaken the same task and obtained the values of the surface area for a series of neutral and polar probes. These authors used neutral reference surfaces like polytetrafluoroethylene (PTFE) or polyethylene and used Eq. (6) to calculate a:

$$a = \frac{RT \ln V_N}{2N \sqrt{\gamma_S^D \gamma_L^D}} \qquad (6)$$

where γ_S^D is the dispersive contribution to the energy of the reference solid surface determined by CA and V_N the net retention time of a given probe (see below). Two major limitations can be ascribed to this approach, i.e., (1) a universal reference surface is needed and (2) the surface energy of PTFE used in Eq. (6), was obtained from CA measurements which always give lower values of γ_S^D than those obtained from IGC.

Table 3 summarizes the values of the surface area of probe molecules as obtained from models or other different approaches.

It is worth noting that the molecular area of a probe remains a key parameter in IGC measurements because it can vary with the nature of adsorbent. Thus, a fluctuation of 20% in the values of this parameter induces an error of 56% in the calculated γ_S^D value.

C. Acid–Base Properties

The acid–base properties of the probes used for IGC surface investigations are based mostly on Gutmann's approach [43]. Gutmann determined acceptor (AN) and donor (DN) numbers of polar molecules, but his approach has been the subject of discussion because AN and DN values are determined by two completely different methods.

AN concerns the acidity (or the electron acceptor capacity) of a substance (mostly solvents). It is determined from the ^{31}P NMR chemical shift of triethyl phosphine oxide $(C_2H_5)_3 - P = O$ (TPO) in its presence according to the following interaction:

$$
\begin{array}{c}
\text{Et} \\
| \quad \frown \\
\text{Et---P---\overset{..}{O}:} \to \quad \text{acceptor} \\
| \\
\text{Et}
\end{array}
\tag{7}
$$

TPO is taken as a base (or electron donor) reference. When its oxygen atom interacts with an acceptor, the electron density at the phosphorus atom decreases and induces a downfield chemical shift (δ) of ^{31}P in the corresponding NMR spectrum. Gutmann has extrapolated the values of δ to infinite dilution and used hexane as a reference solvent to which he assigned an AN value of 0. The δ values of the other solvents were corrected for the difference in volume susceptibilities with respect to hexane. Gutmann used antimony pentachloride in 1,2-dichloroethane as the strongest Lewis acid, to which the value of 100 was attributed. The other solvents were related to this arbitrary scale following Eq. (8):

$$
AN \equiv \frac{\delta_{corr} * 100}{\delta_{corr(TPO.SbCl_5)}} = 2.348 * \delta_{corr}
\tag{8}
$$

The Gutmann acid number concept is difficult to interpret from the point of view of absolute values of AN, since, for example, pyridine, which is well known as a rather strong Lewis base, has an AN value of 14.2. Recently, Rieddle and Fowkes [44] raised some problems related to the AN scale and proposed corrected values of AN, noted AN^*. This new parameter takes into account the dispersive effect of the investigated acid. In fact, they showed that liquids with only dispersive properties, such as hexane, induce a significant shift in the ^{31}P NMR spectrum of TPO. With this correction, the AN value for pyridine fell from 14.2 to 0.5. Rieddle and Fowkes [44] also showed that the corrected values of AN can be correlated to the enthalpy of formation of the TPO–acid adduct. Thus, AN^* and DN can now be expressed with the same units (see below). They also proposed a correlation between the two scales:

$$AN^* = \Delta H_{(A-TPO)} = 0.288(AN - AN_d) \tag{9}$$

where AN is the original Gutmann number and AN^d that related to the dispersion contribution and reported by Rieddle and Fowkes [44].

The values of AN^* for probes used in IGC are reported in Table 4. The acid–

TABLE 4   Values of the Acid–Base Numbers of Molecules Commonly Used as Probes in IGC According to Gutmann's [43], Fowkes' [44] and Drago's [49] Approaches

| Probes | Gutmann | | Fowkes | Drago | | | |
	AN	DN	AN^*	C_A	E_A	C_B	E_B
Methylene chloride	20.4	0	3.9	—	—	—	—
Chloroform	23.1	0	5.4	0.15	3.31	—	—
Carbon tetrachloride	8.6	0	0.7	—	—	—	—
Benzene	8.2	0.1	0.17	—	—	0.707	0.486
p-Xylene	—	—	—	—	—	1.78	0.574
Nitromethane	20.5	2.7	4.3	—	—	—	—
Methanol	41.5	19.0	12.0	—	—	—	—
Ethanol	37.9	20.0	10.3	—	—	—	—
n-Butanol	36.8	—	9.1	1.12	−0.05	—	—
Tetrahydrofuran	8.0	20.0	0.5	—	—	4.27	0.978
Diethyl oxide	3.9	19.2	1.4	—	—	3.25	0.963
Triethylamine	—	61.0	—	—	—	11.09	0.991
Pyridine	14.2	33.1	0.14	—	—	6.4	1.17
Acetonitrile	19.3	14.1	4.7	—	—	1.34	0.886
1,4-Dioxane	10.8	14.8	0	—	—	2.38	1.09
Acetone	12.5	17.0	2.5	—	—	2.33	0.987
Ethyl acetate	9.3	17.1	1.5	—	—	1.74	0.975
Water	54.8	18.0	15.1	2.45	0.33	—	—

base concept, applied to the solid surfaces, were reviewed recently in a special book edited to celebrate the 75th birthday of Professor Frederick M. Fowkes [45]. In the same book, Fowkes [46] recommended the use of chloroform as an acid probe and of diethyl ether or triethylamine as basic probes because they possess the weakest energy of self-association.

DN, which reflects the basicity (or electron donor) ability, is determined as the molar enthalpy for the reaction of the given electron donor probe in a 10^{-3} M solution in 1,2-dichloroethane [43] with the very strong acceptor $SbCl_5$ taken as a reference:

$$D + SbCl_5 \rightleftharpoons D \cdot SbCl_5 (-\Delta H_{D \cdot SbCl_5} = DN) \qquad (10)$$

The acceptor and donor numbers are dimensionless parameters even if DN should be expressed in J/mol. For the systematic comparison between a reference material and its modified forms, this approach can provide interesting information.

Some authors [47,48] have used the Drago [49] approach to select the acid and basic probes. The four-parameter equation proposed by Drago predicts the enthalpy of formation of the one-to-one molecular adduct in the gas phase or in noncoordinating solvents:

$$-\Delta H_{AB} = C_A C_B + E_A E_B \qquad (11)$$

where the C parameters represent the covalent contributions and the E ones those arising from the electrostatic interactions between the acid and the basic components of the adduct.

In reality many molecules have both acceptor and donor sites, but amphoteric probes have not been taken into account in this formalism. Table 4 gives the values of C and E for different probe molecules used in IGC measurements to establish the acid–base characteristics of a solid surface.

VIII. CALCULATION PROCEDURES

A. Determination of Retention Time

For perfectly symmetrical peaks, the simplest method for retention time (t_r) determination consists of taking the time corresponding to their maximum intensity. For asymmetrical peaks, Conder et al. [50] have introduced the skew ratio η, which is a simple measure of asymmetry. This parameter is the ratio of the slope of the trailing (last eluted) edge of the peak to that of the leading (first eluted) edge, both edges being measured at the respective points of inflexion. The skew ratio is therefore equal to a/b as shown in Fig. 1. For symmetrical peaks η is obviously equal to unity. Conder and Young [51] recommended the use of the peak maximum method when the skew ratio varies from 0.7 to 1.3. The retention

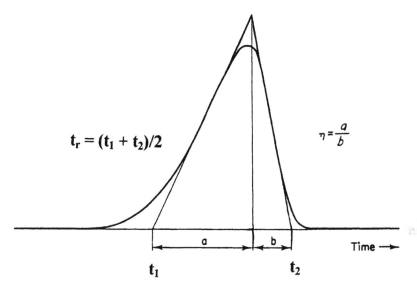

$$t_r = (t_1 + t_2)/2 \qquad\qquad \eta = \frac{a}{b}$$

FIG. 1 Determination of the skew ratio (η) and the retention time (t_r) from assymetrical peaks. (From Ref. 51.)

time of asymmetrical peaks with a skew ratio lower than 0.7 or higher than 1.3 is best given by [51]:

$$t_r = \frac{t_1 + t_2}{2} \tag{12}$$

where t_1 and t_2 are the times at which the tangents to the leading and trailing side of the peak intercept the baseline, respectively, as shown in the typical chromatogram of Fig. 1.

Recently, Kamdem and Riedl [52] carried out a comparative investigation using peak maxima and Conder and Young's method and calculated thermodynamic parameters of lignocellulosic fibers coated with thermosetting polymers. The results of this work are reported in appropriate Sec. IX.A.2 below.

B. Determination of Retention Volume

The IGC chromatograms provide the following information: (1) the retention time, t_r, of a given probe; (2) the retention time, t_o, of the noninteracting marker; (3) the carrier gas flow rate, F; and (4) the inlet (P_i) and outlet (P_o) pressure of the column. These experimental data allow the calculation of the net retention volume, V_N, which is the key thermodynamic parameter of IGC measurements [53]:

$$V_N = FD(t_r - t_o) \tag{13}$$

where D is the James-Martin correction factor [54] for gas compressibility, which can be calculated from Eq. (14):

$$D = \frac{3\left[\left(\dfrac{P_i}{P_o}\right)^2 - 1\right]}{2\left[\left(\dfrac{P_i}{P_o}\right)^3 - 1\right]} \tag{14}$$

C. Determination of Dispersive Properties

Two approaches have been elaborated to calculate the dispersive component of the surface energy of solids investigated by IGC at infinite dilution conditions. Both are based on the fundamental Fowkes' formalism, which considers that the surface energy of any substance can be written as a sum of dispersive and nondispersive (or specific) contributions [55]:

$$\gamma_S = \gamma_S^D + \gamma_S^{SP} \tag{15}$$

The same approach has been extended to the work of adhesion (W_A) between two phases [55,56]:

$$W_A = W_A^D + W_A^{SP} \tag{16}$$

Dorris and Gray [57] have assumed that the free energy of adsorption of a methylene group ($\Delta G_A^{(CH_2)}$) can be correlated to its work of adhesion with the surface of the solid under investigation as follows:

$$\Delta G_A^{(CH_2)} = N a_{(CH_2)} W_A^{(CH_2)} \tag{17}$$

where $a_{(CH_2)}$ is the surface area of a methylene group and N Avogadro's number.

According to Girifalco and Good [58], the dispersive work of adhesion can be expressed as:

$$W_A^D = 2\sqrt{\gamma_S^D \gamma_L^D} \tag{18}$$

Combining Eqs. (17) and (18) leads to

$$\frac{-\Delta G_A^{(CH_2)}}{N a_{(CH_2)}} = 2\sqrt{\gamma_{(CH_2)}^D \gamma_S^D} \tag{19}$$

or

$$\gamma_S^D = \frac{(\Delta G_A^{(CH_2)})^2}{4N^2(a_{CH_2})^2\gamma_{(CH_2)}} \tag{20}$$

where $\gamma_{(CH_2)}$ is the surface energy of polyethylene-type polymers with a finite molecular weight, given by Eq. (21) [59]:

$$\gamma_{(CH_2)} = 34.0 - 0.058T(mJ/m^2) \tag{21}$$

where T is expressed in °C.

The free energy ΔG_A was correlated either to the surface partition coefficient, K_s, which characterizes the adsorbate–adsorbent interaction, or as a function of the net retention volume, V_N, as follows:

$$\Delta G_A = -RT \ln\left(\frac{K_s p_{s,g}}{\Pi_s}\right) \tag{22}$$

or:

$$\Delta G_D = -\Delta G_A = RT \ln\left(\frac{V_N P_0}{Sm\pi_0}\right) \tag{23}$$

where ΔG is the free energy of desorption (or adsorption) of 1 mol of solute from a reference state, defined by the bidimensional spreading pressure π_0 of the adsorbed film, to a reference gas phase state, defined by the partial pressure P_0 of the solute; $p_{s,g}$ is the adsorbate vapor pressure in the gas phase; Π_s is the spreading or surface pressure of the adsorbed gas in the standard adsorption state (see below); S is the specific area and m the weight of the substrate in the column.

Two reference states are generally considered, i.e., that defined by Kemball and Ridel [60], where $P_0 = 1.013 \times 10^5$ Pa and $\pi_0 = 6.08 \times 10^{-5}$ N/m, and that proposed by De Boer [61], where $P_0 = 1.013 \times 10^5$ Pa and $\pi_0 = 3.38 \times 10^{-4}$ N/m.

Equations (22) and (23) each contain five constants and they can be easily simplified to give:

$$\Delta G_A = RT \ln K_s + C_1 \tag{24}$$

$$\Delta G_A = RT \ln V_N + C_2 \tag{25}$$

Dorris and Gray [57] introduced the free energy corresponding to a methylene group, $\Delta G_A^{(CH_2)}$, as:

$$\Delta G_A^{(CH_2)} = \Delta G_A^{(C_{n+1}H_{2n+4})} - \Delta G_A^{(C_nH_{2n+2})} \tag{26}$$

and from Eqs. (24) and (25) they deduced the free energy corresponding to a methylene group, $\Delta G_A^{(CH_2)}$, as:

$$\Delta G_A^{(CH_2)} = -RT \ln \left[\frac{K_s^{(C_{n+1}H_{2n+4})}}{K_s^{(C_nH_{2n+2})}} \right] \tag{27}$$

or:

$$\Delta G_A^{(CH_2)} = RT \ln \left[\frac{V_N^{(C_{n+1}H_{2n+4})}}{V_N^{(C_nH_{2n+2})}} \right] \tag{28}$$

Combining Eqs. (20) and (27) on the one hand and Eqs. (20) and (28) on the other, the London component of the surface energy can be written following Eqs. (29) and (30), respectively, as follows:

$$\gamma_D^S = \frac{RT \ln \left[\frac{K_s^{(C_{n+1}H_{2n+4})}}{K_s^{(C_nH_{2n+2})}} \right]^2}{4N^2 (a_{CH_2})^2 \gamma_{(CH_2)}} \tag{29}$$

or

$$\gamma_S^D = \frac{\left[RT \ln \frac{V_N^{(C_{n+1}H_{2n+4})}}{V_N^{(C_nH_{2n+2})}} \right]^2}{4N^2 (a_{CH_2})^2 \gamma_{(CH_2)}} \tag{30}$$

Figure 2 shows a schematic representation of the approach proposed by Dorris and Gray [57].

The value of 6 Å^2 is the most widely used for the cross-sectional area of CH_2, but other values have been also put forward. Thus, according to Groszek's model [62], based on the calculation of the surface area of the hexagon in the graphite basal plane, the surface area of a methylene unit is 5.2 Å^2. Two years later, Clint [63] studied the adsorption of hydrocarbons on graphitized carbon and found an increment per CH_2 group of 5.5 Å^2. It is worth mentioning that the use of Groszek's or Clint's value of the surface area of a methylene group induces an increase of the dispersive component γ_S^D of 30% and 20%, respectively.

Schultz et al. [40–42] have proposed a second method for the determination of the dispersive component of the surface free energy of the adsorbate under investigation. They used the assumption of Dorris and Gray [see Eq. (17)] and correlated it to the free energy of adsorption-desorption determined by Eq. (25):

$$RT \ln V_N = 2Na \sqrt{\gamma_S^D \gamma_L^D} + C \tag{31}$$

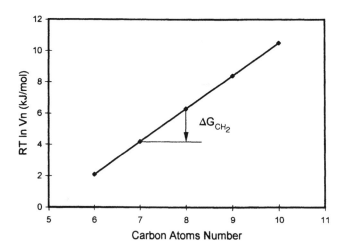

FIG. 2 Example of the determination of the free energy of adsorption related to a methylene group, according to the Dorris and Gray approach [57].

where N is Avogadro's number, a the surface area of the probe molecule, and γ_L^D the dispersive component of the surface tension of the probe molecule in the liquid state.

As can be seen from Eq. (31), the use of the n-alkane series as probes that interact with the substrates under investigation only through dispersive forces, γ_S^D can be obtained from the slope of the plot $RT \ln V_n$ versus $a(\gamma_L^D)^{1/2}$ which is in principle a linear function usually called the "reference line." Figure 3 shows a typical example of this approach.

Schultz et al. [40–42] compared the values of γ_S^D determined by both procedures [graphic approach on the basis of Eq. (31), shown in Figure 3, and Gray's approach following Eq. (30)] and found that they were in very good agreement when the cross-sectional surface area of CH_2 group of 6 Å^2 was used for the calculations.

D. Determination of Thermodynamic Parameters

The net retention volumes have been also correlated to the surface (K_s) and bulk (K_b) partition coefficients of the stationary phase according to Eq. (32);

$$V_N = K_s A + K_b V \qquad (32)$$

where A and V are the total surface area and volume of the stationary phase, respectively.

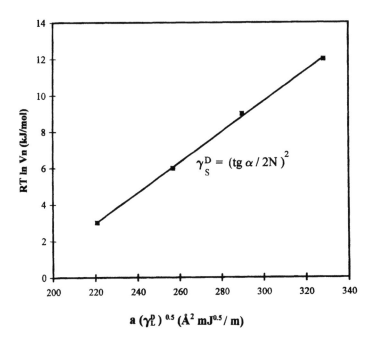

$$\gamma_S^D = (tg\,\alpha\,/\,2N\,)^2$$

a $(\gamma_L^D)^{0.5}$ (Å^2 mJ$^{0.5}$/ m)

FIG. 3   Example of the determination of the dispersive contribution to the surface energy, according to Schultz et al. [41].

In IGC conditions the solution of the probe in the stationary phase is negligible and Eq. (32) reduces to:

$$V_N = K_s A \tag{33}$$

The surface partition coefficient that characterizes the adsorbate–adsorbent interaction may therefore be determined for different probes if A has been previously determined.

The three fundamental thermodynamic parameters related to the adsorption–desorption processes are often used to evaluate the physicochemical affinity between the probe and the sorbent. These parameters are (1) the standard free energy change, ΔG_A, (2) the enthalpy change, ΔH_A, and (3) the standard entropy change, ΔS_A which can be determined from the partition coefficient K_s and/or the net retention volume of adsorption–desorption. We report below the experimental approaches and the calculation procedures used to obtain these parameters and their use to determine the acid–base properties of solid substrates.

1. Determination from the Partition Coefficient

The partition coefficient K_s can be correlated to the thermodynamic parameters of the solid surface [64] using Eqs. (34)–(36):

$$-\Delta G_A/RT = \ln K_s + C \tag{34}$$

$$\frac{d \ln (K_s)}{d\left(\dfrac{1}{T}\right)} = -\left(\frac{\Delta H_A}{R}\right) \tag{35}$$

$$\Delta S_A = -\left(\frac{\Delta H_A + \Delta G_A}{T}\right) \tag{36}$$

2. Determination from the Net Retention Volume

As mentioned above, the retention volume of a given probe, V_N, is related to the free energy of adsorption by Eq. (25). Carrying out experiments at different temperatures, the free enthalpy ΔH_A and the free entropy of desorption ΔS_A can be obtained by applying Eq. (37):

$$\Delta G_A = \Delta H_A - T\Delta S_A \tag{37}$$

Plotting ΔG_A versus T (Fig. 4), the values of ΔH_A and ΔS_A can be deduced from the intercept and the slope, respectively, according to Eq. (37). Equation (38) can also be used for this purpose:

$$\frac{\Delta G_A}{T} = \frac{\Delta H_A}{T} - \Delta S_A \tag{38}$$

E. Determination of Acid–Base Properties

In addition to dispersive interactions, polar probes are injected to study possible donor–acceptor interactions with the solid surfaces studied. These interactions show up as a deviation from the straight line (reference alkane line) behavior and can be quantified by the free energy, ΔG_A, of adsorption–desorption of the specific polar probe [40]:

$$\Delta G_A = \Delta G_A^D + \Delta G_A^{SP} = Na W_A^D + Na W_A^{SP} \tag{39}$$

The dispersive free energy of adsorption can be expressed as:

$$\Delta G_A^D = Na W_A^D = RT \ln V_{Nref} \tag{40}$$

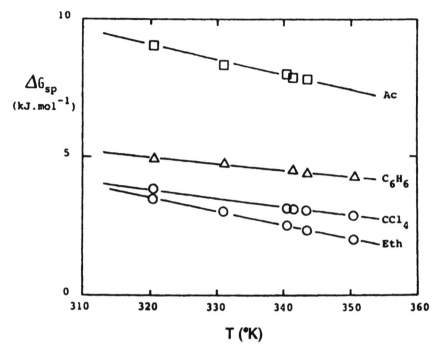

FIG. 4 Example of the determination of the enthalpies of adsorption ΔH_A for different polar probes, according to Schultz et al. [41]. Ac, acetone; Eth, diethyloxide. (From Ref. 41.)

The combination of Eqs. (25), (39), and (40) gives:

$$RT \ln V_N = RT \ln V_{Nref} + Na W_A^{SP} \tag{41}$$

or

$$\Delta G_A^{SP} = Na W_A^{SP} = RT \left(\frac{\ln V_N}{\ln V_{Nref}} \right) \tag{42}$$

where V_N is the retention volume of the corresponding polar probe and V_{Nref} that derived from the n-akanes reference line, at the value of $a(\gamma_L^D)^{1/2}$ corresponding to the polar probe used, as shown in Fig. 5.

In order to assess the acid–base characteristics of solid surfaces by IGC different approaches have been proposed in the literature. The simplest one consists in studying the specific interactions between the surface investigated and two reference polar molecules, i.e., a donor probe and an acceptor one. The quantity

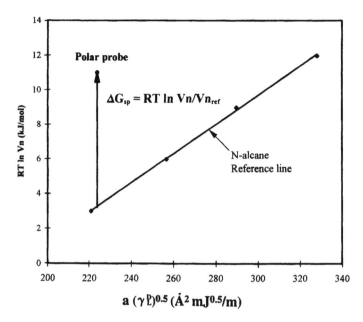

FIG. 5 Schematic determination of the specific free energy of adsorption, ΔG_{sp} of polar probes according to Schultz et al. [41].

ΔG_A determined for each probe is related to the corresponding AN_s and DN_s values for the solid [eqs. (43) and (44)]. It is assumed that these parameters describe the acid and the basic character of the surface [65] respectively, and thus the ratio AN_s/DN_s gives an indication of their relative importance [eq. (45)].

$$\Delta G^{SP}_{A\ donor} \equiv AN_s \qquad (43)$$

$$\Delta G^{SP}_{A\ acceptor} \equiv DN_s \qquad (44)$$

$$\left[\frac{\Delta G^{SP}_{A\ donor}}{\Delta G^{SP}_{A\ acceptor}} \right] \equiv \left[\frac{AN_s}{DN_s} \right] \qquad (45)$$

The following arbitrary acid–base scale has been proposed [65]:

$AN_s/DN_s \geq 1.1$ acid surface   (46)

$AN_s/DN_s \leq 0.9$ basic surface   (47)

$0.9 < AN_s/DN_s < 1.1$ amphoteric surface   (48)

$AN_s \approx DN_s \approx 0$ neutral (nonpolar) surface   (49)

The donor and acceptor couples most often used are tetrahydrofuran/chloroform [32,65,66] and diethyloxide/methylene chloride [67].

Although the above approach is somewhat arbitrary, it gives interesting information. Thus, for example, an increase in the AN_s/DN_s ratio after a given surface modification means that the applied treatment increased its acidic character.

The second approach was proposed by Schultz et al. [40–42]. It was deduced from the acid–base theory introduced by Gutmann [43] and applied to the polar probes. In fact, these authors elaborated a semiquantitative approach to characterize the acid–base properties of solid surfaces based on the correlation of ΔH_{sp} with the acid–base properties of polar probes (DN, AN) and those of the investigated surface, following Saint Flour and Papirer [68]:

$$\Delta H_{sp} = K_A DN + K_B AN \tag{50}$$

where K_A and K_B describe, respectively, the acid and base characteristics of the solid surface and are the slope and the intercept of the plot $\Delta H_{sp}/AN$ versus AN/DN (Fig. 6) according to:

$$\frac{\Delta H_{sp}}{AN} = K_A \frac{DN}{AN} + K_B \tag{51}$$

More recent investigations have reported results on the determination of acid and base numbers of solid surfaces using the corrected AN^* parameters [44].

The third approach was developed by Boluk and Schreiber [47] on the basis of Drago's acid–base concept [49]. These authors determined the interaction parameter, Ω, of homologous series of alcohols, used as acidic probes, and amines, used as basic ones. This interaction parameter reflects the acid–base properties of the solid surface and is defined as follows: for acidic surfaces, the specific retention volume (V_g^0), of the basic probe exceeds that of the acidic alcohol (n-butylamine and n-butyl alcohol are commonly used to calculate Ω) and thus Ω is negative:

$$\Omega = 1 - \frac{(V_g^0)_b}{(V_g^0)_a} < 0 \tag{52}$$

For basic surfaces, the retention volume of the acidic probe exceeds that of the basic amine, yielding a positive value of Ω:

$$\Omega = 1 - \frac{(V_g^0)_a}{(V_g^0)_b} - 1 > 0 \tag{53}$$

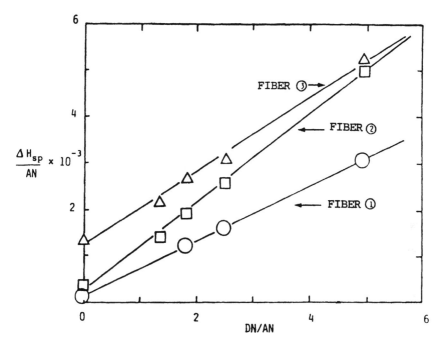

FIG. 6 Determination of the donor and acceptor numbers of solid surfaces according to Schultz et al. (From Ref. 41.)

where $(V_g^0)_a$ and $(V_g^0)_b$ are the specific retention volumes for the acidic and basic probe, respectively, which can be obtained from the following expression:

$$V_g^0 = \frac{273 V_N}{mT} \tag{54}$$

where V_N is the net retention volume, m the mass of stationary phase (g), and T the column temperature (K).

This approach is very similar to the first one, based on Tetrahydrofuran (THF) and chloroform as polar probes, but its originality resides in the fact that the main skeleton of both polar probes used is the same, i.e., n-butyl. As already pointed out for the first approach, the present procedure is qualitative and the absolute value of Ω does not bear any physical meaning, but remains useful, e.g., for the comparison of the surface properties before and after modification. Moreover, the use of the net retention volume, V_N, instead of specific retention volume V_g^0, gives the same results since m and T have constant values for the same column under investigation.

The fourth approach was proposed by St. Flour and Papirer [69] and consists of the calculation of a specific interaction parameter (I_{sp}) between a given polar probe and the substrate surface. This parameter is determined from the plot of the free energy change associated with the polar probe against the logarithm of the vapor pressure under standard conditions (ΔG_A versus log p_0), as shown in Fig. 7. The comparison between the values of I_{sp} of an acceptor and a donor probe gives an indication of the nature of the surface under investigation.

Very recently, Vidal et al. [70] proposed another approach to calculate the specific interaction parameter (I_{sp}), which differs only by the fact that these authors plotted ΔG_A versus the surface area of the adsorbed molecule. They obtained the n-alkane reference line and used Eq. (42) to calculate the specific interaction parameter I_{sp} (Fig. 8).

Tiburcio and Manson [48,71] also proposed a novel way to determine the acid–base properties of solid surfaces based on Drago's four constants [Eq. (11)]. They determined the enthalpies of adsorption associated with the acid–base interactions between polar probes and the solid surface using polar probes and neutral model ones with very similar molecular sizes:

$$\Delta H^{A-B} = (\Delta H_{exp} - \Delta H^{D})_{probe} - (\Delta H_{exp} - \Delta H^{D})_{model} \tag{55}$$

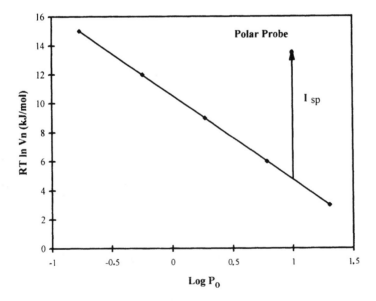

FIG. 7 Determination of the specific interaction parameter I_{sp} of polar probes according to St. Flour and Papirer [69].

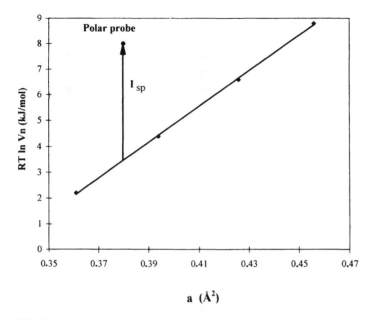

a (Å²)

FIG. 8 Determination of the specific interaction parameter I_{sp} of polar probes according to Vidal et al. [70].

where ΔH_{exp} is the enthalpy of adsorption determined experimentally [eqs. (37) and (38)] and ΔH^D the enthalpy of vaporization of the probe taken from the literature [72] or calculated following Truton's rule [73] or Riedel's equation [74]:

$$V_N = k\left[\exp\left(-\frac{\Delta H_{exp}}{RT}\right)\right] \tag{56}$$

The model probe was carbon tetrachloride for chloroform and methylene chloride (acidic) and n-propane and n-pentane for the acetone and diethyl ether (basic) [48,71].

F. Determination of the Fiber–Matrix Interaction Parameter

Knowing the acceptor and the donor numbers for both fibers and matrix of a given composite material, it is easy to deduce, by analogy with the St. Flour and

Papirer approach [68] [eq. (50)], the interaction parameter A, which describes the acid–base interactions between the fibers (f) and the matrix (m) [40–42]:

$$A = K_{A(f)}K_{B(m)} + K_{B(f)}K_{A(m)} \tag{57}$$

where $K_{A(f)}$ and $K_{A(m)}$ are the acid numbers of the fibers and matrix, respectively, and $K_{B(f)}$ and $K_{B(m)}$ are their corresponding basic numbers.

As expected, Schultz and Lavielle [40–42] have shown that the mechanical properties based on epoxy resin–containing carbon fibers could be improved by increasing the interaction parameter A (see also Sec. XI).

In the context of paints and inks, Lara and Schreiber [65] proposed an empirical expression to calculate the interaction parameter of a given pigment (p) with the surrounding resin (r):

$$I_{sp} = \sqrt{AN_rDN_p} + \sqrt{AN_pDN_r} \tag{58}$$

where AN and DN are, respectively, the acceptor and donor numbers, as determined above by Eqs. (43) and (44).

IX. DISPERSIVE PROPERTIES

A. Characterization of Fibers

1. Cellulose Fibers

Cellulosic fibers are used extensively in papermaking, textiles, productions etc. In many cases, these fibers are submitted to a specific surface treatment, e.g., in paper coating where a thin layer of filled polymer emulsion is deposited on the sheet. These processes are in part conditioned by the surface properties of cellulosics. Moreover, recently, cellulose fibers have received much attention as reinforcing elements in different types of polymeric matrices in order to obtain composites with good mechanical properties. The reasons that justify the use of these natural fibers are, among others, the widespread availability of cheap fibers with different morphologies, their low density, and the possibility of burning the corresponding composites without the obvious problems associated to the same operation applied to glass fibers-based composites. In addition, these fibers possess many other advantages such as good mechanical properties and ease of isolation directly from any plant material (i.e., major renewable resources associated with the biological activity which produces through photosynthesis about 3×10^{11} tonnes/year of vegetal biomass containing more than 45% fibers).

As cellulose is a highly hydrophilic natural polymer, its incorporation into hydrophobic matrices must be carried out in such a way as to protect the ensuing composite from water uptake. Thus, compatibilization between the surfaces of these fibers and that of synthetic matrices like polyethylene, polypropylene, or polystyrene is essential to optimize the performances of these novel materials. Corona [75,76] or plasma [77] treatment or specific chemical modifications [32,78,79] of either the fiber surface or the matrix have been applied to improve the quality of the cellulose–polyolefin interface.

Dorris and Gray [57] were the first to investigate cellulose fibers by IGC. They studied cotton fibers and found a dispersive component of surface energy of about 50 mJ/m^2 at 40°C. This value was in good agreement with that obtained from measurements at finite concentration and slightly higher than those deduced from CA measurements.

Lee and Luner [80] studied the surface properties of "confetti" from Whatman No. 1 chromatographic paper extracted with THF and sized with a solution of alkylketene dimer (AKD) (Structure I). The dispersive contribution to the surface energy of the pristine cellulose (Table 5) was about 48 mJ/m^2, which is very close to the value obtained by Dorris and Gray [57]. This parameter increased substantially to 60 mJ/m^2 after the AKD treatment and subsequent extraction by THF, but decreased when the THF extraction was not applied after sizing. The authors explained these results through a mechanism involving the interaction of the alkane chains among grafted AKD molecules giving a weak brush-type network. However, we feel that the extent of the grafted AKD molecules after extraction was far too low (0.02–0.04%) to justify the proposed mechanism. Moreover, the authors did not mention whether they analyzed the shape of the peaks, which should have become asymmetrical (through mechanical retention of the probes) if such an interaction had indeed occurred.

$$R{=\!=}\!\!\left\langle\begin{array}{c}O\\|\\R\end{array}\right\rangle\!\!{=\!=}O$$

I

R = C$_{14}$-C$_{16}$ aliphatic saturated chain

The same authors [81] investigated the same cellulosic substrate after different cycles of solvent treatment. The initial confetti were swollen in water for 2 days, after which the water was exchanged with ethanol which was in turn exchanged successively with acetone, toluene, and n-heptane. The confetti were kept in each solvent for 2 days and each solvent was replaced twice by the next fresh solvent. The final confetti were dried under vacuum at 105°C for 3 days. The value of the dispersive energy was almost three times as high as that obtained in the previous studies discussed above [57,80] (Table 5). These authors excluded artefacts related to diffusion of the probes into microcapillarity of the substrate since the net retention time remained constant when they varied the flow rate of the carrier gas from 11.5 to 16.5 ml/min. They attributed these high values to the fact that

TABLE 5 Values of the Dispersive Component of the Surface Energy of Cellulosics

Material	Trade name or notation and supplier	Treatment or calculation method	γ_S^D at (T °C) (mJ/m²)	$-d\gamma_S^D/dT$ (mJ/m² °C)	Ref.
Cotton	Whatman no. 1	None	49.9 (40°C)	—	57
Cellulose (cell)	Whatman no. 1	None	48.4 (25°C)	0.21	80
		Sizing with AKD (AKD/cell = 0.026%)	60.6 (30°C)	—	80
		Sizing with AKD (AKD/cell = 0.043%)	51.2 (30°C)	—	80
		Sizing with AKD, unextracted	38.6 (30°C)	—	80
		Solvent exchanged	111.9 (30°C)	0.85	81
Amorphous cellulose beads	Whatman CF-11	Surface area a constant	71.5 (50°C)	0.36	82–84
		a from liquid density	70.9 (50°C)	0.37	82–84
		a as perfect gas	62.8 (50°C)	0.48	82–84
		a as real gas	44.2 (50°C)	0.82	82–84
		None	70.5 (50°C)	—	85
		Trifluoroethoxyacetate cell (DS = 1.0)	32.4 (50°C)	—	85
		Fluorination # 1 (DS = 0.5)	39.0 (50°C)	—	85
		Fluorination # 2 (DS = 0.1)	43.6 (50°C)	—	85
		Ethoxyacetate cell. (DS = 0.2)	42.8 (50°C)	—	85
Condenser paper		None	39.0 (40°C)	0.20	87
		+ Al₂O₃	78.5 (40°C)	0.08	87
		+ TiO₂	41.5 (40°C)	0.12	87
		Grafting with EG-SiO₂	40.0 (40°C)	0.14	87

α-Cellulose	Sigma	None	31.9 (40°C)	0.18	30,66
		Corona (15 mA)	40.2 (40°C)	0.17	66
		Corona (32 mA)	42.2 (40°C)	0.20	66
		Corona (40 mA)	46.3 (40°C)	0.27	30,66
		Extraction with acetone (AcE)	47.4 (40°C)	0.34	30,66
		AcE + Corona (15 mA)	44.2 (40°C)	0.32	66
		AcE + Corona (32 mA)	45.2 (40°C)	0.32	66
		AcE + corona (40 mA)	46.3 (40°C)	0.27	66
		Extraction with cold water	31.9 (50°C)	—	30
		Extraction with hot water	32.2 (50°C)	—	30
		Extraction with ether	43.3 (50°C)	—	30
		Grafting with styrene	37.9 (50°C)	—	30,32
Cellulose	Avicell, Sigma	None	42.5 (50°C)	—	32
		Grafting with II	40.9 (50°C)	—	32
		Grafting with IV	37.3 (50°C)	—	32
		Grafting with Va	37.6 (50°C)	—	32
		Grafting with Vb	39.1 (50°C)	—	32

AKD, alkylketene dimer; DS, degree of substitution; EG-SiO$_2$, silicon oxide grafted with ethylene glycol.

in the micropores the contact area for methylene groups may be higher than on a flat surface. We disagree with this interpretation because 6 Å^2 is the highest area that a CH_2 group can cover based on the fact that it is calculated from bonding length. Moreover, the authors evoked a slow diffusion of the probes into micropores, which contradicts their macroscopic findings excluding the possibility of diffusion artefacts.

Garnier and Glasser [82–85] also studied the surface energy of dried amorphous cellulose. They measured the dispersive contribution to the surface energy of the substrate at different temperatures (Table 5) and found that the temperature dependence was five times higher than that observed with conventional techniques such as contact angle or surface tension of polymers in the melt [7,86]. The authors correlated this strong dependence to the fact that the assumption that the surface area of the adsorbed molecule is independent of the temperature is not always verified [82–84]. They calculated the surface energy of the substrate from the corresponding retention volumes using an a parameter (1) calculated considering that the probes are alternatively liquids, perfect gases, and real gases, or (2) obtained in a previous investigation [41]. Through this comparison they reached the conclusion that the most correct values of the dispersive energy are obtained when the surface areas are those determined by Schultz et al. [41]. In a very recent work [85], the same authors investigated the surface properties of pristine cellulose and four different derivatives namely: (1) fluorinated, (2) trifluoroethoxyacetate, (3) ethoxyacetate, and (4) laurate [85]. They showed that γ_S^D decreased from 70 mJ/m^2 to about 30 mJ/m^2 for fluorinated cellulose and to 40 mJ/m^2 for the other derivatives.

Ignatova et al. [87] investigated papers for condensers as such and with three different fillers, namely, Al_2O_3, TiO_2, and SiO_2 "grafted" with diethylene glycol. Whereas the first filler produced a substantial increase in the surface energy of the paper from 39 to 48 mJ/m^2, the other two did not induce any appreciable change.

More recently, Belgacem et al. [30,32,66] studied the surface properties of different types of cellulose fibers as such, purified by extraction with different solvents, grafted with different coupling agents, or corona-treated. The dispersive components of the surface energy of these materials are summarized in Table 5. As can be seen from these data, γ_S^D of α-hardwood cellulose fibers increased from about 35 to 50 mJ/m^2 when they had been acetone-extracted or corona-treated at a discharge level of 40 mA [66]. In the second publication [30], the authors showed that the efficiency of purification of α-hardwood cellulose fibers by diethyl ether is slightly lower than that attained with acetone. They also reported that water (cold or hot) did not remove the impurities from the fiber surface. In another study [32], microgranular cellulose (MGC) was also investigated as such and after grafting with four different coupling agents (structures **II–V**):

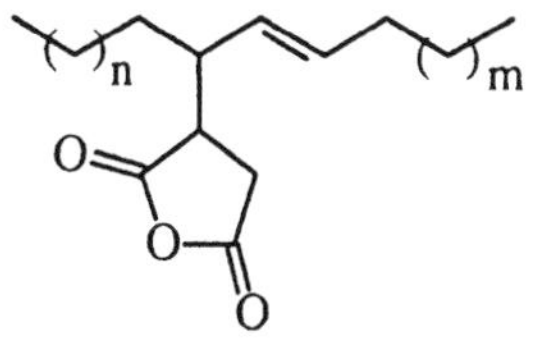

II

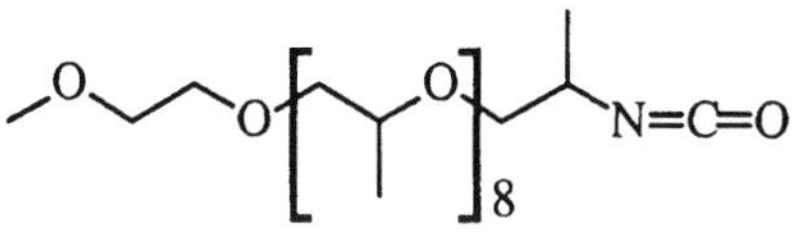

III

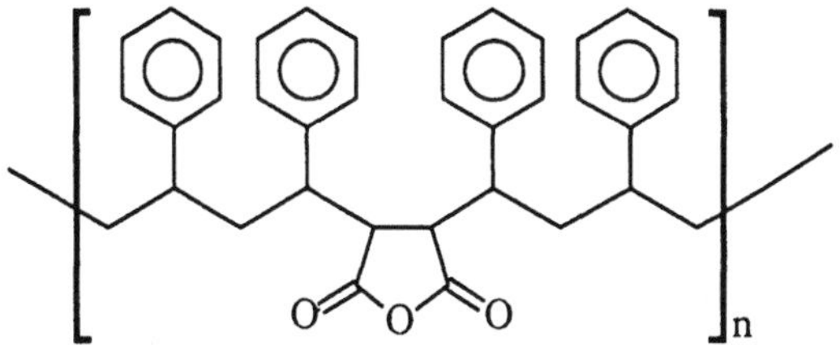

IV

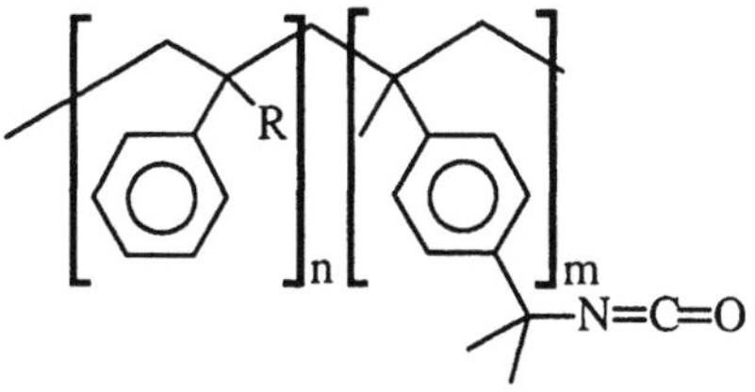

Va, R = H
Vb, R = CH$_3$

The dispersive properties of MGC and its grafted derivatives remained practically unchanged (Table 5).

Finally, Borch [88] reviewed the use of IGC at zero coverage conditions in the characterization of cellulosic materials.

2. Lignocellulosic Fibers

Lignocellulosic fibers have also attracted the attention of researchers. They have been the subject of different investigations dealing with their surface properties in view of their use as reinforcing elements as an alternative to pure cellulose, but keeping the same basic advantages.

Dorris and Gray studied the adsorbtion behavior of thermomechanical pulps (TMPs) [57] and found that these materials have lower dispersive energy (38.8 mJ/m^2) than that of pure cellulose, as reported in Table 6.

Lignocellulosic fibers, as obtained from wood delignification and after modification (bleaching, beating, grafting, etc.), have also been characterized [52,89–100]. Garnagul and Gray [89] studied the surface properties of bleached kraft pulps from black spruce before and after mechanical beating at four different beating energies and found that the dispersive surface energy of the pulps was not sensitive to this parameter (see Table 6). Kamdem and Riedl determined the dispersive component of the surface energy of chemithermomechanical pulps (CTMPs) obtained from hardwood and softwood (50% spruce and 50% balsam fir), as received or treated by different phenol-formaldehyde resins (PFR) or polymethyl methacrylate (PMMA) [90,91]. The values of γ_S^D as a function of temperature are given in Table 6. These data indicate that the dispersive component of the treated fibers tended to reach those of the corresponding polymer. The same authors reported data on the effect of grafting CTMP fibers with PMMA [90]. Hydrogen peroxide–treated CTMP fibers showed a higher dispersive energy (40.1 mJ/m^2) compared with that of the untreated substrate (36.4 mJ/m^2) because this reagent extracted some lignin and hemicellulose fragments from their surface (Table 6). Grafting with PMMA tended to increase γ_S^D, which reached a maximum value of 46.0 mJ/m^2 when grafting attained 50% w/w, and decreased thereafter [91].

These authors [52] applied the peak maximum [50] and Conder and Young's

TABLE 6 Values of the Dispersive Component of the Surface Energy of Wood Fibers

Material	Treatment or calculation method	γ_S^D at (T °C) (mJ/m^2)	$-d\gamma_S^D/dT$ (mJ/m^2 °C)	Ref.
TPM	None	38.8 (30°C)	—	57
CTMP	None	25.2 (40°C)	0.33	90
	None	36.4 (25°C)	—	91
	None	46.2 (23°C)	—	98
	Peak maxima method (PM)	37.0 (20°C)	—	52
	Conder-Young method (CY)	37.9 (20°C)	—	52
	Cleaning with H_2O_2	27.6 (40°C)	—	90
	Cleaning with H_2O_2	40.1 (25°C)	—	91
	Coating with PFR (PM)	38.6 (20°C)	—	52
	Coating with PFR (CY)	39.5 (20°C)	—	52

TABLE 6 Continued

Material	Treatment or calculation method	γ_S^D at $(T\ °C)$ (mJ/m^2)	$-d\gamma_S^D/dT$ (mJ/m^2 °C)	Ref.
	Coating with 8% PFR-A	31.2 (40°C)	0.45	90
	Coating with 8% PFR-B	35.3 (40°C)	—	90
	Coating with 20% PFR-A	38.3 (40°C)	0.49	90
	Coating with 20% PFR-B	42.5 (40°C)	0.63	90
	Grafting with PMMA (30%)	42.3 (25°C)	0.57	91
	Grafting with PMMA (50%)	46.0 (25°C)	0.70	91
	Grafting with PMMA (60%)	34.2 (25°C)	0.76	91
	Grafting with PMMA (100%)	46.0 (25°C)	—	92
	Grafting with PMMA (130%)	33.0 (25°C)	—	92
	a from liquid density	100 (50°C)	—	96
	a from Dorris correction	71.5 (50°C)	—	96
	a from BET fit	45.9 (50°C)	—	96
	Using of Eq. (30)	39.1 (50°C)	—	96
Kraft pulps (KP)	None	45.1 (20°C)	—	89
	None	6.8 (50°C)	0.07	97
	Beating for 3000 revolutions	45.1 (20°C)	—	89
	Beating for 6000 revolutions	46.0 (20°C)	—	89
	Beating for 9000 revolutions	46.7 (20°C)	—	89
	Beating for 11000 revolutions	46.7 (20°C)	—	89
Wood birch	None	43.8 (20°C)	—	93
Bleached sulfite	None	44.0 (40°C)	—	94
pulps (BSP)	None	45.1 (60°C)	—	95
	ASA	33.0 (40°C)	—	94
	Plasma, NH$_3$, 5 mn	43.2 (60°C)	—	95
	Plasma, NH$_3$, 15 mn	42.5 (60°C)	—	95
	Plasma, NH$_3$, 60 mn	40.7 (60°C)	—	95
	Plasma, N$_2$, 5 mn	43.7 (60°C)	—	95
	Plasma, N$_2$, 15 mn	43.1 (60°C)	—	95
	Plasma, N$_2$, 60 mn	41.0 (60°C)	—	95
	Plasma, MMA, 5 mn	40.1 (60°C)	—	95
	Plasma, MMA, 15 mn	36.9 (60°C)	—	95
	Plasma, MMA, 60 mn	38.2 (60°C)	—	95
Explosion pulps	None	9.8 (50°C)	0.25	97
Softwood KP	Bleaching with ClO$_2$	47.8 (50°C)	—	98
	Bleaching with H$_2$O$_2$	44.5 (50°C)	—	98
	Bleaching with O$_3$	43.8 (50°C)	—	98
Hardwood KP	Bleaching with O$_3$	47.6 (50°C)	—	98

TMP, thermomechanical pulps; CTMP, chemithermomechanical pulps; PFR, phenol-formaldehyde resin; PFR-A, phenol-formaldehyde resin A, low molecular weight PFR; PFR-B, phenol-formaldehyde resin B, high molecular weight PFR; ASA, alkenylsuccinic anhydride; PMMA, polymethyl methacrylate; MMA, methyl methacrylate.

[51] methods to calculate the surface properties of the CTMP fibers and found that there was little difference in the ensuing γ_S^D values (see Table 6). Kamdem et al. [92,93] determined the dispersive and acid–base properties of the surface of white birch wood meal. As shown in Table 6, these authors found results very similar to those obtained for CTMP materials.

Felix et al. investigated the dispersive and acid–base properties of the surface of bleached sulfite pulps obtained from 60% beechwood and 40% birchwood [94,95]. The lignocellulosic fibers were purified by Sohxlet extraction with toluene and treated with alkenylsuccinic anhydride (ASA, structure **II**) [94], or treated by plasma under different atmospheres and for different times [95]. They studied the divergence between the DCA and IGC measurements and showed that the latter technique gives higher values of the dispersive contribution to the surface energy [94]. They correlated this behavior to the specificity of the IGC data obtained at infinite dilution. In fact, at zero coverage conditions the detection of high-energy sites is more efficient. As expected, the treatment of these fibers by ASA was found to decrease the dispersive energy of the fibers as shown in Table 6. The different plasma treatments of bleached sulfite pulp fibers produced a very slight change in their γ_S^D as shown by the values reported in Table 6.

Jacob and Berg [96] characterized the surface energy of CTMPs and softwood bleached kraft pulp (SWKPs) fibers. They calculated the corresponding γ^P_S using different surface areas of the adsorbed molecules. They concluded that the most satisfactory values were obtained when the surface areas of the adsorbed molecules were determined from a fit of the BET model to adsorption of n-alkanes onto the surface of CTMP fibers. They also obtained good results when they used Dorris and Gray's approach with a value of the surface area of a methylene group of 6 Å^2.

Chtourou et al. [97] have studied kraft and steam explosion pulps and found extremely low values of the dispersive energy (about 10 and 7 mJ/m^2, respectively, at 50°C). These authors did not give an explanation for this surprising result. As reported in Table 6, the extrapolation of their data to 20°C gives γ_S^D of 20 and 10 mJ/m^2, respectively, which are clearly unrealistic.

Lunqvist et al. [98] determined γ_S^D of softwood and hardwood kraft pulps bleached with different bleaching agents (ClO$_2$, H$_2$O$_2$, O$_3$). The values obtained and reported in Table 6 showed very similar and quite high surface energies, which indicates that the use of less conventional bleaching agents is as efficient (at least in terms of the resulting surface energy) as working with chlorine (the commonly used bleaching agent that is now being proscribed). Jacob and Berg [99] also studied CTMP fibers and found results very similar to those reported in the literature [89,93–95] (see Table 6).

Finally, Simonsen et al. [100] studied the surface energy of wood flour from

Douglas fir (T14) and TMP fibers from western hemlock as such and after deposition of different amounts of polystyrene. They showed that the dispersive surface energy of T14 increased from about 35 to 60 mJ/m^2 with increased deposited polystyrene, whereas that of TMP decreased very slightly from about 34 to 30 mJ/m^2, as shown in Fig. 9. Since the surface energy of polystyrene at room temperature is known to be about 35 mJ/m^2, the growing values of γ_S^D with polystyrene-coated T14 do not seem coherent.

3. Carbon Fibers

Carbon fibers have found extensive use in the elaboration of composite materials, mostly based on epoxy resins. As for all composites, the final properties strongly depend on the surface properties of the constituents and their interfacial compatibility. To reach highly resistant materials, carbon fibers are usually subjected to chemical or physical treatment. IGC was found to be a good complementary

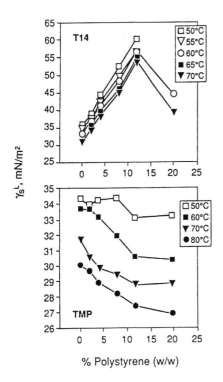

% Polystyrene (w/w)

FIG. 9 London component of the surface energy for Douglas fir and TMP fibers as a function of polystyrene loading as reported by Simonsen et al. [100].

technique to characterize their surface properties before and after such modifications.

Vukov and Gray [64,101,102] have studied the adsorption behavior of two commercial polyacrylonitrile-based carbon fibers. They examined two different untreated and unsized fibers as received and after cleaning by thermal treatment under nitrogen for 100–120 h. γ_S^D was found to increase substantially after cleaning with both type of fibers, as shown by the data reported in Table 7.

Schultz et al. [40–42,103] studied the surface properties of high-resistance polyacrylonitrile-derived carbon fibers as such (designed as untreated fibers), oxidized, oxidized and sized, electrolytically oxidized, coated, and subjected to enzymatic treatment. The authors investigated these fibers in order to optimize composite materials with epoxy resin as a matrix. The resin was based on diglycidyl ether of bisphenol A and two types of hardener: (1) 35 parts by weight diaminodiphenyl sulfone (resin I) and (2) 55 parts by weight polyamine amide (resin II). The resins were used as powders (200–300 μm) for IGC measurements. The dispersive contribution to the surface energy according to their own approach [Eq. (31)] was compared to that obtained with Gray's formalism [Eq. (30)]. The ensuing values of γ_S^D are summarized in Table 7. From these data, two main conclusions can be drawn: (1) the values of the dispersive properties according to Gray's and Schultz's approaches are very close, and (2) wetting measurements also gave very similar results compared to IGC. The untreated fibers had a dispersive energy of 50 mJ/m^2, whereas most treatments induced a decrease in γ_S^D, except the electrolytic oxidation which produced an increase in surface energy.

Bolvari and Ward [104,105] studied the effect of grafting of carbon fibers by silane-type coupling agents (**VI**). They investigated three commercial PAN-derived carbon fibers, namely) untreated fibers and two commercial varieties of

VI

surface-modified fibers. One of them, along with the untreated sample, was submitted to grafting with coupling agent **VI**. The dispersive component of the surface energy was calculated from measurements at zero coverage conditions according to Eq. (30) (Dorris and Gray's approach [57]). The values of this parameter, which are reported in Table 7, showed that the surface grafting was efficient because γ_S^D decreased from about 60 to 26 mJ/m^2.

Wilkinson and Ward [106] studied the surface properties of seven types of intermediate-modulus carbon fibers from two different suppliers as such and after a heat treatment, and found very similar results to those obtained by Vukov and

TABLE 7 Values of the Dispersive Component of Surface Energy of Carbon Fibers

Fiber type	Treatment or calculation method	γ_S^D at (T °C) (mJ/m²)	$-d\gamma_S^D/dT$ (mJ/m² °C)	Ref.
PAN-based	None	38.8 (50°C)	—	64
	None	36.2 (50°C)	—	102
	None	36.3 (50°C)	—	101
	None	50.0 (50°C)	—	40–42,103
	Heating (ht) (160°C)	76.8 (50°C)	—	64
	Cleaning	70.9 (50°C)	—	102
	Cleaning	74.9 (50°C)	—	101
	Ht (105°C)	52.0 (47°C)	—	109
	Ht (150°C)	72.0 (47°C)	—	109
	Ht (200°C)	75.0 (47°C)	—	109
	Oxidation (ox)	49.0 (25°C)	—	42
	Electrochemical ox + ht (105°C)	56.0 (47°C)	—	109
	Electrochemical ox + ht (150°C)	81.0 (47°C)	—	109
	Ox + sizing (si)	36.0 (25°C)	—	41,42,103
	Coating (co)	29.0 (25°C)	—	42
	Ox + co + ht (105°C)	36.0 (47°C)	—	109
	Ox + co + ht (150°C)	32.0 (47°C)	—	109
	Electrolytic ox (elox)	59.0 (25°C)	—	42
	Enzymatic	36.0 (25°C)	—	40
	None	54.7–65.1	—	104,105
	Not reported	39.3–47.5	—	104,105
	Grafting with silane agent VI	23.3–26.3	—	104,105
	Heating (sample A)	85.0 (40°C)	—	106
	A + 50% surface treatment (ST)	45.0 (40°C)	—	106
	A + 50% ST + heating	56.0 (40°C)	—	106

TABLE 7 Continued

Fiber type	Treatment or calculation method	γ_S^D at (T °C) (mJ/m²)		$-d\gamma_S^D/dT$ (mJ/m² °C)	Ref.
	A + 100% ST	57.0	(40°C)	—	106
	A + 100% ST + heating	78.0	(40°C)	—	106
	A + 50% ST	54.0	(40°C)	—	106
	50% ST + heating	65.0	(40°C)	—	106
	None (sample B)	78.0	(40°C)	—	106
	B + 10% ST	43.0	(40°C)	—	106
	B + 50% ST	39.0	(40°C)	—	106
	B + 100 ST	36.0	(40°C)	—	106
	Graphitization	100	(29°C)	—	107
	Carbonization	46.5	(29°C)	—	108
	Stabilization	44.8	(29°C)	—	108
	Ht 1280 (B)	67	(25°C)	—	112
	B + ox	68	(25°C)	—	112
	B + ox with HNO₃	56	(25°C)	—	112
	B + ht (2800°C)	91	(25°C)	—	112
	G-210, North American, none	540	(200°C)	—	134
	AC-1000, Ashland Co., none (C)	455	(200°C)	—	134
	C + ox with 15 M HNO₃ solution	475	(200°C)	—	134

Pitch-based	Heating at 160°C	54.4 (50°C)	—	64
	None	37.7 (50°C)	—	101
	Cleaning	50.8 (50°C)	—	101
	None	37.0 (50°C)	—	102
	Cleaning	50.1 (50°C)	—	102
	Heating at (1400–1500°C)	136 (20°C)	0.37	110,111
	Heating at (1650–1750°C)	127 (20°C)	0.21	110,111
	Heating at (1900–2000°C)	122 (20°C)	0.20	110,111
	Heating at (2200–2350°C)	101 (20°C)	0.19	110,111
	Heating at (2500–2700°C) (D)	100 (20°C)	0.15	110,111
	D + anodic (Io)	117 (20°C)	0.29	110,111
	D + anodic (3*Io)	114 (20°C)	0.13	110,111
	D + anodic (10*Io)	110 (20°C)	0.15	110,111
	D + anodic (30*Io)	109 (20°C)	0.27	110,111
	Ht (2600°C) (E)	87 (25°C)	—	112
	E + elox in HNO$_3$	83 (25°C)	—	112
	Ht (900°C)	281 (25°C)	—	114
Experimental	Ht (105°C)	53.0 (47°C)	—	109
	Ht (150°C)	71.0 (47°C)	—	109
	Ht (200°C)	90.0 (47°C)	—	109
Not reported	None	84 (40°C)	—	113
	Oxidation	66 (40°C)	—	113
	Sizing	34 (40°C)	—	113

PAN, polyacrylonitrile.

Gray [64,101,102]. Dong et al. [107] determined the dispersive energy of graphitized carbon fibers and found a value close to 100 mJ/m². Nardin et al. [108] studied carbonized and stabilized carbon fibers and found a value of about 45 mJ/m⁻², which is close to that obtained by Lavielle and Schultz [109] for PAN-based fibers before thermal treatment (see Table 7). These authors [109] showed that the dispersive component of the surface energy of thermally treated fibers increased with the treatment temperature. Donnet and coworkers [110,111] also studied pitch-base carbon fibers before and after thermal and anodic oxidation treatments and found exactly the inverse trend than that of Lavielle and Schultz [109]. In fact, the dispersive contribution to the surface energy of the fibers decreased after heating. They also found that the increase of the current intensity of anodic treatment induced a corresponding decrease in the dispersion energy of the surface. More recently, Tsutsumi and Ban [112] studied the same material with the same type of treatment and found results very similar to those reported by several laboratories and in opposition to those published by Donnet et al. [110,111]. These values are also summarized in Table 7.

Very recently, Jacobasch et al. [113] and Park and Brendlé [114] also studied the surface energetics of carbon fibers. In the first study it was shown that the dispersive energy of fibers before and after sizing were in very good agreement with those obtained by almost all previous reports except again for that of Donnet and coworkers [110,111]. In the second investigation [114], the authors showed that the thermal treatment of carbon fibers can induce an increase in their dispersive properties up to a value of 280 mJ/m². These data are resumed in Table 7, which gives all of the values found in the literature concerning the surface energy of carbon fibers as determined by IGC at infinite dilution.

4. Glass Fibers

Glass fibers are also used as a reinforcing element in different polymeric matrices. The most common matrix in this context is made with unsaturated glycerophthalic polyesters that are crosslinked by their radical copolymerization with styrene. Again, the mechanical properties of the composite will depend on the properties of the glass fibers and polyester matrices and the quality of the interface that they can establish. These fibers are usually submitted to a specific surface treatment before they are incorporated into the matrix. By far the most common surface modification involves the use of silanes as OH coupling agents as shown below [115].

Papirer et al. [116–118] studied the surface characteristics of glass fibers before different treatments, namely, (1) heating, (2) washing with water or aqueous acidic (HCl, AlCl₃), basic (NaOH), and EDTA solutions, and (3) grafting with styrene-butadiene (SB), γ-aminopropyltriethoxysilane (**VII**) and isopropyltris (tridecylbenzenesulfonyl) (**VIIIa**) and isopropyltris(dodecylbenzenesulfonyl) titanate (**VIIIb**) (see Table 8).

TABLE 8 Values of the Dispersive Component of the Surface Energy of Glass
Fibers

Fiber type	Supplier	Treatment or calculation method	γ_S^D at (T °C) (mJ/m^2)		Ref.
Short	Rockwool	None	50	(60°C)	116,117,118
		Washing (ws) with water	51	(100°C)	116
		Ws with 0.01 N HCl for 1 h	36	(100°C)	116
		Ws with 0.1 N NaOH for 1 h	40	(100°C)	116
		Ws with 100 ppm AlCl$_3$, 6 h	30	(100°C)	116
		Ws with EDTA	45	(100°C)	116
		Heating (ht) at 54°C	38	(100°C)	116
		Ht at 59°C	30	(100°C)	116
		Ht at 68°C	18	(100°C)	116
		Ht at 71°C	17	(100°C)	116
		Ht at 85°C	42	(100°C)	116
		Ht at 97°C	29	(100°C)	116
		Ht at 500°C	21	(100°C)	116
		Grafting by SBR in cyclo-hexane	42.9	(40°C)	116
E glass	Ashai	None	49	(25°C)	119
		Grafting by VI	48	(25°C)	119
		Grafting by VII	40	(25°C)	119
		Grafting by IX	38	(25°C)	119
		Grafting by X	39	(25°C)	119
		Grafting by XI	69	(25°C)	119
		Grafting by XII	24	(25°C)	119

SBR, styrene-butadiene resin; EDTA, ethylenediaminetetraacetic acid.

VII

VIIIa: R = $C_{12}H_{25}$
VIIIb: R = $C_{13}H_{27}$

The authors showed that the dispersive energy of the fibers' surface increased
when they washed it with water and HCl solution; decreased by washing with
an EDTA solution, grafting with SB, and heating at 500°C for 24 h; but remained
unchanged when washed with NaOH and AlCl$_3$ or grafted by **VII** and **VIII**. The
variation of γ_S^D with temperature [116] is shown in Fig. 10.

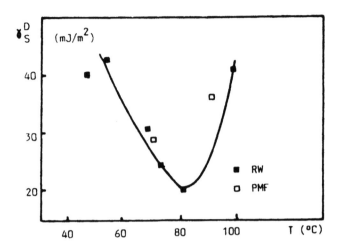

FIG. 10   Effect of the temperature on the dispersive contribution to the surface energy
of glass fibers. (From Ref. 116.)

Tsutsumi and Oshuga [119] also studied glass fibers before and after grafting
with six different silane coupling agents (**VI, VII**, and **IX–XII**).

In this work, the London component of the surface energy decreased from 40
mJ/m^2 to about 30 mJ/m^2 for grafted materials. Two exceptions were noted,
however, as expected: grafting with coupling agents **XI** and **XII** gave, respectively, 55 and 19 mJ/m^2.

5. Polyamide Fibers

Thanks to its very high modulus and low density, aramid fibers have found a
large success in the elaboration of composite materials for aeronautics. The most

abundant aramid fibers are made with poly (p-phenylene terephthalamide) (PPTA, Kevlar). The surface energetics of this material have been studied essentially by Williams and coworkers [120–124] who found that the dispersive contribution to the surface energy of pristine uncleaned fibers was about 35 mJ/m^2 while cleaning by solvent extraction (acetone, water, haxane, etc.) raised the energy to about 55 mJ/m^2.

Very recently, Tate et al. [125] studied nylon-66 fibers spun with different draw ratios in dry and humid atmospheres and found that these fibers had an average value of the dispersive component of their surface energy of about 40 mJ/m^2 (see Table 9).

6. Polyethylene Terephthalate Fibers

To the best of our knowledge, only Anhang and Gray [126,127] and Yang and Ward [128] have studied the surface properties of polyethylene terephthalate (PET) fibers by IGC at zero coverage conditions. These papers reported that the dispersive energy of PET fibers is very close to that found for nylon-66, i.e., about 40 mJ/m^2 as shown in Table 9.

7. Polyethylene Fibers

The surface properties of polyethylene fibers have been studied by Chtourou et al. [97,129]. They examined commercial low-density PE fibers as received and after modification by ozone in aqueous solution for 2 and 3 h and by fluorine gas. Surprisingly, the dispersive component of the surface energy of these materials remained constant even after the fluorine modification, as shown in Table 9. The authors emphasized that the most significant change in the surface properties should be reflected on the acid–base interactions, as discussed below.

8. Cellulose Acetate Fibers

Cellulose acetate fibers and fibrils have been investigated by Yang and Ward [128] who found a γ_S^D values of 30 and 20 mJ/m^2, respectively (Table 9). The latter value seems unusually low for that polymer structure.

9. Cellophane Film

To the best of our knowledge, only Katz and Gray [130] have studied the dispersive energy of cellophane by IGC at zero coverage conditions. They found a value of 42 mJ/m^2 at 20°C (Table 9), which is very reasonable.

B. Characterization of Organic Fillers and Powders

1. Activated Carbons

Carbon black is constituted by very porous solid particles (varying in size from 5 to 50 nm) that possess very interesting adsorption properties [131] because they contain a wide range of pores of different sizes and shapes that give rise to specific characteristics such as molecular sieve behavior. Additionally, these

TABLE 9 Values of the Dispersive Component of the Surface Energy of Polyamides, Polyethylene Terephthalate and Poly(ethylene) Fibers

Fiber type	Description or supplier	Treatment or calculation method	γ_S^D at (T °C) (mJ/m²)	$-d\gamma_S^D/dT$ (mJ/m² °C)	Ref.
Kevlar 49	Roving type 968	Drying	33.3 (30°C)	—	120
	Du Pont de Nemours	None	32.0 (20°C)	—	121,124
	Roving type 968-7100 denier	Drying	32.0 (20°C)	—	122
	Roving type 968	Solvent extraction	54.1 (30°C)	—	120
	Du Pont de Nemours	Solvent extraction	42.0 (20°C)	—	121
	Roving type 968-7100 denier	Extraction with hexane + drying	43.4 (20°C)	—	122
	Roving type 968-7100 denier	Extraction with water + drying	52.5 (20°C)	—	122,124
	Roving type 968-7100 denier	Extraction with acetone + drying	65.1 (20°C)	—	122,124
	Fabric type 353-CS800	Drying	36.3 (30°C)	—	120
	Type 353 cloth, CS800, 1150 denier	Drying	37.5 (20°C)	—	122
	Fabric type 755-CS816	Drying	35.2 (30°C)	—	120
	Fabric type 755-CS800	Drying	37.7 (30°C)	—	120
	Fabric type 755-CS800	Solvent extraction	55.2 (30°C)	—	120

Material	Style/Company	Treatment	Value (°C)		Ref.
	Style 710, CS800	None	37.0 (20°C)	—	123
	Style 755, CS800	None	37.2 (20°C)	—	123
	Style 755	Solvent extraction	35.7 (20°C)	—	123
	Style 755, CS815	None	34.1 (20°C)	—	123
	Style 755, CS816	None	37.4 (20°C)	—	123
	Style 755, CS800SS	None	57.7 (20°C)	—	123
Nylon-66	Monsanto Company	Drawn (1 time) at 0% RH	38.3 (30°C)	—	125
		Drawn (1 time) at 60% RH	42.2 (30°C)	—	125
		Drawn (4 times) at 0% RH	34.2 (30°C)	—	125
		Drawn (4 times) at 60% RH	37.8 (30°C)	—	125
PET	Du Pont Mylar	wa. (MeOH-Water), ext. acetone	42.6 (15°C)	0.18	126,127
	Hoechst Celanese Co.	None	41.0 (0°C)	0.25	128
PE	Du Pont	None	32.7 (50°C)	0.34	97,129
		Aqueous ozonation (2 h)	37.1 (50°C)	0.48	97
		Aqueous ozonation (3 h)	28.9 (50°C)	0.27	97,129
		Fluorination #1	32.8 (50°C)	0.31	97
		Fluorination #2	31.3 (50°C)	0.33	97,129
Cellulose acetate	Hoechst Celanese Co.	In the form of fibers	30 (0°C)	—	128
		In the form of fibers	20 (0°C)	—	128
Cellophane	British Cellophane Ltd.	Film cut in the form of confetti	42 (20°C)	—	130

PE, polyethylene; PET, polyethylene terephthalate; wa., washing.

particles bear different polar groups at their surface (OH, $C=O$, COOH, SH, NH_2, etc.), depending on the way in which they were prepared and/or treated after isolation. They find a large spectrum of applications as fillers, radical traps (e.g., in tires formulations), reinforcing materials, printing ink pigments, etc. These applications require a good knowledge of the adsorption features and surface energy of carbon black because in all cases its particles interact with the surrounding matrix; thus, the quality of the ensuing dispersion and/or interface will be conditioned by the compatibility of the two phases.

The most important investigations in the surface characterization of graphitic fillers were carried out by a reseach team led by Donnet, Vidal, and Jagiello [70, 114,132–146]. These authors have published a series of papers in which they investigated the surface energy of carbon black from different sources, before and after different types of modification, namely, (1) heating, (2) microwave plasma treatment, (3) nitric acid oxidation, (4) solvent extraction, (5) mechanical compression, and (6) grafting with bis(3-triethoxysilylpropyl)tetrasulfane (TSPTS).

The values of the dispersive surface energy of these pristine and treated materials are summarized in Table 10. These data suggest the following general remarks:

1. The dispersive surface properties of initial carbonaceous materials can vary from about 100 to 500 mJ/m^2, depending on their source [114,134,137,146].
2. Thermal treatment induces a substantial increase in the dispersive surface energy of carbon black [132,133,137,140]. This is roughly linear up to 500°C, after which it tends to be less pronounced [140].
3. Microwave plasma treatment seems to be effective only when it is preceded by a thermal treatment or when it is applied under oxidative atmosphere (air) [114,135,141] because of residual adsorbed hydrocarbons on the surface of the carbon particles.
4. Nitric acid oxidation produces a slight decrease (about 10–15%) in the dispersive energy of the surface of carbon black [134,135,144].
5. Acetone, toluene, 1,2-dichlorobenzene, and acetophenone were found to be the most efficient solvents for the surface purification of carbon particles [70], whereas water and n-alcanes were ineffective [70].
6. The mechanical compression of carbon black, previously extracted with toluene, induces an increase in its dispersive energy from 140 to 475 mJ/m^2 [142]. The compression of the pristine material appeared to be ineffective since the dispersive component gained only about 30% after being compressed under maximum load [142].
7. Grafting with silane coupling agent TSPTS induced a decrease in the dispersive energy from 140 to about 120 mJ/m^2 [143].

2. Lignin Powders

Recently, Belgacem et al. [30] studied the dispersive properties of different lignins obtained by conventional processes, such as kraft or sulfite pulping and new delignification refineries like steam explosion and organosolv cooking. They

TABLE 10 Values of the Dispersive Component of Surface Energy of Carbon Black

Filler type	Description or supplier	Treatment	γ_S^D at (T °C) (mJ/m²)	$-d\gamma_S^D/dT$ (mJ/m² °C)	Ref.
Graphite	Natural	None	134–138 (25°C)	0.28	144,132,136
		Heating at 800°C	151–155 (25°C)	0.30–0.40	114,132
		Heating (Ht)	165 (0°C)	0.41	132,136
		Plasma (n-BuOH)	138 (25°C)	0.51	136
		Plasma (ammonia)	138 (25°C)	0.44	136
		Plasma (n-BuNH₂)	132 (25°C)	0.23	136
		Ht + plasma (n-BuOH)	176 (25°C)	0.28	136
		Ht + plasma (ammonia)	241 (25°C)	1.74	136
		Ht + plasma (n-BuNH₂)	131 (25°C)	0.42	136
	Lonza Company	None	135 (0°C)	0.29	114
	Silicon Carbide	None	99 (0°C)	0.53	114
Carbon black	Sheron 6, Cabot Co.	None	340 (170°C)	—	133
		Ht at 800°C	600 (170°C)	—	133
	N326, Degussa Co.	None	380 (170°C)	—	133
	N347, Cabot Co.	None	300 (160°C)	—	133
		Ht at 800°C	410 (170°C)	—	133
		Ht at 800°C	550 (180°C)	—	133
	Westvaco	None	110 (30°C)	0.09	134
	Westvaco	None	313 (200°C)	—	135
		Ox. with 15 M HNO₃ solution (25°C, 2.5 h)	292 (250°C)	—	135
	Norit	None	575 (250°C)	—	135
		Ox with 15 M HNO₃ solution (25°C, 2.5 h)	475 (250°C)	—	135
	Calgon	None	580 (250°C)	—	135
		Ox with 15 M HNO₃ solution (25°C, 2.5 h)	549 (250°C)	—	135
	North American	None	513 (250°C)	—	135
		Ox with 15 M HNO₃ solution (25°C, 2.5 h)	474 (250°C)	—	135

TABLE 10 Continued

Filler type	Description or supplier	Treatment	γ_S^D at (T °C) (mJ/m²)	$-d\gamma_S^D/dT$ (mJ/m² °C)	Ref.
	N 110, Degussa	None	404 (200°C)	—	137
	N 220, Degussa	None, column 1	447 (200°C)	—	137
	N 220, Degussa	None, column 2	505 (200°C)	—	137
	N 330, Degussa	None	241 (200°C)	—	137
	N 330 A, Degussa	None	300 (200°C)	—	137
	N 330 B, Degussa	None	241 (200°C)	—	137
	N 330 C, Degussa	None	213 (200°C)	—	137
	N 660, Degussa	None, column 1	115 (200°C)	—	137
	N 660, Degussa	None, column 2	118 (200°C)	—	137
	N 774, Degussa	None	95 (200°C)	—	137
	N 772, Degussa	None	140 (80°C)	—	70,142
		None	147 (80°C)	0.6	143
		Extraction with n-alkanes	143 (80°C)	—	70
		Extraction with ether	130 (80°C)	—	70
		Extraction with chloroform	146 (80°C)	—	70
		Extraction with acetone	162 (80°C)	—	70
		Extraction with toluene (ExTl)	165 (80°C)	—	70
		Extraction with 1,2-dichlorbenzene	166 (80°C)	—	70
		Extraction with ethylbenzene	160 (80°C)	—	70
		Extraction with acetophenone	173 (80°C)	—	70
		Ht at 180°C	150 (120°C)	—	140
		Ht at 300°C	210 (120°C)	—	140
		Ht at 500°C	305 (120°C)	—	140

Active	Treatment				
N550G, Degussa	Ht at 700°C	420	(120°C)	—	140
	Ht at 900°C	460	(120°C)	—	140
	Pressing at 600 MPa (Pr600M)	185	(80°C)	—	142
	ExTl + Pr600M	475	(80°C)	—	142
	ExTl + pressing at 100 MPa	225	(80°C)	—	142
	ExTl + pressing at 200 MPa	375	(80°C)	—	142
	ExTl + pressing at 400 MPa	425	(80°C)	—	142
	Grafting with TSPTS	121	(80°C)	0.55	143
	None	136	(200°C)	—	137
	None	150	(100°C)	—	141
	Ht at 900°C for 4 h (ht9-4)	250	(100°C)	—	137
	Plasma (hydrogen)	152	(100°C)	—	141
	Plasma (air)	305	(100°C)	—	141
	Plasma (ammonia)	112	(100°C)	—	141
	Plasma (argon)	152	(100°C)	—	141
	Ht9-4 + plasma (hydrogen)	275	(100°C)	—	141
	Ht9-4 + plasma (air)	403	(100°C)	—	141
	Ht9-4 + plasma (ammonia)	455	(100°C)	—	141
	Ht9-4 + plasma (argon)	440	(100°C)	—	141
	none	47.9	(90°C)	—	144
	Ox with 25% HNO_3 solution (3 h)	48.9	(90°C)	—	144
	Ox with 25% HNO_3 solution (3 h)	49.4	(90°C)	—	144
Carbonization of furfuryl alcohol	Formulation 1	560.5	(25°C)	1.83	146
	Formulation 2	448.7	(25°C)	2.03	146
	Formulation 3	405.4	(25°C)	1.21	146
	Formulation 4	419.5	(25°C)	1.72	146

TSPTS, bis(3-triethoxysilylpropyl)tetrasulfane; Ox, oxidation.

TABLE 11   Values of the Dispersive Component of the Surface Energy of Lignins,
Starch, Chitin, Chitosans, and Cork

Material	Description and/or supplier	γ_S^D at (T °C) (mJ/m^2)	$-d\gamma_S^D/dT$ (mJ/m^2 °C)	Ref.
Lignins	Organosolv, Repap Technologies	44.7 (50°C)	—	30
	NaOH-anthraquinone	115.6 (50°C)	—	30
	Kraft	46.6 (50°C)	—	30
	Steam explosion	49.0 (50°C)	—	30
	Lingnosulfonate	66.7 (50°C)	—	30
Starch	Prolabo	27.4 (50°C)	—	30
Chitin	Aber Technologies	38.3 (50°C)	—	30
Chitosans	(50% NH$_2$), Aber Technologies	45.2 (50°C)	—	30
	(80% NH$_2$), Aber Technologies	55.7 (50°C)	—	30
Cork	*Ouercus suber* Champcork Co.	41.1 (25°C)	0.22	29

showed that except for the soda–anthraquinone lignin and lignosulfonates, the other lignins investigated—namely (1) kraft, (2) organosolv, and (3) steam explosion—showed γ_S^D values of 45–50 mJ/m^2. Soda antraquinone lignins appeared to have a very high surface dispersive energy (116 mJ/m^2) because of phenolate groups (Ph-O$^-$) arising from the residual basic medium. Lignosulfonates gave an intermediate dispersive component (67 mJ/m^2) because of the presence of some sulfonate moieties. The values of γ_S^D are presented in Table 11.

3.   Starch, Chitin, and Chitosan Powders

Belgacem et al. [30] investigated the surface dispersive properties of starch, chitin, and chitosans with different degrees of residual acetyl groups. They showed that starch has a low γ_S^D in comparison with that of cellulose (see Table 11). This difference was surprising considering that these two natural polysaccharides have very similar chemical structures. The main difference between these two polymers resides in the fact that cellulose has a linear semicrystalline structure, whereas starch is a highly branched and totally amorphous macromolecule. Indeed, the crystallinity was found to be at the origin of the different surface energy since the authors found a good agreement of their results with the empirical Eq. (59), which correlates the surface energy with bulk and surface crystallinity.

$$\log \gamma_{sc} - \log \gamma_a = \frac{1}{3} F_{c,s}^* x_{s,\,max} \tag{59}$$

where γ_{cs} and γ_a are the dispersive components of surface energy of the semicrystalline and amorphous polymers, respectively; $F_{c,s}$ the surface crystallinity of the semicrystalline polymer (0–1) fraction; and $x_{s,\,max}$ the maximum bulk polymer crystallinity.

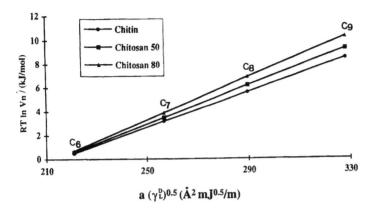

$$a \ (\gamma_L^D)^{0.5} \ (\mathring{A}^2 \, mJ^{0.5}/m)$$

FIG. 11   RT ln V_n versus $a(\gamma^D{}_S)_{1/2}$ plots for chitin and chitosans at 50°C. (From Ref. 30.)

The same authors [30] reported results related to the surface energy of chitin and two chitosans (see Table 11). As expected, they found that chitin showed a slightly lower value of γ_S^D (38.3 mJ/m²) compared with that of cellulose (**XIII**). In fact, chitin (**XIV**) and cellulose have again very similar structures except for the fact that one of the secondary hydroxy groups of cellulose has been transformed into a secondary amide function in chitin. The lower intermolecular cohesive energy of the amide group (lack of mobile hydrogens) with respect to that of the hydroxy function inevitably induces a decrease in the value of γ_S^D. The surface energy of chitosans (**XV**), which are a product of the partial deacetylation of chitin, increased by increasing the degree of deacetylation to reach 55.7 mJ/ m² for chitosans with only 20% of residual amide groups ($x = 0.8$), i.e., a value higher than that of cellulose. Figure 11 shows the results of these determinations. This results from the fact that the primary amino groups give rise to a higher cohesive energy than that induced by OH functions. In other words, the intermolecular hydrogen bonding in cellulose is somewhat less energetic than that occurring in highly deacetylated chitosans.

XIII

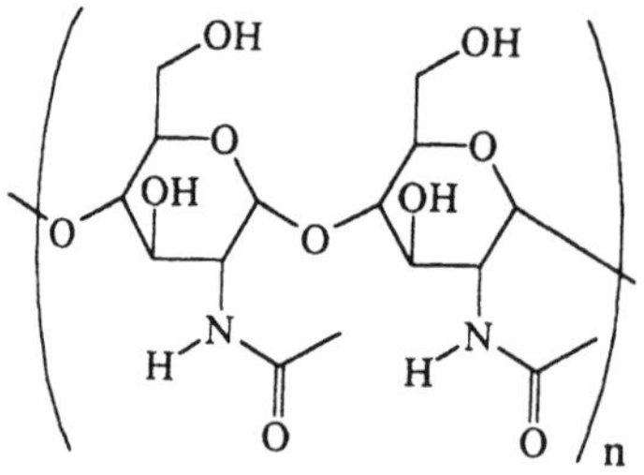

XIV

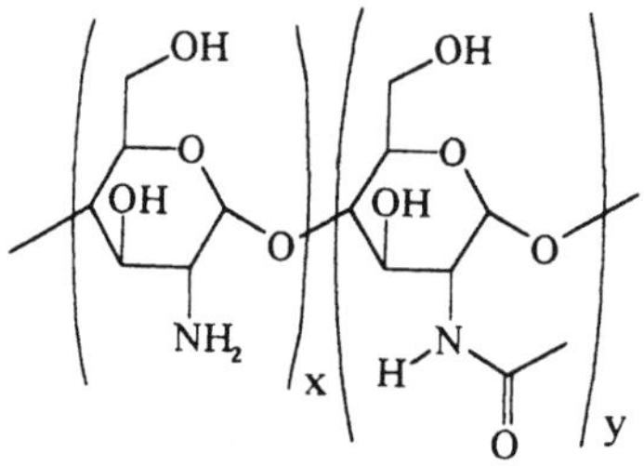

x = 0.8, y = 0.2

XV

4. Cork Powders

Cordeiro et al. [29] studied the dispersive properties of cork from *Quercus suber L.* The values of γ_S^D at different temperature are reported in Table 11. As seen from these data, the dispersive contribution to the surface energy of cork extrapolated to 25°C is about 40 mJ/m².

5. Synthetic Polymers

In the literature, the majority of studies dealing with the surface characterization of polymers has been carried out by adsorbing them onto commercial chromosorbs. This type of column falls outside the present scope and these investigations will not be discussed here. Nevertheless, IGC can be a very suitable technique to investigate polymers that do not display film-forming properties because of their insoluble and infusible character, as well as those that can be degraded when submitted to film processing (e.g., thermal degradation under extrusion).

The most thoroughly studied powderous polymers are polyconjugated structures reported by Chehimi and coworkers [147–151]. Only a recent publication by Voelkel et al. [23] has been devoted to more common macromolecular materials such as polymethyl methacrylate (PMMA). The dispersive component of these polymers as such and after different modifications (doping with chloride salts, p-toluenesulfonate etc., annealing under helium or air) were determined by IGC at zero coverage conditions, as summarized in Table 12. From these data the following remarks can be given:

1. Polypyrrole (PPy) doped by an aqueous solution of $FeCl_3$ (PPyCl) showed a decrease in its dispersive properties from 55 to 35 mJ/m^2 [147,148]. The charge generation associated with doping casts serious doubts about the reliability of this result. Indeed, 3 years later, the same group of researchers communicated that the dispersive contribution to the surface energy of PPyCl is about 140 mJ/m^2 [150,151] without any comment concerning the enormous discrepancy with their previous value of 35 mJ/m^2!

2. The doping of PPy with sodium p-toluenesulfonate was more efficient because it induced an increase in the dispersive component up to about 90 mJ/m^2 [149].

3. The coating of PPy with PMMA decreased its dispersive component of the surface energy. At a high ratio of the adsorbed layer of PMMA, the dispersive component tended to reach that of pure PMMA, i.e., 39 mJ/m^2 [23,150,151].

4. The annealing under helium or air of six different acrylic polymers and one copolymer (see Table 12) produced very little difference in their surface energetics [23].

C. Characterization of Inorganic Fillers and Powders

1. Silicas and Other Oxides

Oxide surfaces have been used for a long time as adsorbents in chromatographic processes, catalysts or catalyst supports, fillers for polymeric matrices, etc. At first, they were used as such and later they were subjected progressively to more and more specific treatments in order to either (1) improve their surface compatibility with the matrix and thus enhance the adhesion strength at the interface or (2) disperse homogeneously the particles in such applications as paints or coating compositions. In all cases, their surface properties must be assessed in order to predict the mechanical, optical, and other properties of the materials in which they are to be added [152–154]. The benefits of using IGC in the surface characterization of pigments have been recently reviewed [155–158].

The most significant work in the area of the characterization of the energetics of oxide surfaces has been carried out by Papirer and coworkers who investigated silicas, clays, kaolinites, illites, mica, talc, etc. [26,157–177]. In addition, other studies on the same topic have appeared in the literature [178–193]. Since Dr. Papirer also contributes to this book, we will give here only a very brief outline of the application of IGC to these materials. The interested reader will find more details in the appropriate chapter. The surface properties of different untreated silicas from various sources have been examined before subjecting them to sev-

TABLE 12 Values of the Dispersive Component of the Surface Energy of Some Synthetic Polymers

Polymer	Treatment	γ_S^D at (T °C) (mJ/m^2)	$-d\gamma_S^D/dT$ (mJ/m^2 °C)	Ref.
Polypyrrole	None	58 (48°C)	0.01	147
	Doping by FeCl$_3$	37 (48°C)	0.15	147
	Doping by FeCl$_3$	144 (47.6°C)	—	150,151
	Doping by p-toluenesulfonate	88.5 (24.6°C)	0.16	149
	Coating with a solution of PMMA in:			
	Tetrahydrofurane	69.4 (47.6°C)	—	150,151
	1,4-Dioxane	39.2 (47.6°C)	—	150,151
	Toluene	45.4 (47.6°C)	—	150,151
	Tetrachloromethane	32.7 (47.6°C)	—	150,151
	1,2-Dichloroethane	54.1 (47.6°C)	—	150,151
	Methylene chloride	70.4 (47.6°C)	—	150,151
	Chloroform	55.0 (47.6°C)	—	150,151
PBDM	None	27.0 (50°C)	0.03	23
	Annealing under He at 80°C	27.6 (50°C)	0.07	23
	Annealing under He at 160°C	28.0 (50°C)	0.04	23
	Annealing under air at 80°C	26.2 (50°C)	0.03	23
	Annealing under air at 160°C	28.4 (50°C)	0.03	23
PPDM	None	30.1 (50°C)	0.06	23
	Annealing under He at 80°C	30.3 (50°C)	0.02	23
	Annealing under He at 160°C	31.5 (50°C)	0.07	23
	Annealing under air at 80°C	28.4 (50°C)	0.015	23
	Annealing under air at 160°C	32.9 (50°C)	0.095	23

POEDM	None	26.5 (50°C)	0.075	23
	Annealing under He at 80°C	25.7 (50°C)	0.085	23
	Annealing under He at 160°C	24.8 (50°C)	0.055	23
	Annealing under air at 80°C	25.6 (50°C)	0.085	23
	Annealing under air at 160°C	25.0 (50°C)	0.05	23
PTEDM	None	25.1 (50°C)	0.05	23
	Annealing under He at 80°C	26.9 (50°C)	0.08	23
	Annealing under He at 160°C	31.7 (50°C)	0.07	23
	Annealing under air at 80°C	26.8 (50°C)	0.07	23
	Annealing under air at 160°C	31.1 (50°C)	0.07	23
PNDM	None	39.8 (50°C)	0.065	23
	Annealing under He at 80°C	40.1 (50°C)	0.06	23
	Annealing under He at 160°C	41.4 (50°C)	0.08	23
	Annealing under air at 80°C	40.4 (50°C)	0.02	23
	Annealing under air at 160°C	44.7 (50°C)	0.155	23
P(BDM-co-MA)	None	24.8 (50°C)	0.025	23
	Annealing under He at 80°C	26.5 (50°C)	0.05	23
	Annealing under He at 160°C	27.4 (50°C)	0.05	23
	Annealing under air at 80°C	25.8 (50°C)	0.045	23
	Annealing under air at 160°C	26.3 (50°C)	0.04	23
PMMA	None	36.1 (50°C)	0.03	23
	Annealing under He at 80°C	37.2 (50°C)	0.045	23
	Annealing under He at 160°C	37.6 (50°C)	0.065	23
	Annealing under air at 80°C	37.0 (50°C)	0.02	23
	Annealing under air at 160°C	37.4 (50°C)	0.025	23

PBDM, poly(butane-1,4-diol dimethacrylate); PPDM, poly(pentane-1,5-diol dimethacrylate); POEDM, poly(2,2′-oxybisethanol dimethacrylate); PTEDM, poly(2,2′-thiobisethanol dimethacrylate); PNDM, poly(N-methyldiethanolamine dimethacrylate); P(BDM-co-MA), BDM-co-maleic anhydride copolymer; PMMA, poly(methyl methacrylate).

eral modifications, namely, (1) heat treatment, (2) grinding, and (3) grafting with polyethylene glycols of different lengths or with perfluorosilane agents.

From these results (see Table 13), the following general remarks can be drawn:

1. The London component of the surface energy of illites from different sources varies from about 140 to 185 mJ/m^2 [26].
2. The dispersive surface energy of various kaolinites was found to fall in the range of 155–240 mJ/m^2 [26].
3. The surface energy of titanium oxide varied from 25 to 55 mJ/m^2 [30,65,183,188,192,193]. The plasma treatment induced a decrease of the dispersive surface energy (about 35 mJ/m^2) when the treatment was carried out under CH$_4$ or C$_2$F$_4$ but not when Ar or ammonia was used [192,193].
4. The dispersive contribution to the surface energy of aluminas from various sources varied from about 40 to 165 mJ/m^2 [67,156,169,187].
5. The dispersive component of the surface energy of untreated mica (muscovite) is about 30 mJ/m^2 [159,177]. This parameter increased up to about 100 mJ/m^2 when the material was ground for 75 h in different solvents and solutions, as shown in Fig. 12 [159,177].
6. Untreated amorphous silica, from different sources, has a dispersive energy between 30 and 100 mJ/m^2 [160,161,165,168,175,179,180,184–186, 189,191].
7. Crystalline silicas gave a γ_S^D between 175 and 220 mJ/m^2 [165,175].
8. The dispersive contribution to the surface energy of γ-alumina was about 92 mJ/m^2 [167].
9. Magnesium and zinc oxides were found to have a dispersive energy of about 97 and 85 mJ/m^2, respectively [170].
10. Grafting of silicas with perfluorinated agents reduced considerably their dispersive properties, all the way down to 21 mJ/m^2 [171].
11. Talc has a dispersive component of surface energy of between 150 and 200 mJ/m^2 [172,174,176]. Its grafting with methanol increased γ_S^D slightly, whereas its coupling with long-chain alcohols (pentanol and dodecanol), as well as its surface treatment with sodium carbonate and O$_2$/SiH$_4$, decreased this parameter down to about 45 mJ/m^2 [172,179]. The thermal treatment of talc decreased its γ_S^D to 74 mJ/m^2.
12. The dispersive contribution to the surface energy of α-alumina was found to be about 45 mJ/m^2. The grinding of this material increased it up to about 160 mJ/m^2 [173].
13. The surface energy of ceramic superconductor powder of YBa$_2$Cu$_3$O$_x$ was found to be extremely low compared with the values commonly reported for oxides. These values were in the range 14–24 mJ/m^2 [178].
14. The dispersive surface energy of natural clays (smectites from bentonites) was in the range of 160 mJ/m^2 [181,182]. The adsorbtion of furfuryl alcohol

TABLE 13 Values of γ_S^D of Different Silicas and Other Oxides

Material	Supplier	Treatment	γ_S^D at (T °C) (mJ/m²)	$-d\gamma_S^D/dT$ (mJ/m² °C)	Ref.
Micas	Bihar, India	None	30 (25°C)	—	159,177
		None	49.2 (100°C)	—	156
Silicas	Aerosil A300, Degussa (Germany)	None	76 (60°C)	0.30	161
		Grafting by 5% of PEG-2000	43 (60°C)	0.20	161
		Grafting by 7% of PEG-4000	36 (60°C)	0.08	161
		Grafting by 17% of PEG-2000	38 (60°C)	0.15	161
		Grafting by 23% of PEG-2000	36 (60°C)	0.08	161
		Grafting by 56% of PEG-2000	30 (60°C)	0.10	161
		Grafting by 57% of PEG-10000	30 (60°C)	0.10	161
		Grafting by 72% of PEG-4000	30 (60°C)	0.10	161
		Grafting by 75% of PEG-10000	32 (60°C)	0.13	161
		Grafting by monoethylene glycol	51 (60°C)	—	161
		Grafting by diethylene glycol	60 (60°C)	—	161
		Grafting by triethylene glycol	40 (60°C)	—	161
		Grafting by tetraethylene glycol	39 (60°C)	—	161
		Grafting by 1,4-butanediol	63 (60°C)	—	161
		Grafting by 1,6-butanediol	40 (60°C)	—	161
		Grafting by 1,4-hexanediol	40 (60°C)	—	161
		Grafting by 1,8-octanediol	44 (60°C)	—	161
		Grafting by 1,10-decanediol	49 (60°C)	—	161
	Aerosil A130, Degussa	None	40 (80°C)	—	175
		None	75.3 (20°C)	0.32	160,179
		Grafting with methanol	70.2 (20°C)	0.32	160,179
		Grafting with hexadecanol	38.7 (20°C)	0.10	160,179
	Rhône Poulenc	None	98.2 (20°C)	0.41	160,179

TABLE 13 Continued

Material	Supplier	Treatment	γ_S^p at (T °C) (mJ/m²)	$-d\gamma_S^p/dT$ (mJ/m² °C)	Ref.
		Grafting with methanol	68.2 (20°C)	0.29	160,179
		Grafting with hexadecanol	35.6 (20°C)	0.06	160,179
		None	54.3 (65°C)	0.24	168
	Aerosil A150, Degussa	Grafting with TMEDOMSI	45.2 (65°C)	0.21	168
		None	50.6 (110°C)	—	186
	Aerosil A200, Degussa	Grafting by hexamethydisoloxane	36.2 (110°C)	—	186
		None	41 (200°C)	—	180
	Davidson 952	Heating at 600°C	38 (200°C)	—	180
		Heating at 800°C	31 (200°C)	—	180
		Heating at 600°C	87 (200°C)	—	180
	Aldrich	None	29.6 (100°C)	0.15	191
	Silica C18	None	30.4 (100°C)	0.07	191
	Lichroprep RP18	None	31.0 (100°C)	0.16	191
	Lichroprep Si 100	None	80 (20°C)	—	171
	XOB 75	Grafting with TDFOMDS	30 (20°C)	—	171
	Rhône Poulenc	Grafting with PDFDMCS	24 (20°C)	—	171
Smectites	Bentonites	Saturation with 1 N NaCl solution (M)	159 (180°C)	—	181
		M + ws with Al(OH)₃ solution (MI)	171 (180°C)	—	181
		MI + heating at 400°C (MA)	160 (180°C)	—	181
		M	155 (150°C)	—	182
		MA	153 (150°C)	—	182
		MA + FA + polymerization, POL (MAP)	122 (150°C)	—	182
		M + FA + POL + heating at 700°C (MC)	232 (150°C)	—	182
		MAP + ht at 700°C (MAC)	37 (150°C)	—	182
	Hydrocalcites	Mg-Al-CO₃ Mg/Al = 2 (H₂)	65 (150°C)	—	182
		Mg-Al-CO₃, Mg/Al = 3 (H₃)	51 (150°C)	—	182

Category	Source	Treatment	Value (temp)	Value	Ref.
Aluminas	Pechiney aluminum	H₂ + FA + POL (H2P)	141 (150°C)	—	182
	Degussa	H₃ + FA + POL (H3P)	142 (150°C)	—	182
		H₂ + FA + POL + ht at 700°C (H₂C)	231 (150°C)	—	182
		H₃ + FA + POL + ht at 700°C (H₂C)	165 (150°C)	—	182
		None	65–162	—	67
		γ-Aluminas, none	92 (100°C)	—	167
		Ht at 300°C	125 (100°C)	—	167
		Ht at 300°C	85 (100°C)	—	167
		None	92 (100°C)	—	169
	Pechiney aluminum	α-aluminas, none	50 (100°C)	—	173
		ht at 450°C	70 (100°C)	—	173
		Grinding	up to 160	—	173
	Aldrich	None	72 (200°C)	—	180
		Ht at 600°C	89 (200°C)	—	180
Illites	Vosges, France	None	177 (80°C)	0.10	26
	Szabadsag, Hungary	None	174 (80°C)	0.15	26
	Brive, France	None	138 (80°C)	0.03	26
Kaolinites	Charente, France	None	211 (80°C)	0.28	26
	Ploemeur, France	None	198 (80°C)	0.34	26
	Provins, France	None	154 (80°C)	0.01	26
	SPS International	Coating with 10 parts SB-latex + 1 part CMC	28 (50°C)	—	203
	SPS International	Coating with 10 parts starch	55 (50°C)	—	203
TiO₂	Sicpa	None	39.7 (50°C)	—	30
		None	25–55	—	65,183,188,192,193
		None	50–124	—	155
ZnO	Rhône Poulenc	None	85.3 (25°C)	0.4	170
MgO	Merck	None	95.6 (25°C)	0.2	170

PEG, polyethylene glycol; TMEDOMSI, tetramethylethylenedioxydimethylsilane; TDFOMDS, tridecafluorooctylmethyldichlorosilane; PDFDMCS, pentade-cafluorodecylmethyldichlorosilane; FA, furfuryl alcohol; SB, styrene-butadiene; CMC, carboxymethylcellulose.

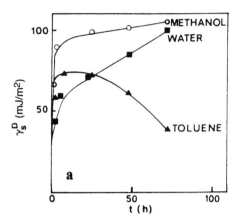

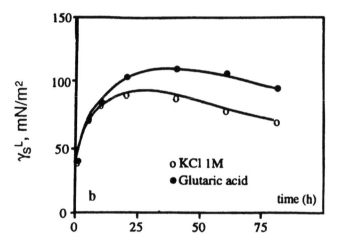

FIG. 12 γ_S^D of muscovite surface as a function of grinding time in different solvents. (From Ref. 159 for top and Ref. 177 for bottom.)

(FA) on the surface of these materials and its subsequent polymerization lowered the surface energy to 37 mJ/m². The carbonization of these particles restored a high surface energy of 120–230 mJ/m² [182]. Synthetic hydrotalcites were also investigated and showed different results from those obtained on natural smectites since the initial materials had a γ_S^D of 50–65 mJ/m² and their surface coating by a layer of polyFA yielded a γ_S^D of 140 mJ/m². The carbonization of these particles again produced an increase of γ_S^D [182].

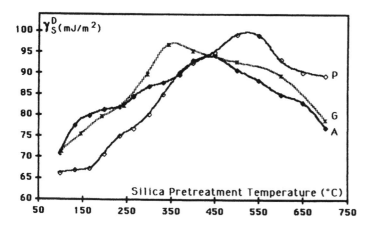

FIG. 13 Variation of the dispersive contribution to the surface energy of amorphous silicas heat-treated at increasing temperature. (A) aerosil, (G) gel, and (P) precipitated. (From Ref. 165.)

15. The dispersive component of the surface energy of fumed silica decreased from 51 to 36 mJ/m^2 when the materials were grafted with hexamethyldisilazane as coupling agents [184–186]. γ_S^D increased when these particles were submitted to heat treatment up to 600°C, as shown in Fig. 13 [165]. The same trend was observed with other types of silica (Fig. 14) [186].

16. The dispersive surface energy of grafted silica was found to be a linear function of the surface concentration of silane [190].

17. Grafting silicas with alcohols bearing different n-alkyl chains resulted in the behavior shown in Fig. 15 which emphasizes the role of the specific (short) length of the carbon chain on the actual value of γ_S^D.

2. Aluminum

There are only two papers that have reported the surface energy of aluminum as determined by IGC at infinte dilution conditions. The values of γ_S^D obtained were about 120 mJ/m^2 for base-treated samples and 140 mJ/m^2 for acid-treated ones [194,195].

3. Calcium Carbonate and Calcium Oxide

The fact that calcium carbonate is frequently used in paper coating and as a filler for thermoplastics and composite materials justifies the need of a good knowledge of its surface properties, which have indeed been investigated [157,158,196,197]. As for most fillers, before its incorporation into plastics, calcium carbonate is usually subjected to a specific surface treatment in order to facilitate its disper-

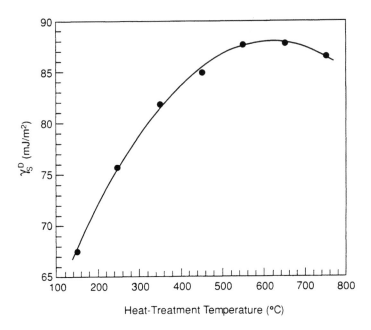

FIG. 14   Variation of γ_S^D of fumed silica under heating. (From Ref. 186.)

sion. The most commonly used coupling agent for calcium carbonate is stearic acid [198]. IGC has been used for the determination of the surface energy of calcium carbonate [192,193,199–203].

From these investigations one can extract the following:

1. Untreated calcium carbonate has a surface energy in the range 45–70 mJ/m^2 [30,157,158,192,193,199,203].
2. Plasma-treated calcium carbonate yields particles with lower dispersive surface energy (about 30 mJ/m^2) when the treatment is carried out under CH_4 or C_2F_4 whereas when Ar or ammonia is used no significant change is observed [192,193].
3. The treatment of calcium carbonate with stearic acid decreases its dispersive component of the surface energy down to 25–35 mJ/m^2 [199,200,202].
4. The thermal treatment of calcium carbonate leads to an increase of its surface energy. In fact, for a sample conditioned at 300°C the γ_S^D reached 250 mJ/m^2 [201].
5. The adsorption of styrene-butadiene latex as a binder on calcium carbonate particles induces a decrease of their γ_S^D to 25 mJ/m^2, whereas their coating with starch does not bring about any substantial change [203].

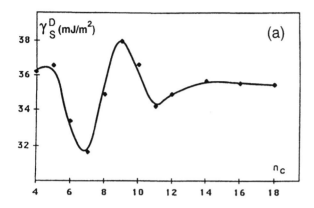

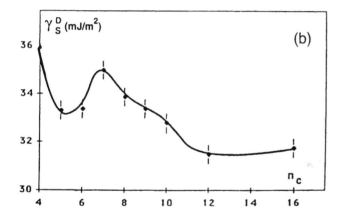

FIG. 15 Variation of γ_S^D of silica grafted with n-alkyl chains of growing size. (From Ref. 163 for 15a and Ref. 164 for 15b.)

6. The surface energy of calcium oxide was found to increase with grinding time (200 h) from 42 to 60 mJ/m^2 [204].

These data are summarized in Table 14.

4. Printing Ink Pigments

The quality of the dispersion of printing ink pigments into organic vehicles depends on their energetic and acid–base surface properties [205]. There is little information published about these parameters except for very recent papers [30,65] which reported that the dispersion of common printing ink pigments (magenta, cyan, violet, green, yellow) was about 35 mJ/m^2, as shown in Table 14.

TABLE 14 Values of γ_s^p of Different Fillers, Pigments, and Miscellaneous Materials

Material	Supplier	Treatment	γ_s^p (mJ/m²)	Ref.
CaCO₃	Solvay	None	52.0 (50°C)	157,158
	Omya	None	48.0 (50°C)	157,158
	Laboratory	None	58.0 (20°C)	199
		Grafting by stearic acid		
		50% coverage	33.0 (20°C)	199
		75% coverage	28.0 (20°C)	199
		100% coverage	27.0 (20°C)	199
		None	44.4 (70°C)	200
		Grafting by stearic acid	29.0 (70°C)	200
	Laboratory	Precipitation (P)	50 (100°C)	201
		P and heating at 300°C	250 (100°C)	201
	Marble	Heating at 100°C	50 (100°C)	201
		Heating at 300°C	95 (100°C)	201
	Chalk I	Heating at 100°C	180 (100°C)	201
		Heating at 300°C	200 (100°C)	201
	Chalk II	Heating at 100°C	135 (100°C)	201
		Heating at 300°C	150 (100°C)	201
	Laboratory	None (A)	50.3 (60°C)	202
		A + 1 monolayer AKD	35.6 (60°C)	202
		None (B)	52.6 (60°C)	202
		B + 1/4 monolayer AKD	48.9 (60°C)	202
		B + 1 monolayer AKD	37.3 (60°C)	202
	Omya	none (C)	68 (50°C)	203
		C + 10 parts SB-latex	25 (50°C)	203
		C + 5 parts SB-lat. + 1 part CMC	44 (50°C)	203
		C + 10 parts SB-lat. + 1 part CMC	27 (50°C)	203

Material	Source	Treatment	Value	Temp	Ref
		C + 10 parts SB-lat. + 1 part CMC	25	(50°C)	203
		C + 10 parts starch	58	(50°C)	203
	Prolabo	None	21.3	(50°C)	30
	Sicpa	Unknown	68.6	(50°C)	30
CaO		None	42.4	(20°C)	204
		Grinding for 10 h	48.6	(20°C)	204
		Grinding for 20 h	49.5	(20°C)	204
		Grinding for 100 h	54.5	(20°C)	204
		Grinding for 200 h	60.8	(20°C)	204
		Heating at 400°C	50.0	(20°C)	204
		Heating at 600°C	45.2	(20°C)	204
		Heating at 800°C	40.8	(20°C)	204
		Heating at 1000°C	38.5	(20°C)	204
Magenta		None	26.5	(60°C)	65
Violet		None	29.8	(60°C)	65
Green		None	43.0	(60°C)	65
Magenta	Sicpa	None	40.3	(50°C)	30
Cyan	Sicpa	None	38.4	(50°C)	30
Yellow	Sicpa	None	34.4	(50°C)	30
Theophylline	Sigma	None	50.8	(35°C)	206
Caffeine	Sigma	None	39.9	(35°C)	206
α-Lactose	Lactochem	None	42	(29°C)	207
		Heating at 40°C	42	(29°C)	207
Stainless		Passivation	105	(40°C)	208
		Electropolishing (coiled)	49	(40°C)	208
		Electropolishing (linear)	43	(40°C)	208
	Alltech	None	33	(40°C)	208
Salbutanol	Glaxo	Batch 1, heating at 40°C	83	(29°C)	209
sulfate		Batch 2, heating at 40°C	38	(29°C)	209

AKD, alkylketene dimer; CMC, carboxymethylcellulose; SB-latex; Styrene-butadiene latex.

5. Miscellaneous

In addition to the above substances, IGC has been very recently applied to less common products. Thus, the dispersive properties of theophylline, caffeine, α-lactose monohydrate, stainless steel, submatol sulfate, and petroleum pitches have been reported [206–210]. The values of the corresponding γ_S^D are summarized in Table 14.

X. ACID–BASE PROPERTIES

The acid–base properties of fibers and fillers are also of great interest in the elaboration of composites and blends. IGC is undoubtedly a good technique to evaluate these properties starting from classical acid–base concepts as discussed in Sec. VIII.E. In this survey we will report only acid–base numbers as obtained using St. Flour and Papirer's approach [68] [Eq. (50)] and the interaction parameter Ω according to Schreiber's formalism [47] [Eqs. (52) and (53)]. This choice is motivated by the fact that K_A and K_B are temperature-independent and that Ω is always obtained from the same acid–base probes, i.e., n-BuOH as an acid and n-BuNH$_2$ as a base [47]. These considerations facilitate the comparison between the data reported in the literature. The other acid–base parameters are less homogeneous because they are temperature- and probe-dependent, which renders any quantitative comparison impossible.

A. Fibers

1. Cellulose and Pulp Fibers

Kamdem et al. [93] studied the acid–base properties of extractive-free white birch wood meal and found that this material had an acidic surface ($K_A/K_B = 1.5$) as shown in Table 15, which collects the K_A/K_B values for these and other fibers.

Felix and Gatenholm reported the acid–base properties of bleached softwood pulp fibers before and after grafting them with ASA and detected an increase in the acidity component [94] (see Table 15). The authors did not give a rational explanation of this behavior because they simply evoked the fact that K_A and K_B decreased by 44% and 83%, respectively, due to the coverage of hydroxy and glucosidic groups on the surface of the fibers. From our point of view, a more logical interpretation, while accepting that the decrease of these parameters is related to the surface grafting, explains the important quantitative evidence that K_A/K_B increased form 1.29 to 4.25 in terms of the presence of the carboxylic acid groups arising from the esterification between the anhydride functions of ASA (Structure **II**) and the OH groups of cellulose. The same research team investigated bleached softwood pulp fibers before and after treatment with plasma under different atmospheres [95]. They found, as expected, that the plasma treatment in the presence of ammonia and nitrogen increased the surface basicity,

TABLE 15 Acid–Base Properties of Different Modified and Untreated Fibers

Material	Treatment	K_A	K_B	K_A/K_B	Ref.
Birch wood meal	None	4.40	3.00	1.47	93
Bleached softwood	None	0.31	0.24	1.29	94
pulp	Grafted by ASA	0.17	0.04	4.25	94
	None	0.25	0.23	1.09	95
	NH_3 plasma for:				
	5 s	0.14	0.28	0.50	95
	15 s	0.09	0.31	0.29	95
	60 s	0.16	0.22	0.73	95
	N_2 plasma for				
	5 s	0.22	0.27	0.81	95
	15 s	0.16	0.30	0.53	95
	60 s	0.19	0.22	0.86	95
	MMAc plasma for				
	5 s	0.30	0.20	1.5	95
	15 s	0.33	0.15	2.2	95
	60 s	0.39	0.10	3.9	95
α-Hardwood	None	—	—	3.0	66
cellulose	Acetone extraction (AcE)	—	—	2.4	66
	Corona treatment at:				
	15 mA	—	—	4.8	66
	32 mA	—	—	3.9	66
	40 mA	—	—	3.3	66
	AcE + corona treated at:				
	15 mA	—	—	4.1	66
	32 mA	—	—	3.3	66
	40 mA	—	—	2.5	66
Explosion pulp	None	2.4	12.1	0.2	97
Kraft pulp	None	5.5	5.0	1.1	97
Carbon fibers	None	6.5	1.5	4.3	40–42,103
	Oxidation	10.0	3.2	3.1	40–42,103
	Enzymation	8.6	13.0	0.7	40
	Coating	8.6	13.0	0.7	40–42,103
	Coating	9.1	9.3	1.0	41–42,103
	Carbonization	1.0	1.4	0.7	108
	Stabilization	1.5	1.3	1.2	108
	Heating at 1500°C (A)	6.5	8.0	0.8	110
	Heating at 1800°C (B)	5.4	6.6	0.8	110
	Heating at 2100°C (C)	4.6	6.2	0.75	110
	Heating at 2400°C (D)	4.4	6.2	0.7	110
	Heating at 2700°C (E)	3.6	5.2	0.7	110
	E + anodic oxidation				
	at I_o current	8.6	9.0	0.75	110

TABLE 15 Continued

Material	Treatment	K_A	K_B	K_A/K_B	Ref.
	at 3 * I_0 current	9.9	10.0	1.0	110
	at 10 * I_0 current	12.1	7.6	1.6	110
	at 30 * I_0 current	8.6	4.6	1.9	110
	Carbonization at 1280°C (F)	343	84	4.1	112
	F + oxidation with O_2 (G)	359	292	1.2	112
	G + oxidation with HNO_3	908	277	3.3	112
	Carbonization at 2600°C (H)	0	0	—	112
	H + oxidation with HNO_3	192	255	0.85	112
	None	0.1	0.7	0.1	113
	Oxidation	0.5	0.7	0.7	113
	Sizing	1.1	0.8	1.4	113
	Sizing	0.9	2.2	0.4	113
Polyethylene fibers	None	0	0	—	97
	Ozonation for 2 h	3.5	0.8	4.4	97
	Ozonation for 3 h	3.3	1.0	3.3	97
	Fluorination 1	7.3	2.5	2.9	97
	Fluorination 2	10.3	9.2	1.1	97

ASA, alkylsuccinic anhydride; MMAc, methyl methacrylic acid.

whereas the surface acidity of the fibers was found to increase when the treatment was carried out in a methacrylic acid medium [95] (see Table 15).

More recently, Belgacem et al. [66] reported the acid–base surface properties of different types of cellulosic fibers as such, purified by acetone extraction and corona-treated. The acid–base properties changed significantly as summarized in Table 15.

Finally, Chtourou et al. [97] investigated unbleached steam-exploded and kraft pulps and found that they had K_A/K_B ratios of 0.2 and 1.1, respectively. The strong basic character of the steam-exploded pulps was not interpreted.

2. Carbon Fibers

Schultz et al. [40–42,103] were the first to investigate the acid-base properties of carbon fibers in the form of commercial polyacrylonitrile-based fibers corresponding to different stages of manufacture. The values of K_A and K_B thus obtained are resumed in Table 15. From these values it can be concluded that the surface of carbon fibers before and after an oxidizing treatment has an acidic character (K_A/K_B = 3–4), whereas coating with epoxy resins gives a neutral surface (K_A/K_B = 1). The enzymatic treatment of these fibers yielded a surface of high basicity (K_A/K_B = 0.66).

Nardin et al. have carried out investigations on carbonized and stabilized carbon fibers. The first were obtained from carbonization at 1100°C, whereas the others stem from an intermediate stage of the process [108]. They found that, contrary to the surface of carbonized fibers which showed a basic character ($K_A/K_B = 0.7$), that of stabilized homologues displayed slightly acidic properties ($K_A/K_B = 1.1$).

Donnet and Park have investigated pitch-based carbon fibers [110] and found that thermal treatments did not affect their surface acid–base character ($K_A/K_B = 0.7$–0.8), but anodic oxidation increased significantly the acidity of their surface (K_A/K_B increased from 0.7 to 1.9).

Recently, Tsutsumi and Ban [112] also studied both polyacrylonitrile- and pitch-based carbon fibers as such and heated or electrolytically oxidized. The authors found surface characteristics that only partially agreed with previous investigations [40–42,103]. In fact, they detected an acidic surface for the untreated fibers ($K_A/K_B = 4$) but, surprisingly, a more amphoteric character ($K_A/K_B = 1.2$) for the oxygen-oxidized material. The other disagreement with previous data concerned the fibers electrolytically oxidized with HNO_3, which were less acidic and indeed more basic than the surface of the starting material. The corresponding values of the donor and acceptor numbers are summarized in Table 15.

Finally, Jacobasch et al. [113] studied the surface properties of carbon fibers before and after oxidation or sizing. However, since the origin of these fibers was not specified, it is not easy to establish any comparison of their acid–base properties with those related to previous studies. Moreover, the authors found that the initial material had a strong basic character ($K_A/K_B = 0.1$), which is in contradiction with most previous work on carbon fibers [40–42,103,112]. They also found that oxidation and sizing increased the surface acidity of fibers, an observation that is commonly reported (Table 15).

3. Glass Fibers

Osmont and Schreiber have reported the values of Ω related to glass fibers before and after treatment with silane coupling agents [211]. They found that the surface of the starting fibers was acidic ($\Omega = -0.14$) and its treatment with 3-chloropropyltrimethoxylsilane increased its acidity further ($\Omega = -2.33$) but became basic when treated with hexadimethoxysilane ($\Omega = 0.83$) and amphoteric with aminopropylmethoxysilane ($\Omega = 0.16$).

4. Polyethylene Fibers

Polyethylene (PE) fibers were studied as such and after ozonation and fluorination with different treatment times. As shown in Table 15, whereas the surface of untreated PE fibers was found neutral with $K_A = K_B \approx 0$, both ozonation and fluorination substantially increased its acidic character [97].

B.   Fillers and Powders

1.   Activated Carbon

Donnet et al. studied the donor–acceptor properties of natural graphite as such, heated at 800°C and plasma-treated under vapors of nBuOH, ammonia, and nBuNH$_2$ [136]. The main points that can be drawn from these results, given in Table 16, are:

1. The surface of virgin natural graphite is acidic (K_A/K_B = 1.5).
2. Heating this material gives rise to a basic surface (K_A/K_B = 0.6).
3. Plasma treatment in the presence of vapors of nBuOH did not affect the surface of untreated carbon (K_A/K_B = 1.5) but increased the surface acidity of heated samples (K_A/K_B = 1.2).
4. Plasma treatment in the presence of vapors of nBuNH$_2$ increased the basic character of the surface of both heated and unheated samples.

2.   Calcium Carbonate

Schreiber et al. have reported values of the interaction parameter Ω of CaCO$_3$ as received and plasma-treated [212–214], and showed that the initial surface of this filler possessed a basic character (Ω = 0.30). This parameter increased when the filler was plasma-treated in the presence on nBuNH$_2$ and ammonia (Ω =

TABLE 16   Acid–Base Properties of Different Materials

Material	Treatment	K_A	K_B	K_A/K_B	Ref.
Natural	None (A)	1.24	0.83	1.49	136
graphite	A + nBu-OH plasma	1.37	0.89	1.54	136
	A + NH$_3$ plasma	0.49	0.90	0.54	136
	A + nBu-NH$_2$ plasma	0.38	0.87	0.44	136
	Heating at 800°C (B)	0.59	0.94	0.63	136
	B + nBu-OH plasma	1.18	0.96	1.23	136
	B + NH$_3$ plasma	0.79	1.72	0.46	136
	B + nBu-NH$_2$ plasma	0.47	1.02	0.45	136
Polypyrrole	Doping by p-toluenesulfonate	0.273	0.026	10.5	149
	Doping by aq. sol. of FeCl$_3$	0.261	0.436	0.6	149
Polyoctylthio-	None	0.0138	0.298	0.46	149
phene					
α-Alumina	Pure acidic (Bayer process)	12.2	6.3	1.9	157,158
		20.5	8.1	2.5	157,158
Cork	None	0.32	0.29	1.1	29
Theophylline	None	0.072	0.105	0.68	206
Caffeine	None	0.051	0.034	1.55	206

aq. sol., aqueous solution.

1.75 and 0.70, respectively) and decreased when nBuOH and ethylene were used as gas for plasma treatment ($\Omega < 0$ with both). The use of argon gave a more amphoteric surface ($\Omega = 0.19$).

3. Polymeric Powders

The surface properties of p-toluenesulfonate-doped polypyrrole was found to bear a very acidic character ($K_A/K_B = 10.5$), whereas those of chloride-doped polypyrrole and poly(3-octylthiophene) were basic ($K_A/K_B = 0.6$ and 0.5, respectively) [149] (see Table 16).

4. Cork

Cordeiro et al. [29] studied the acid–base properties of cork powders and found that cork has a slightly acidic surface ($K_A/K_B = 1.3$), as shown in Table 16.

5. Miscellaneous

Whereas the surface of theophylline powder was found to be basic ($K_A/K_B = 0.7$), that of caffeine appeared to be acidic ($K_A/K_B = 1.6$) [206] (see Table 16).

XI. INTERACTION PARAMETERS IN COMPOSITES AND BLENDS

Schultz et al. [40–42,103] have also characterized the acid–base properties of different carbon fibers according to Eq. (50). The values of K_A and K_B thus obtained were used to calculate the interaction parameter A [Eq. (57)] and it was shown that the interfacial shear strength (τ) increased when A increased as illustrated in Fig. 16.

Lara and Schreiber [65] have also correlated the interaction parameter I_{sp}, [Eq. (58)] between pigments and vehicles (liquid matrix) to the adsorption capacity of various resins and found a linear relationship between these two parameters, as shown in Fig. 17.

These two examples show the importance of IGC in the search for correlations between the physical chemistry of the surface of both matrix and filler or fiber and the macroscopic properties of the resulting blends or composites.

XII. CONCLUSIONS

By writing this survey, we wished to emphasize the importance of IGC as a technique for the surface characterization of fibers and fillers. We pointed out the fact that IGC is suitable for materials that have no film forming ability and/ or those that lose their original surface properties where they are cast. Of course, there are some peculiarities of this method and the values obtained from IGC measurements are not always readily comparable with those obtained from other

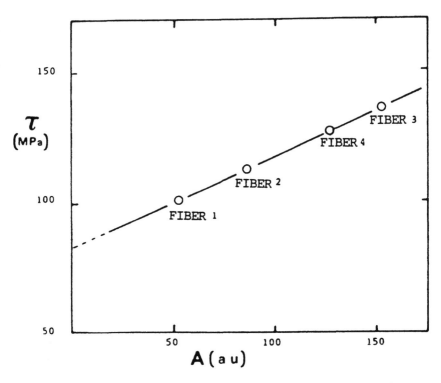

FIG. 16 Interfacial shear resistance versus specific interaction parameter (*A*) between an epoxy resin matrix and carbon fibers as defined by Schultz et al. [41].

techniques. Nonetheless, the microscopic character of IGC (adsorption at infinitely dilute conditions) provides much information that more conventional techniques are not able to detect. The main conclusion that we want to convey is that the reader should keep in mind that IGC is not only a useful complementary technique, but in some instances it is the *only* method that can be applied to study the surface properties of a material. If more laboratories are stimulated to adapt their conventional GC equipment to IGC, we will consider that we have reached the goal that we outlined when we accepted the invitation to contribute to this book.

FIG. 17 Rationalizing adsorption of different resins onto the surface of various pigments using the interaction parameter I_{sp} as defined by Lara and Schreiber [65]. (a) S1 is a polyester with M_n of 1800 and an acid value (*AV*) of 8.0. (b) A3 is a polyester with M_n of 1200 and an *AV* of 0.7. A3 was modified by amine group backbones. (c) *A*1 and (d) *A*2 are very similar to A3, except that they have an *AV* of 7.7 and 7.6, respectively.

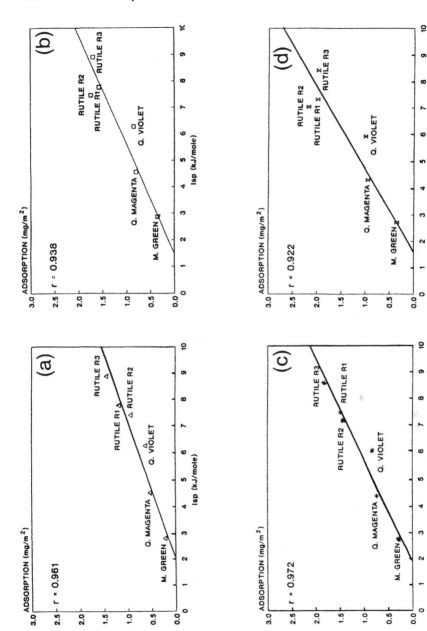

ADDENDUM

Since this review was sent to the editor, some papers relevant to its aim have appeared in the literature [215–217]. Thus, data concerning fully bleached softwood kraft pulps [215] and chemicothermomechanical pulps (CTMP) [216] and other, different modifications have been reported. More recently, Botaro and Gandini [217] proposed an original approach for preparing composite materials reinforced by cellulosic fibers, based on the idea of introducing polymerizable alkenyl moieties at the surface of the fibers. The aim of this operation was to promote the subsequent participation of these unsaturations in polymerizations involving conventional monomers i.e., the in situ synthesis of a matrix bearing covalent links with the fibers. From the data of this report, the main information relevant to the present chapter is the fact that the γ_S^D of pristine material was slightly higher than that measured after modification and that the acidic character of the initial fibers was reduced considerably.

SYMBOLS AND ABBREVIATIONS
Symbols

γ_S^D	dispersive component of the surface energy of solid surface.
γ_L^D	dispersive component of the surface energy of the probes in the liquid state.
AN	acid or acceptor number of the probe molecules according to Gutmann [43].
DN	basic or donor number of the solid probe molecules (Gutmann [43]).
AN_s	acid or acceptor number of the solid surface (Lara and Schreiber [65]).
DN_s	basic or donor number of the solid surface (Lara and Schreiber [65]).
K_A	acid or acceptor number of the solid surface (St. Flour and Papirer [68]).
K_B	basic or donor number of the solid surface (St. Flour and Papirer [68]).
K_s	surface partition coefficient.
Ω	acid–base interaction parameter according to Lara and Schreiber [65].
A	interaction parameter between matrix and fibers (Schultz et al. [40–42,103]).
I_{sp}	interaction parameter between matrix and filler (Lara and Schreiber [65]).
t_r	net retention time of a given probe molecule.
V_g^0	specific retention volume of a given probe molecule.

V_n net retention volume of a given probe molecule.

ΔG_A free energy of adsorption of a given probe molecule.

$\Delta G_A^{(CH_2)}$ free energy of adsorption of a methylene group.

ΔG_A^{sp} specific (polar) free energy of adsorption of a given probe molecule.

a_{CH_2} surface area of a given probe molecule.

a_{CH2} surface area of a methylene group

ΔH_A free enthalpy of adsorption of a given probe molecule

ΔS_A free entropy of adsorption of a given probe molecule

Abbreviations

AKD	alkylketene dimer
ASA	alkenylsuccinic anhydride
CMC	carboxymethylcellulose
CTMP	chemithermomechanical pulp
DS	degree of substitution
EDTA	ethylenediaminetetraacetic acid
EG-SiO$_2$	ethylene glycol–modified silicon oxide
FA	furfuryl alcohol
MMAc	methyl methacrylic acid
PAN	polyacrylonitrile
PBDM	polybutane-1,4-diol dimethacrylate
P(BDM-co-MA)	BDM-co-maleic anhydride copolymer
PE	polyethylene
PDFDMCS	pentadecafluorodecylmethyldichlorosilane
PEG	polyethylene glycol
PET	polyethylene terephthalate
PFR	phenol-formaldehyde resin
PFR-A	low molecular weight PFR
PFR-B	high molecular weight PFR
PMMA	polymethyl methacrylate
PNDM	poly$N$-methyldiethanolamine dimethacrylate
POEDM	poly(2,2′-oxybisethanol dimethacrylate)
POT	polyoctylthiophene
PPDM	polypentane-1,5-diol dimethacrylate
PTEDM	poly-2,2′-thiobisethanol dimethacrylate
SBR	styrene-butadiene copolymer
TDFOMDS	tridecafluorooctylmethyldichlorosilane
TMP	thermomechanical pulp
TMEDOMSI	tetramethylethylenedioxydimethylsilane
TSPTS	bis(3-triethoxysilylpropyl)tetrasulfane

REFERENCES

1. W. J. Feast, H. S. Munro, and R. W. Richard, *Polymer Surface and Interfaces*, Vol. 2, John Wiley and Sons, Chichester, 1993.
2. F. Garbassi, M. Morra, and E. Occhiello, *Polymer Surfaces, From Physics to Technology*, John Wiley and Sons, Chichester, 1994.
3. W. Neagle and D. R. Randell, *Surface Analysis Techniques and Applications*, Royal Society of Chemistry, U.M.I.S.T., Manchester, 1989.
4. D. Briggs, *Polymer 25*:1379 (1984).
5. S. Brunauer, P. H. Emmet, and E. Teller, *J. Amer. Chem. Soc. 60*:309 (1938).
6. M. E. Schrader and G. I. Loeb, *Modern Approaches to Wettability: Theory and Applications*, Plenum Press, London, 1992.
7. S. Wu, *Polymer Interface and Adhesion*, Marcel Dekker, New York, 1982.
8. F. M. Fowkes, *Adv. Chem. Ser. 43*:99 (1964).
9. C. J. Van Oss, in Ref 1. Chap. 11, pp. 267–290.
10. W. A. Zisman, *Adv. Chem. Ser. 43*:1 (1964).
11. D. K. Owens and R. C. Wendt, *J. Appl. Polym. Sci. 13*:1741 (1969).
12. J. M. Braun and J. E. Guillet, *Adv. Polym. Sci. 21*:108 (1976).
13. D. G. Gray, *Prog. Polym. Sci. 5*:1 (1977).
14. D. R. Lloyd, T. C. Ward and H. P. Schreiber, *Inverse Gas Chromatography*, ACS Symp. Series No. 391, American Chemical Society, Washington, D.C., 1989.
15. Z. Y. Al-Saigh, *Polym. News 19*:269 (1994).
16. A. V. Kiselev, in *Advances in Chromatography* (J. C. Gidding and R. A. Keller, eds.), Marcel Dekker, New York, 1967.
17. O. Smidsrød and J. E. Guillet, *Macromolecules 2*:272 (1969).
18. D. Patterson, Y. B. Tewari, H. P. Schreiber, and J. E. Guillet, *Macromolecules 4*: 356 (1971).
19. W. R. Summers, Y. B. Tewari, and H. P. Schreiber, *Macromolecules 5*:12 (1972).
20. H. P. Schreiber, Y. B. Tewari, and D. Patterson, *J. Polym. Sci., Polym. Chem. Ed. 11*:15 (1973).
21. H. P. Schreiber and D. R. Lloyd, in Ref. 14, Chap. 1, pp. 1–10.
22. A. E. Bolvari, T. C. Ward, P. A. Koning, and D. P. Sheehy, in Ref. 14, Chap. 2, pp. 12–19.
23. A. Voelkel, E. Andrzejewska, R. Maga, and M. Andrzejewski, *Polymer 37*:455 (1996) and *39*:3499 (1998).
24. G. M. Dorris and D. G. Gray, *J. Colloid Interface Sci. 71*:93 (1979).
25. H. P. Schreiber, M. R. Wertheimer, and M. Lambla, *J. Appl. Polym. Sci. 27*:2269 (1982).
26. A. Saada, E. Papirer, H. Balard, and B. Shiffert, *J. Colloid Interface Sci. 175*:212 (1995).
27. C.-T. Chen and Z. Y. Al-Saigh, *Macromolecules 22*:2974 (1989).
28. M. F. Grenier-Loustalot, Y. Barthomieu, and P. Grenier, *Surf. Interface Anal. 14*: 187 (1989).
29. N. Cordeiro, C. Pascoal Neto, A. Gandini, and M. N. Belgacem, *J. Colloid Interface Sci. 174*:246 (1995).

30. M. N. Belgacem, A. Blayo, and A. Gandini, *J. Colloid Interface Sci. 182*:431 (1996).
31. N. Cordeiro, P. Aurenty, M. N. Belgacem, A. Gandini, and C. Pascoal Neto, *J. Colloid Interface Sci. 187*:498 (1997).
32. J. A. Trejo O'Reilly, J.-Y. Cavaillet M. N. Belgacem, and A. Gandini, *J. Adhesion 67*:359 (1998).
33. P. H. Emmet and S. Brunauer, *J. Am. Chem. Soc. 59*:1553 (1937).
34. G. Dorris and D. G. Gray, *J. Chem. Soc., Faraday Trans. 177*:413 (1981).
35. G. Dorris and D. G. Gray, *J. Chem. Soc., Faraday Trans. 1:77*:725 (1981).
36. T. L. Hill, *J. Chem. Phys. 16*:181 (1948).
37. N. N. Avgul, G. I. Berezin, A. V. Kiselev, and I. A. Lygina, *Izu. Akad. Nauk. SSSR, Otd. Khi. Nauk* 1021 (1957).
38. A. L. McClellan and H. F. Harnsberger, *J. Colloid Interface Sci. 23*:577 (1967).
39. L. R. Snyder, *Principle of Adsorption Chromatography*, Marcel Dekker, New York, 1968, pp. 194–195.
40. J. Schultz, L. Lavielle, and C. Martin, *J. Chimie Phys. 84*:231 (1987).
41. J. Schultz, L. Lavielle, and C. Martin, *J. Adhesion 23*:45 (1987).
42. J. Schultz and L. Lavielle, in Ref. 9, Chap. 14, pp. 186–202.
43. V. Gutmann, *The Donor–Acceptor Approach to Molecular Interactions*, Plenum Press, New York, 1978.
44. F. L. Riddle and F. M. Fowkes, *J. A. Chem. Soc. 112*:3259 (1990).
45. K. L. Mittal and H. R. Anderson, *Acid–Base Interactions: Relevance to Adhesion Science and Technology*, VSP, Utrecht, The Netherlands, 1991.
46. F. M. Fowkes, in Ref. 45, pp. 93–115.
47. Y. M. Boluk and H. P. Schreiber, *Polym. Comp. 7*:295 (1986).
48. A. C. Tiburcio and J. A. Manson, *J. Appl. Polym. Sci. 42*:427 (1991).
49. R. S. Drago, G. G. Vogel, and T. E. Needham, *J Am. Chem. Soc. 93*:6014 (1971).
50. J. R. Conder, S. McHale, and M. A. Jones, *Anal. Chem. 58*:2663 (1989).
51. J. R. Conder and C. L. Young, *Physicochemical Measurements in Gas Chromatography*, John Wiley and Sons, New York, 1979.
52. D. P. Kamdem and B. Riedl, *J. Colloid Interface Sci. 150*:507 (1992).
53. J. E. Lipson and J. E. Guillet, in *Development in Polymer Characterization* (J. V. Dawkins, ed.), 3rd ed. Vol. 3, Chap. 2, Applied Science, New York, 1982, pp. 33–74.
54. A. T. James and A. J. P. Martin, *Biochem. J. 50*:679 (1952).
55. F. M. Fowkes, *Ind. Eng. Chem. 56*:40 (1964).
56. F. M. Fowkes, *J. Phys. Chem. 6*:382 (1982).
57. G. M. Dorris and D. G. Gray, *J. Colloid Interface Sci. 77*:353 (1980).
58. L. A. Girifalco and R. J. Good, *J. Phys. Chem. 61*:904 (1957).
59. R. J. Aveyard, *J. Colloid Interface Sci. 52*:621 (1975).
60. C. Kemball and E. K. Rideal, *Proc. Roy. Soc. A 187*:53 (1946).
61. J. H. De Boer and S. Kruyer, *Proc. K. Ned. Akad. Wet. B 55*:451 (1952).
62. A. J. Groszek, *Proc. Roy. Soc. A 314*:473 (1970).
63. H. J. Clint, *Trans. Faraday Soc. 68*:2239 (1972).
64. A. J. Vukov and D. G. Gray, *Langmuir 4*:743 (1988).

65. J. Lara and H. P. Schreiber, *J. Coating Technol.* 63:81 (1991).
66. M. N. Belgacem, G. Czeremuszkin, S. Sapieha, and A. Gandini, *Cellulose* 2:145 (1996).
67. E. Papirer, J. M. Perrin, B. Siffert, and G. Philipponeau, *J. Colloid Interface Sci.* 144:263 (1991).
68. C. Saint-Flour and E. Papirer, *Ind. Eng. Chem. Prod. Res. Dev.* 21:666 (1982).
69. C. Saint-Flour and E. Papirer, *J. Colloid Interface Sci.* 91:69 (1983).
70. A. Vidal, W. Wang, and J. B. Donnet, *Kautschuk Gummi Kunststoffe* 46:770 (1993).
71. A. C. Tiburcio, *Polym. Ma. Sci. Eng., 70,* 462 (1993).
72. D. R. Little (ed.), *CRC Handbook of Chemistry and Physics,* 76th ed., CRC Press, Boca Raton, 1995.
73. W. J. Moore, *Physical Chemistry,* 4th ed., Printice-Hall, Englewood Cliffs, NJ, 1972, p. 213.
74. L. Riedel, *Chem. Ing. Tech.* 26:679 (1954).
75. C. Y. Kim, G. Surani, and D. A. I. Goring, *J. Polym. Sci., Part C* 30:533 (1992).
76. M. N. Belgacem, P. Bataille, and S. Sapieha, *J. Appl. Polym. Sci.* 53:379 (1994).
77. X. Tu, A. Young, and F. Denes, *Cellulose* 1:87 (1994).
78. J. M. Felix and P. Gatenholm, *J. Appl. Polym. Sci.* 42:609 (1991).
79. R. G. Raj and B. V. Kokta, in *Viscoelasticity of Biomaterials* (W. Glasser and H. Hatakeyama, eds.), ACS Symp. Series No. 489, American Chemical Society, Washington, D.C., 1992, pp. 99–117.
80. H. L. Lee and P. Luner, *Nordic Pulp Paper Res. J.* 9:164 (1989).
81. H. L. Lee and P. Luner, *J. Wood Chem. Technol.* 13:127 (1993).
82. G. Garnier and W. G. Glasser, *Polym. Mat. Sci. Eng.* 70:507 (1993).
83. G. Garnier and W. G. Glasser, *J. Adhesion* 46:165 (1994).
84. G. Garnier and W. G. Glasser, *Proc. International Symposium on the Interface, 16th Annual Meeting of the Adhesion Society,* Williamsburg, VA, February 21–26, 1993, pp. 135–137.
85. G. Garnier and W. G. Glasser, *Polym. Eng. Sci.* 36:885 (1996).
86. J. Brandrup and E. H. Immergut, *Polymer Handbook,* 3rd ed., John Wiley and Sons, New York, 1989.
87. T. D. Ignatova, V. V. Gorichko, L. A. Kaptioukh, A. L. Komarovski, and A. E. Hesterov, *Sb. Nauch. Trud. Ukr. Bum. Prom.* 99 (1988).
88. J. Borch, *Proc. Papermaking 10th. Fund. Res. Symp.,* Oxford, UK, September 20–24, 1993, pp. 209–236.
89. N. Gurnagul and D. G. Gray, *Can. J. Chem.* 65:1935 (1987).
90. D. P. Kamdem and B. Riedl, *Colloid Polym. Sci.* 269:595 (1991).
91. D. P. Kamdem and B. Riedl, *J. Wood Chem. Technol.* 11:57 (1991).
92. B. Riedl and D. P. Kamdem, *J. Adhesion Sci. Technol.* 6:1053 (1992).
93. D. P. Kamdem, S. K. Bose, and P. Luner, *Langmuir* 9:3039 (1993).
94. J. M. Felix and P. Gatenholm, *Nordic Pulp and Paper Res. J.* 8:200 (1993).
95. J. M. Felix, P. Gatenholm, and H. P. Schreiber, *J. Appl. Polym. Sci.* 51:285 (1994).
96. P. N. Jacob and J. C. Berg, *Langmuir* 10:3086 (1994).
97. H. Chtourou, B. Riedl, and V. K. Kokta, *J. Adhesion Sci. Technol.* 9:551 (1995).
98. Å. Lundqvist, L. Ödberg, and J. C. Berg, *Tappi J.* 78:139 (1995).

99. P. N. Jacob and J. C. Berg, *J. Adhesion* 54:115 (1995).
100. J. Simonsen, Z. Hong, and T. G. Rials, *Wood Fiber Sci.* 29:75 (1997).
101. A. J. Vukov and D. G. Gray, *Proc. 4th Int. Carbon Conf.*, Baden-Baden, 1986, pp. 394–396.
102. A. J. Vukov and D. G. Gray, in Ref. 14, Chap. 13, pp. 168–184.
103. J. Schultz, L. Lavielle, and C. Martin, *Proc. 10th Annual Meeting of the Adhesion Society*, Williamsburg, VA, February 22–27, 1987, pp. 513–528.
104. A. E. Bolvari and T. C. Ward, *Polym. Mat. Sci. Eng.* 58:655 (1988).
105. A. E. Bolvari and T. C. Ward, in Ref. 14, Chap. 16, pp. 217–229.
106. S. P. Wilkinson and T. C. Ward, *Proc. 35th International SAMPE Symposium*, April 2–5, 1990, pp. 1180–1192.
107. S. Dong, M. Brendlé, and J. B. Donnet, *Chromatographia* 28:469 (1989).
108. M. Nardin, H. Balard, and E. Papirer, *Carbon* 28:43 (1990).
109. L. Lavielle and J. Schultz, *Langmuir* 7:978 (1991).
110. J.-B. Donnet and S.-J. Park, *Carbon* 29:955 (1991).
111. R.-Y. Qin and J.-B. Donnet, *Carbon* 32:165 (1994).
112. K. Tsutsumi and K. Ban, *Surf. Sci. Catal.*, 80:679 (1993).
113. H.-J. Jacobasch, K. Grundke, P. Uhlmann, F. Simon, and E. Mäder, *Composite Interfaces 3*, 293 (1996).
114. S.-J. Park and M. Brendlé, *J. Colloid Interface Sci.* 188:336 (1997).
115. E. P. Plueddemann, *Silane Coupling Agents*, Plenum Press, New York, 1982.
116. E. Papirer, in *Composite Interface* (H. Ishida and J. L. Koening, eds.), Elsevier, New York, 1986.
117. E. Papirer, H. Balard and A. Vidal, *Eur. Polym. J.* 24:783 (1988).
118. E. Papirer and H. Balard, in Ref. 44, pp. 191–205.
119. K. Tsutsumi and T. Oshuga, *Colloid Polym. Sci.* 268:38 (1990).
120. P. J. C. Chappell and D. R. Williams, *ICCM/ECCM Conference*, Elsevier, London, July 1987, pp. 346–355.
121. B. J. Briscoe, S. Roach and D. R. Williams, in *Interfacial Phenomena in Composite Materials* (F. Jones, ed.), Butterworth, Guilford Press 1989, pp. 132–137.
122. P. J. C. Chappell and D. R. Williams, *J. Colloid Interface Sci.* 128:336 (1989).
123. P. J. C. Chappell and D. R. Williams, *J. Adhesion Sci. Technol.* 4:7 (1990).
124. D. R. Williams, in *Controlled Interface Composite Materials* (H. Ishida, ed.), Elsevier, New York, 1990, pp. 219–232.
125. M. L. Tate, Y. K. Kamath, S. P. Wesson, and S. B. Ruetsch, *J. Colloid Interface Sci.* 177: 579 (1996).
126. J. Anhang and D. G. Gray, *J. Appl. Polym. Sci.* 27: 71 (1982).
127. J. Anhang and D. G. Gray, *Physico-chemical Aspects of Polymer Surfaces, Proc. Int. Symp.*, 1983, pp. 659–667.
128. Y. Yang and T. C. Ward, *Polym. Mater. Sci. Eng.* 58:237 (1988).
129. H. Chtourou, B. Riedl, and B. V. Kokta, *Polym. Mater. Sci. Eng.* 70:509 (1990).
130. S. Katz and D. G. Gray, *J. Colloid Interface Sci.* 82:318 (1981).
131. S. L. Gregg and K. S. W. Sing, *Adsorption Surface Area and Porosity*, Academic Press, London, 1982.
132. J. B. Donnet, S. J. Park, and H. Balard, *Chromatographia* 31:434 (1991).
133. E. Papirer, S. Li, H. Balard, and J. Jagiello, *Carbon* 29:1135 (1991).

134. J. Jagiello, T. J. Bandosz, and J. A. Schwarz, *Chromatographia 33*:441 (1992).

135. J. Jagiello, T. J. Bandosz, and J. A. Schwarz, *Carbon 30*:63 (1992).

136. J. B. Donnet, S. J. Park, and M. Brendlé, *Carbon 30*:263 (1992).

137. J. B. Donnet and C. M. Lansinger, *Kautsch. Gummi, Kunstst. 45*:459 (1992).

138. H. Balard, W. D. Wang, and E. Papirer, *Polym. Mater. Sci. Eng. 70*:468 (1993).

139. M.-J. Wang and S. Wolff, *Rubber Chem. Technol. 64*:714 (1991).

140. J. B. Donnet, W. Wang, A. Vidal, and M. J. Wang, *Kautsch. Gummi, Kunstst. 46*: 866 (1993).

141. W. Wang, A. Vidal, J. B. Donnet, and M. J. Wang, *Kautsch. Gummi, Kunstst. 46*: 933 (1993).

142. W.-D. Wang, B. Haidar, A. Vidal, and J. B. Donnet, *Kautsch. Gummi, Kunstst. 47*:238 (1994).

143. W.-D. Wang, A. Vidal, G. Nanse, and J. B. Donnet, *Kautsch. Gummi, Kunstst. 47*: 493 (1994).

144. S. Asai, K. Sakata, M. Sumita, K. Miyasaka, and A. Sawatari, *Nippon Kagaku Kaishi 12*:1672 (1991).

145. M.-J. Wang and S. Wolff, *Rubber Chem. Technol. 65*:890 (1992).

146. F. J. Lopez-Garzon, M. Pyda, and M. Domingo-Garcia, *Langmuir 9*:531 (1993).

147. M. M. Chehimi, E. Pigois-Landureau, and M. Delamar, *J. Chim. Phys. 89*:1173 (1992).

148. M. M. Chehimi and E. Pigois-Landureau, *J. Mat. Chem. 4*:741 (1994).

149. M. M. Chehimi, S. Lascelles, and P. Armes, *Chromatographia 41*:671 (1995).

150. M. M. Chehimi, M. L. Abel, and Z. Sahraoui, *J. Adhesion Sci. Technol. 10*:287 (1996).

151. M. M. Chehimi, M. L. Abel, M. Delamar, J. F. Watts, and P. A. Zhdan, *AIP Conference Proceedings 354*:351 American Institute of Physics (1996).

152. G. Kraus, *Reinforcement of Elastomers*, Interscience, New York, 1965.

153. K. K. Unger, *Porous Silicas*, Elsevier, New York, 1979.

154. R. K. Iler, *The Chemistry of Silica*, John Wiley and Sons, New York, 1979.

155. C. R. Hegedus and I. L. Kamel, *J. Coating Technol. 820*:23 (1993).

156. C. R. Hegedus and I. L. Kamel, *J. Coating Technol. 820*:31 (1993).

157. H. Balard and E. Papirer, *Proc. Organic Coatings Science and Technology, 18th International Conference*, Athens, Greece, July 1992, pp. 13–32.

158. H. Balard and E. Papirer, *Prog. Org. Coatings 22*:1 (1993).

159. E. Papirer, P. Roland, M. Nardin, and H. Balard, *J. Colloid Interface Sci. 113*:62 (1986).

160. A. Vidal, E. Papirer, M. J. Wang, and J. B. Donnet, *Chromatographia 23*:121 (1987).

161. E. Papirer, H. Balard, Y. Rahmani, A. P. Legrand, L. Faccini, and H. Hommel, *Chromatographia 23*:639 (1987).

162. H. Balard, M. Sidqi, E. Papirer, J. B. Donnet, A. Tual, H. Hommel, and A. P. Legrand, *Chromatographia 25*:707 (1988).

163. H. Balard, M. Sidqi, E. Papirer, J. B. Donnet, A. Tual, H. Hommel, and A. P. Legrand, *Chromatographia 25*:712 (1988).

164. M. Sidqi, H. Balard, E. Papirer, A. Tual, H. Hommel, and A. P. Legrand, *Chromatographia 27*:311 (1989).

165. G. Linger, A. Vidal, H. Balard, and E. Papirer, *J. Colloid Interface Sci. 133*:200 (1989).
166. G. Linger, A. Vidal, H. Balard, and E. Papirer, *J. Colloid Interface Sci. 134*:486 (1990).
167. E. Papirer, G. Linger, H. Balard, A. Vidal, and F. Mauss, in *Chemically Modified Oxide Surfaces* (D. E. Leyden and W. T. Collins, eds.), Gordon and Breach, New York, 1990, pp. 15–26.
168. F. Joachim, A. Vidal, and E. Papirer, in *Chemically Modified Oxide Surfaces* (D. E. Leyden and W. T. Collins, eds.), Gordon and Breach, New York, 1990, pp. 361–373.
169. E. Papirer, J. M. Perrin, B. Siffert, and G. Philipponneau, *Prog. Colloid Polym. Sci. 84*:257 (1991).
170. J. Kuczynski and E. Papirer, *Eur. Polym. J. 27*:653 (1991).
171. E. Papirer, H. Balard, and M. Sidqi, *J. Colloid Interface Sci. 159*:238 (1993).
172. E. Papirer, H. Balard, J. Jagiello, R. Baeza, and F. Clauss, in *Chemically Modified Surfaces* (H. A. Mottola and J. R. Steinmetz, eds.), Elsevier, New York, 1992, pp. 351–368.
173. E. Papirer, J. M. Perrin, B. Siffert, G. Philipponneau, and J. M. Lamerant, *J. Colloid Interface Sci. 156*:104 (1993).
174. H. Balard, D. Yeates, E. Papirer, M. Gastiger, P. Bouard, F. Clauss, and R. Baeza, *Proc. Mineral and Organic Functional Fillers in Polymers: International Symposium*, Namur, Belgium, April 13–16, 1993.
175. H. Hadjar, H. Balard, and E. Papirer, *Colloid Surfaces A99*:45 (1995).
176. H. Balard, A. Saada, J. Hartmann, O. Aoudj, and E. Papirer, *Makromol. Chem., Macromol. Symp. 108*:63 (1996).
177. H. Balard, O. Aouadj, and E. Papirer, *Langmuir 13*:1251 (1997).
178. J. L. Oteo, J. Rubio, J. Tartaj, J. F. Fernandez, and C. Moure, *Bol. Soc. Esp. Ceram. Vidrio 30*:458 (1991).
179. A. Voelkel, *Crit. Rev. Anal. Chem. 22*:411 (1991).
180. Cr. Contescu, J. Jagiello, and J. A. Schwarz, *J. Catalysis 131*:433 (1991).
181. T. J. Bandoz, J. Jagiello, B. Andersen, and J. A. Schwarz, *Clays and Clay Minerals 40*:306 (1992).
182. T. J. Bandoz, K. Putyera, J. Jagiello, and J. A. Schwarz, *Microporous Materials 1*:73 (1993).
183. F. Bosse, A. Eisenberg, Z. Deng, and H. P. Schreiber, *J. Adhesion Sci. Technol. 7*:1139 (1993).
184. M. A. Zumbrum, *Polym. Mater. Sci. Eng. 70*:500 (1993).
185. M. A. Zumbrum, *Proc. International Symposium on the Interface, 16th Annual Meeting of the Adhesion Society*, Williamsburg, VA, February 21–26, 1993, p. 140.
186. M. A. Zumbrum, *J. Adhesion 46*:181 (1994).
187. H. Ping, W. Y. Fu, and Y. Fushan, *J. Liaoning Normal Univ. 18*:312 (1995).
188. M. Fafard, M. El-Kindi, and H. P. Schreiber, *J. Adhesion Sci. Technol. 7*:1139 (1993).
189. A. B. Scholten, H. G. Janssen, J. W. de Haan, and C. A. Cramers, *J. High Res. Chromatogr. 17*:77 (1994).
190. W. Wasiak, A. Voelkel, and I. Rykowska, *J. Chromatography A690*:83 (1995).

191. G. Liu and Z. Xin, *Chromatographia* 42:290 (1996).
192. H. P. Schreiber and F. St. Germain, *J. Adhesion sci. Technol.* 4:319 (1990).
193. H. P. Schreiber and F. St. Germain, in *Acid–Base Interactions: Relevance to Adhesion Science and Technology*, VSP, Utrecht, The Netherlands, 1991, pp. 273–285.
194. J. H. Burness and J. G. Dillard, *Langmuir* 10:1713 (1991).
195. J. H. Burness and J. G. Dillard, *Langmuir* 10:1894 (1994).
196. J. P. Casey, *Pulp and Paper, Chemistry and Chemical Technology*, 3rd ed., Vol. 4, Wiley-Interscience, New York, 1983.
197. D. H. Solomon and D. G. Hawthorne, *Chemistry of Pigments and Fillers*, John Wiley and Sons, New York, 1983.
198. H. S. Katz and J. V. Milewski, *Handbook of Fillers and Reinforcement for Plastics*, Van Nostrand-Reinhold, New York, 1978.
199. E. Papirer, J. Schultz and C. Turchi, *Eur. Polym. J.* 20:1155 (1984).
200. P. Schmitt, E. Koerper, J. Schultz, and E. Papirer, *Chromatographia* 25:786 (1988).
201. D. S. Keller and P. Luner, *Tappi Proceedings, Coating Conferences*, 1992, pp. 349–359.
202. M. Pyda, M. Sidqi, D. S. Keller, and P. Luner, *Tappi J.* 76:79 (1993).
203. A. Lundqvist and L. Odberg, *Pol. International Paper and Coating Chemistry Symposium*, 1996, pp. 29–34.
204. Z. Fu and S. Wei, *Guisuayan Xyebao* 17:308 (1989).
205. R. H. Leach and R. J. Pierce, *The Printing Ink Manual*, 5th ed., Blueprint, London, 1993.
206. J. W. Dove, G. Buckton, and C. Doherty, *Int. J. Pharmaceut.* 138:199 (1996).
207. M. D. Ticehurst, P. York, R. C. Rowe, and S. K. Dwivedi, *Int. J. Pharamaceut.* 141:93 (1996).
208. E. Papirer, H. Balard, E. Brendle, and J. Lignieres, *J. Adhesion Sci. Technol.* 10: 1401 (1996).
209. M. D. Ticehurst, R. C. Rowe, and P. York, *Int. J. Pharamaceut.* 111:241 (1994).
210. L. C. Matveivich, R. Z. Bakhtisina, F. Kh. Koudacheeva, and G. A. Berg, *Khim. Tverd. Topl.* 2:95 (1989).
211. E. Osmont and H. P. Schreiber, *Polym. Mat. Sci. Eng.* 58, 730 (1988).
212. H. P. Schreiber, M. R. Wertheimer, and M. Lambla, *J. Appl. Polym. Sci.* 17:2269 (1982).
213. H. P. Schreiber, C. Richard, and M. R. Wertheimer, in *Physico-Chemical Aspects of Polymer Surfaces*, Vol. 2 (Mittal ed.), Plenum Press, New York, 1983, pp. 739–748.
214. C. Richard, K. Hing, and H. P. Schreiber, *Polym. Composites* 6:201 (1985).
215. A. Lundqvist and L. Ödberg, in *The Fundamentals of Papermaking Materials* (C. F. Baker ed.), Vol. 2, Pira International, Surrey, UK, 1997, pp. 751–769.
216. L. Börås and P. Gatenholm, *Nordic Pulp Paper Res. J.*, 12:220 (1997).
217. V. Botaro and A. Gandini, *Cellulose* 5:65 (1998).

3

Interactions in Cellulose-Polyethylene Papers as Obtained Through Inverse Gas Chromatography

BERNARD RIEDL and HALIM CHTOUROU* Department of Wood Science, CERSIM, Laval University, Quebec City, Quebec, Canada

I.	Introduction	126
II.	Literature Review	126
	A. Early work	126
	B. Dispersive interactions	127
	C. Acid–base interactions	127
	D. Results on fiber, especially cellulose-based fiber	128
III.	Theory and Fundamentals	128
	A. Inverse gas chromatography	128
	B. Contact angle measurements	131
IV.	Experimental	131
	A. IGC measurement details	131
	B. Contact angle measurements	132
	C. Materials	132
V.	Results and Discussion	133
	A. IGC	133
	B. Dispersive component	133
	C. Donor–acceptor component	135

---

* *Current affiliation*: Rétec Inc., Granby, Quebec, Canada.

D. Comparison with wetting characteristics   138
E. Correlation with paper physical characteristics   139

VI. Conclusion   140

Symbols and Acronyms   141

References   142

I. INTRODUCTION

In chromatography, one is concerned with the eluting or mobile phase, which contains the substance to be analyzed. It is also possible to characterize the nonmobile phase, using the mobile phase and molecular probes it carries; thus the name *inverse gas chromatography* (IGC). Variation in surface chemistry of the nonmobile phase will change the elution characteristics of the probes. IGC can be used to measure dispersive and acceptor–donor interactions on surfaces, with subsequent correlation with adhesion between surfaces, as obtained through mechanical properties of composites. It is particularly suited to fibrous or powdery substrates, which may be difficult to characterize by other surface characterization techniques such as contact angle measurements, because of a rough surface but also due to wicking phenomena. The temperature variable is easy to vary in IGC and thus enthalpies of adsorption and related parameters are not difficult to obtain.

In this chapter, we will illustrate how it is possible to compare dispersive and acid–base contributions to adhesion in various lignocellulosic (LG) and polyethylene (PE) fibers, both virgin and chemically modified, as well as in composite papers. These interactions are measured by IGC.

II. LITERATURE REVIEW

A. Early Work

We will not try to encompass all previous work on IGC. There are several good reviews on the subject [1–4]. We will limit ourselves to measurements done at infinite dilution of the molecular probes although it is possible to do measurements at finite dilution.

Guillet [1] was among the first to use gas chromatography to detect the glass transition in polymers. Olabisi [5] applied IGC to the study of polymer–polymer interactions as had done Patterson and coworkers [6]. In these studies, the polymer or blend is coated on a so-called inert support. Munk [7] did extensive studies on whether the support, probes, and general technique did influence results. In studies of polymer blends, the elution behavior of volatile organic compounds, called "molecular probes" and injected in very small quantities in the carrier gas stream, is determined for the two or more pure polymers and compared to that of the polymer blend. If the retention time of a probe is less than the linear combination of retention times for the respective pure polymers, with respect to volume ratio, then there is a negative or favorable interaction between the polymers. An excess retention can thus be interpreted in terms of interpolymer interactions.

B. Dispersive Interactions

Generally interactions between surfaces are either of the dispersive (also called Lifshitz–van der Waals, or nonpolar) type and polar (or acid–base, or acceptor–donor). Some authors differentiate polar and donor–acceptor interactions. However, Fowkes [8] showed that polar interactions are small or nil compared to donor–acceptor interactions.

C. Acid–Base Interactions

Donor–acceptor interactions in IGC are obtained by subtracting the dispersive interaction from the total interaction. The dispersive component is obtained with n-alkanes. For instance, the dispersive component of the heat of adsorption on a substrate is obtained from the temperature dependence of the elution volumes of a series of alkanes. These values are subtracted from the heats of adsorption of different polar probes with different acceptor, donor, or amphoteric properties. What remains is deemed only to be due to donor–acceptor interactions between the polar probes and the polar surface. Volumes of retention for a series of polar probes on PE fiber follow a regular progression with increasing size of the n-alkane probe. It was observed that polar probes were only retained as to the proportion of the dispersive component of their surface tension [9,10]. On polar surfaces, probes with donor characteristics are retained more on surfaces with acceptor surface properties and vice versa. The donor–acceptor characteristics of the probes are known, whereas those of the surface are not. There are several scales or ways of classifying the donor–acceptor numbers of probes. Drago [11] and Gutmann [12] defined scales for such use. Furthermore, donor and acceptor characteristics of the substrate surface can be obtained from a correlation of the polar probe retention characteristics and their donor–acceptor number. Thus the true surface properties of the surface of the substrate can be related to surface adhesion–enhancing treatments and adhesion with other substrates [2].

D. Results on Fiber, Especially Cellulose-
Based Fiber

Fiber or fiber mixture is a bit of an ideal system for performance of IGC, compared to thermoplastic composites or blends where one has partial interpenetra-
tion of surfaces. In fibrous mixtures, surfaces remain neat and separate. Also,
performing contact angles studies on fibers and papers is difficult because of
irregular surface, small substrate size for fibers, and wicking phenomena. IGC
does not suffer from those limitations. Fiber, chemically modified or not, is
packed into a column and times of retention of various probes are studied, as for
any other polymeric substrate. There are several instances in the literature of IGC
studies on paper fiber, especially in terms of the dispersive components [13–15].

   Lundqvist et al. [16] showed that the surface properties of various bleached
pulp fiber could be broken down in dispersive and donor–acceptor contributions.
The dispersive component was found to be approximately 45–48 mJ/m^2 and the
paper had an acid surface. Birch wood meal has a predominantly acid (acceptor)
surface but does have basic sites. The acceptor number (K_A) was 4.40 and the
donor number (K_B) 3.00 [17]. Other work will be mentionned in the discussion
section.

III. THEORY AND FUNDAMENTALS

A. Inverse Gas Chromatography

In this technique a small amount of a gas, referred to as a (molecular) probe, is
injected into a column packed with the substrate of interest, called the nonmobile
phase, while an inert carrier gas sweeps away the probe. A long retention time
indicates that the probe is in strong interaction with the nonmobile phase, whereas
a short retention time indicates low interaction. In general, retention time increases with probe and nonmobile phase polarities. It is supposed that probe vapor
molecules interact with the nonmobile phase surface winhout interacting with
other probe molecules, thus the term "infinite dilution." To ensure flash vaporization, the temperature of the injection ports of the chromatograph is around
300°C, which is 50°C above the boiling point of the probes used. The probes
used to evaluate the dispersive component of the surface energy of the stationary
phases are n-alkanes ranging from n-hexane to n-decane, whereas a series of
polar probes are used to evaluate donor–acceptor interactions.

   Calculations are as follows [2]: By assuming that the retention mechanism is
only due to surface adsorption, the net retention volume, V_N, the fundamental
variable in IGC, which is the volume of the carrier gas required to elute the probe
from the column, is given by:

$$V_N = K_s A = Q(t_r - t_m) \tag{1}$$

where K_s is the surface partition coefficient of the given probe between the stationary and the mobile phases. A is the surface area (m^2) of the stationary phase, equal to the specific surface multiplied by the sample weight, t_r is the retention time of the injected probe through the column, t_m is the retention time of methane, Q is the corrected flow rate of the mobile phase at column temperature and 760 mm Hg obtained as

$$Q = Q_0 j \frac{T_c}{T_a} \left(1 - \frac{P_w}{P_a} \right) \tag{2}$$

where Q_0 is the measured flow rate (mL/min), T_c and T_a are the experimental and ambient temperatures (K), respectively, P_a and P_w are the atmospheric and the saturated vapor pressures of water at ambient temperature, and J is the James-Martin compression correction term determined as follows, where P_1 is equal to P_a plus the pressure drop in the column:

$$J = \frac{3}{2} \frac{1 - \left(\frac{P_1}{P_a} \right)^2}{1 - \left(\frac{P_1}{P_a} \right)^3} \tag{3}$$

The specific net retention volume (V_g^0) at) at 0°C and per gram of absorbent is given by the following equation [18], where W is the weight of nonmobile phase:

$$V_g^0 = \frac{273.15}{T_c} \frac{V_N}{W} \tag{4}$$

The standard free energy, the standard enthalpy, and the standard entropy of adsorption, ΔG_A^0, ΔH_A^0, and ΔS_A^0, respectively, are given by the following expressions:

$$-\Delta G_A^0 = RT \ln V_g^0 + C = RT \ln \left(K_s \frac{P_{s,g}}{\Pi_s} \right) \tag{5}$$

$$-\Delta H_A^0 = R \frac{d(\ln V_g^0)}{d\frac{1}{T}} \tag{6}$$

$$-\Delta S_A^0 = \frac{\Delta G_A^0 - \Delta H_A^0}{T} \tag{7}$$

where $P_{s,g}$ is the absorbate vapor pressure in the gaseous standard state, equal to 101 kN/m^2, and Π_s is the spreading pressure of the absorbed film to a reference gas phase state defined by the pressure $P_{s,g}$ of the solute, equal to 0.338 mN/m [19].

A linear variation of $-\Delta G_A^0$ (or $RT \ln V_N$) as a function of the number of carbon atoms in the nonpolar probes (n-alkanes) is a common observation. The free energy of adsorption corresponding to one methylene group, $-\Delta G_A^0(-CH_2-)$, was obtained through the injection of a homologous series of n-alkane probes:

$$-\Delta G_A^0(-CH_2-) = RT \ln \frac{V_N(C_{n+1}H_{2n+4})}{V_N(C_nH_{n+2})} \tag{8}$$

The London dispersive component (γ_S^D) of the surface energy of the stationary phase is calculated by the following equation [20]:

$$\gamma_S^D = \frac{1}{4} \frac{\Delta G_A^0(-CH_2-)^2}{\gamma(CH_2)N^2a^2} \tag{9}$$

where N is Avogadro's number, a is the area of an absorbed methylene group, and $\gamma(CH_2)$ is the surface energy of pure methylene group surface [20], $\gamma(CH_2)$ = 35.6 + 0.058 (293 − T), in mJ/m^2.

St. Flour and Papirer [21] found that linearity is usually obtained when plotting $RT \ln(V_N)$ as a function of $\ln (P_0)$ in the case of n-alkane probes. In addition, Schultz et al. [22] found a linear relation by plotting $RT \ln(V_N)$ as a function of $a(\gamma_L^D)^{1/2}$, where a is the surface area of the probe and γ_L^D is its dispersive component of surface energy in the liquid state. These correlations are observed because n-alkanes at infinite dilution behave nearly as ideal gases.

To determine the contribution of the surface specific interactions to the total surface energy it is necessary to inject polar probes in the column, in addition to n-alkanes. By assuming that alkanes are involved in only dispersive interactions and also assuming that dispersive and polar components of surface energy are additive, the alkane line may be taken as a reference for the determination of the dispersive component for the polar probes. The difference of ordinates between the alkane straight line and the polar probe gives ΔG_A^{SP} or ΔH_A^{SP}, corresponding to the specific interaction.

As noted by Fowkes, the specific interactions are all acid–base. Schultz et al. [22] characterized solid surfaces by an acidic constant K_A and a basic constant K_B, using Gutmann's [12] acid–base concepts, and proposed the following equation:

$$-\Delta H_A^{SP} = K_A DN + K_B AN \tag{10}$$

where ΔH_A^{SP} is the enthalpy of adsorption component corresponding to specific interactions K_A and K_B, which are, respectively, equivalent to a AN (acceptor or acid number) and DN (donor or basic number) of molecules at the surface of the nonmobile phase, in this case paper. According to Lewis, an acid is an electron acceptor and a base an electron donor. The DN expressed in kilojoules per mole is the molar enthalpy of interaction between a base and a reference acceptor $SbCl_5$, in a dilute solution of 1,2-dichloroethane. Unlike the DN, the AN is in arbitrary units set at zero for the NMR shift induced by hexane and at 100 for $SbCl_3$ in dilute solutions of 1,2-dichloroethane [12]. ΔH_A^{SP} is obtained from a subtraction of the dispersive contribution of the probe from the overall heat of adsorption.

Once the donor and acceptor numbers have been determined for each fiber from Eq. (10), a specific interaction parameter can be obtained for the mixture of two fibers in equal proportions as:

$$I_{SP} = K_A^{PE} K_B^{Lig} + K_B^{PE} K_A^{Lig} \tag{11}$$

Note that there is something missing in this last equation: If a fiber has both acceptor and donor sites, there can be interaction between these, and it will contribute to the strength of the paper.

A linear combination of these can be done according to concentration, as detailed in Refs. 10 and 23:

$$I_{SP} = [2P_{PE}^2 \, K_A^{PE} K_B^{PE}] + [2P_{PE}P_{lig}(P_A^{PE} K_B^{Lig} + K_B^{PE} K_A^{Lig})] + [2P_{lig}^2 K_A^{lig} K_B^{Lig}] \tag{12}$$

where P refers to the concentration of each fiber. Thus the first term represents the contribution of contacts between PE fibers, the second term represents the contribution of the PE/LG contacts, i.e., unlike fibers, and the last term represents the contribution of contacts between LG fibers.

B. Contact Angle Measurements

Contact angle measurements can be made indirectly by the Wilhelmy balance technique. Fiber, as such or as paper, is immersed in a probe liquid where the force F, less the buoyancy, exerted on the sample is:

$$F = L\gamma_w \cos \theta \tag{13}$$

where L is the perimeter of the sample, γ_w the surface tension of the probe liquid, and θ the contact angle of the solvent or probe liquid with the paper sample. In the case of n-hexane as probe liquid, the angle is essentially zero and the we can obtain the perimeter of the sample from Eq. (13).

IV. EXPERIMENTAL

A. IGC Measurement Details

The chromatographic measurements were carried out at infinite dilution with a Hewlett-Packard 5700A apparatus equipped with dual flame ionization detectors maintained at 300°C. A 1-μl microsyringe is filled repeatedly with probe vapor and only the content is injected in the column. Peaks are generally symmetrical and their shape is not dependent on the amount injected in this dilution range. The stationary phases were packed into copper tubing of 1.2 m length and 4.0 mm internal diameter. To maintain a constant temperature of the column, which was controlled within ±0.5 °C and monitored with an Omega digital thermometer, a circulating water bath was used (Julabo, Model UC-5B). Nitrogen was used as a carrier gas (also called the mobile phase) and methane was used to determine the dead volume of the column. The columns were packed under vacuum with a weighed amount of the nonmobile phase fiber and conditioned under carrier gas for a few hours prior to measurement.

B. Contact Angle Measurements

Contact angle measurements are made indirectly by the Wilhelmy balance technique: Composite papers are cut into strips of 2 × 3 cm. These are immersed in n-hexane at 26°C as probe liquid, where the contact angle is essentially zero, and we can obtain the perimeter of the sample from Eq. (13). The software automatically corrects for buoyancy effects. The apparatus used was a Kruss K14 Wilhelmy plate microbalance, at 2.7 mm/min immersion rate. Measurements were also done against water as probe liquid at 26°C.

C. Materials

In this work we are interested in characterizing the surface properties of composite papers and their relation with mechanical properties. By composite papers we mean papers made of natural LG fiber and untreated or treated PE fiber.

1. PE Fiber

This material, Pulplus QP 3850, was received from DuPont as thick, low-density sheets. They were defibrated and fiber was used as such or modified by a proprietary oxyfluoration treatment (Air Products and Chemicals) at two different levels of treatment [10,23]. This type of PE fiber contained, in the virgin state, oxygen on the surface, as polyvinyl alcohol and polyvinyl acetate. After treatment, the O/C ratio, from X-ray photoelectron spectroscopy (XPS) measurements, increased from 9.9% for untreated fiber, to 11.2% and 17.1%, for the two treatments, respectively. There was no fluorine on the surface of the untreated PE fiber. After treatment, the F/C ratio was 1.5% and 9.1% for the two treatments, respectively. In precedent work, these were mixed with the natural fiber mentioned below; IGC measurements were done, as well as mechanical characterization of papers, with the same fiber proportions [9,23–24].

2. Lignocellulosic Fiber

Wood pulp was either explosion type or kraft pulp made from Aspen and Birch, respectively [9].

V. RESULTS AND DISCUSSION

In this section, we will illustrate some situations where IGC can be a powerful technique to characterize modified LG and PE surfaces. We will present some results from the literature and some of our own results on both dispersive and donor–acceptor interactions.

A. IGC

More complete data on the following can be found in precedent publications [9,10,23,24]. IGC technique was applied to investigation of chemically modified and unmodified PE fiber and cellulose fiber.

B. Dispersive Component

Pioneering work on measurements of dispersive components of surface tension of cellulose fibers in the presence and absence of water was done through IGC by Dorris and Gray [25,26]. It was found that the surface properties of the water–cellulose system became independent of the amount of water at moisture contents above 20%. Heats of adsorption of alkanes on fibers were higher at high moisture contents.

The standard free energy for the zero-coverage adsorption of n-nonane on various fibers is listed in Table 1, second to last column. Values of PE fiber, treated or not, are very similar, near 28–30 kJ/mol, although they do increase as the concentration of oxygen and fluorine increases at the surface. The chemical treatment done on these fibers increases to a quite large degree their donor–acceptor numbers, as the two preceding columns show. Note that these numbers for untreated PE are near zero, although it does have some oxygen at the surface. The conclusion is that to an increasing extent the polarity of a fiber does not necessarily contribute to its dispersive properties. This leads credence to the assumption that dispersive and polar effects are additive and more or less independent of each other. This can also be seen in the dispersive component of the surface tension (last column of the table), which does not vary much either, for the PE fibers, irrespective of treatment. Again on the natural fibers, kraft and explosion pulps, the value of heats of adsorption, and dispersive components of surface tension do not vary by much. Treatments with some more hydrophobic agents can change the dispersive component in a substantial way [27]. Lee and Luner [15] with treatment with alkylketene dimer, a hydrophobic sizing agent, increased γ_s^d of cellulose fiber from 47.4 to 60.4 mN m^{-1}.

Dispersive interactions can still be very useful in exploring the surface characteristics of composites. For instance, in thermoplastic fiber composites, if the reinforcing fiber and the continuous phase have markedly different values of the dispersive components, one can vary the amount of polymer grafted or coated

TABLE 1   Contact Angles of Water on Composite Papers at 26°C as a Function of
Content in Explosion Pulp, for Papers Made with PE Pulp Using First- and Second-
Degree Oxyfluoration, as Well as Donor (K_B) and Acceptor (K_A) Numbers, Standard
Free Energy for the Zero-Coverage Adsorption (n-Nonane), and Dispersive
Components of Surface Tension for Unmixed Fibers

	Contact angle θ, paper, 26°C, against water	K_A	K_B	$-\Delta G_A^0$ nC$_9$ 50°C (kJ/mol)	γ_s^D50°C (mJ/m^2)
Untreated PE fiber	—	0	0	29.2	32.7
PE, 1st treatment level	(121)	7.3	2.5	28.0	32.8
PE, 2nd treatment level	(100)	10.3	9.2	27.8	31.3
Kraft pulp	—	5.5	5.0	19.4	6.8
Explosion pulp	69	2.4	12.1	20.6	9.8
Mixed paper, %, explosion pulp:					
Treatment 1:					
25	110				
50	88				
75	77				
Treatment 2:					
25	93				
50	86				
75	75				

Values in parentheses were obtained by regression on contact angles versus composition.
Source: IGC data from Refs. 10 and 24.

on the fiber and record the variation of the dispersive components or heats of
adsorption. When the value reaches that of the continuous phase, then we can
infer from the amount of polymer grafted, knowing the specific surface of the
fiber, the thickness of the interfacial layer. At that stage it will also be necessary
to graft more polymer on the fiber surface. In earlier work [28,29], we grafted,
with the free radical mechanism, polymethylmethacrylate (PMMA) onto cellulose fibers, with up to 140% polymer loading on the fiber. The purpose was to
verify at what level of loading, e.g., thickness, the properties of the grafted fiber
at the surface would become like those of the bulk phase. As can be seen in Fig.
1, which shows the dispersive component of the surface tension of paper CTMP
fiber as a function of PMMA fiber loading, this value ceases to vary when the
polymer loading is near 140% (which means the composite contains 50% PMMA
as wt%). At that point, the properties of the grafted PMMA become like those
of the bulk phase (the isolated points on the lower right corner of the figure). At
that loading level, the figure additionally shows the O/C ratio of the surface,
obtained from photoelectron spectroscopy, which also becomes like that of pure

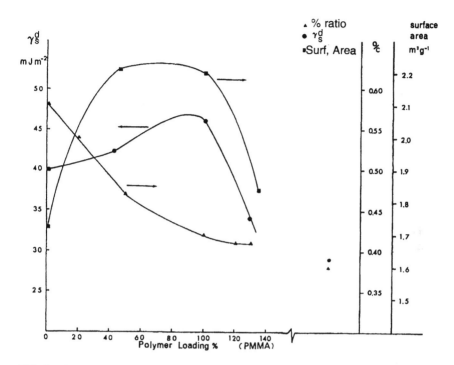

FIG. 1 Comparison between spectroscopic and thermodynamic measurements on poly-methyl methacrylate (PMMA)–grafted chemithermomechanical pulp. Dispersive component of the surface energy, γ_s^d (from IGC), atomic oxygen/carbon (O/C) surface ratio, surface atomic O/C ratio (from photoelectron spectroscopy), and specific surface area (from adsorption isotherms), as a function of polymer loading on the fiber. The two points at the lower right of the figure are for the pure PMMA. (From Ref. 28.)

PMMA. This implies that grafting more polymer onto the surface will not improve adhesion characteristics any more. Thus, from IGC measurements it is possible to optimize surface modification of fibers for the manufacture of composites.

Similar results were obtained by other workers with different substrates and a coating rather than grafting process [30]. The dispersive component of the surface tension of polystyrene (PS) on thermomechanical pulp was obtained (Fig. 2) as a function of PS loading. At 20% loading, values of the dispersive component cease to change. The authors estimate the depth of the interphase at 0.12 μm.

C. Donor–Acceptor Component

Acid–base interactions through IGC and relations with adhesion phenomena have been discussed in the literature [31]. As in Eq. (6), heats of adsorption of the

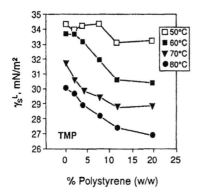

% Polystyrene (w/w)

FIG. 2 Variation of the dispersive component of the surface tension of paper fiber with increasing loading in polystyrene. (From Ref. 30.)

probes can be obtained from the temperature dependence of the volume of retentions. An example of this is given in Fig. 3 [10,24] for kraft pulp. First, n-alkanes show a linear variation with respect to the horizontal axis, which is in $a(\gamma_s^d)^{1/2}$, a measure of the dispersive surface area of the probes. Other horizontal scales have been used by different authors [2]. The other probes as electron donors and acceptors, or both (amphoteric), do show an additional interaction with the fiber surface. With an untreated PE surface, even the untreated PE surface examined here, which does show, in XPS, surface oxygen, all probes nearly line up on the n-alkane reference line [10,24]. With treatment with oxygen and fluorine an additional enthalpic contribution appears, which increases with the degree of treatment. The difference between the n-alkane baseline and the actual value for the polar probes is called the specific contribution to the heat of adsorption and is a function of *both* the donor–acceptor characteristics of the probe and the surface of the nonmobile phase From Eq. (10) the donor and acceptor number of the surface can be ascertained (see Table 1). An electron donor (basic) probe will have a large heat of adsorption on a surface with a high concentration of electron acceptor (acid) sites. For instance, kraft pulp has a high donor number of 5.0, and with the probe in Fig. 3 gives the largest value of the heat of adsorption, CH_3NO_2 has the largest acceptor number, 20.5, of the probes used. Thus, IGC shows (see Table 1) that the surface electron acceptor and donor characteristics of the PE fiber does increase with chemical treatment, and a large contribution of these donor–acceptor interactions to the mechanical properties or adhesion can be expected.

With Eq. (12), this reasoning can be further expanded. Contributions to the I_{sp} can be broken down in several contributions due to PE–PE contacts, PE–LG "unlike" contacts, and LG–LG contacts. In Fig. 4, such a distribution has been

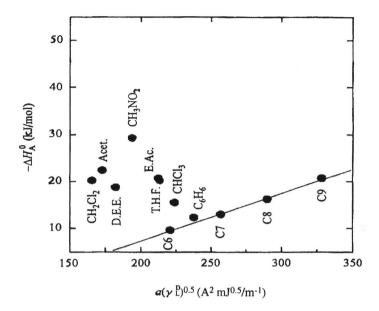

FIG. 3 Plot of $-\Delta H_A^0$ for the *n*-alkanes and the polar probes on kraft pulp fiber. (From Refs. 10 and 24.)

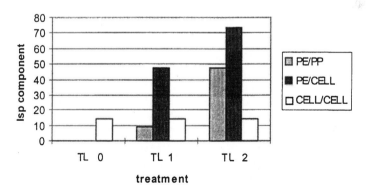

FIG. 4 Variation in donor–acceptor interaction parameter, I_{sp}, as a function of degree of oxyfluoration of PE fiber (T.L. 0 corresponds to no treatment, T.L. 1 and T.L. 2 correspond to level 1 and 2 of treatment, respectively), for 50:50 blend of explosion and PE pulp, and broken down in contributions due to the three different types of fiber–fiber contacts. (Data from Ref. 10.)

done (in the figure "CELL" stands for LG). With no treatment on the fiber, there are no contributions to I_{sp} from the PE fiber, only from the LG–LG contacts, as these fibers have both donor and acceptor surface properties. With treatment 1, there is a dominant contribution (67%) to I_{sp} contacts from unlike interfiber contacts. With further treatment of the PE fiber, this last contribution is still dominant (54%), but the contribution of PE–PE (35%) is stronger. Of course, the contribution of LG–LG contacts is constant because their surface remains the same. Note that if one fiber had only donor sites and the other only acceptor sites, only the unlike contacts contribution will contribute to I_{sp}. Thus, although the PE fiber is less strong mechanically than the LG fiber, as far as the paper is concerned, the mechanical properties can still be increased by increased adhesion between fibers, as will be seen in the last section.

D. Comparison with Wetting Characteristics

There are few comparisons between surface energy measurements by IGC and wetting measurements. Felix and Gatenholm [27] did such a study on cellulose fibers. From IGC, γ_s^D was 44.0 mJ/m², whereas a corresponding value of 25.5 mJ/m² was found with wetting measurements. This difference was attributed to a fundamental difference between IGC and wetting measurements as the former is more sensitive to surface high-energy sites than the latter [32]. Similar differences were seen for treated fibers. Shi et al. [33] also compared, as obtained through IGC and column wicking (capillary rise) measurements, acceptor–donor characteristics of hardwood fiber and automobile fluff (automobile shredder residue). Both techniques showed fluff surface as predominantly acidic, with wood fiber having a dispersive component value of 43–49 mJ/m².

Quillin et al. [34] did an intensive study on Pulpex fiber. They assessed surface properties through wetting (contact angle measurements). They found a positive correlation between donor–acceptor work of adhesion and the internal bond strength of the composite papers. Haidara et al. [35] did a study of the temperature dependence of interfacial tensions with the Wilhelmy plate method. The entropic effects involved were large and results were correlated with acid–base properties. There is also a discussion of donor–acceptor effects in wetting by Berg [36].

With our composite papers, we did not characterize evolution of surface tension by wetting with treatment of PE or proportion of LG fiber. However, we did some angle of contact measurements with water on composite papers, which are shown in Table 1, first column. The angle of contact decreases with a higher proportion of LG fibers in composite papers, being lowest on pure explosion pulp fiber paper. The contact angle also decreases with increasing treatment and polarity of PE, but remains high. This suggests that a high proportion of hydrophobic sites have not been affected by the treatment, although the polarity of the treated fiber, no. 2, is about as high (K_B) or higher (K_A) as that of explosion pulp.

E. Correlation with Paper Physical Characteristics

There are several instances of strong correlations between donor–acceptor characteristics and mechanical properties. One of the best examples is the work of Schultz and Lavielle [22]: The interfacial shear strength for carbon fiber reinforced epoxy is linearly related to an interaction parameter similar to that given in Eq. (11). Other authors have shown such correlations [37]. We have shown in precedent work [23] that tear index and burst strength of composite papers, previously discussed here, are much increased by oxyfluoration treatments, as shown in Fig. 5. These properties are also linearly related with the specific interaction parameter, I_{sp}, as can be seen in Fig. 6 (where degree of treatment progresses from left to right, in both parts of the figure). In this case chemical treatments result in increased interfiber-specific interactions, which end up increasing the mechanical properties of the papers, without change in the mechanical properties of the individual fibers. From the initial values of mechanical properties with

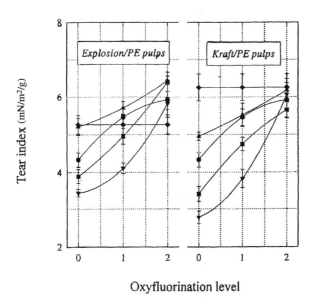

Oxyfluorination level

FIG. 5 Tear index of paper as a function of the oxyfluoration of PE pulp, and LG/PE pulp proportions. ◆, 0% PE; ▲, 25% PE; ■, 50% PE; ▲, 75% PE; ●, 100% PE. (From Refs. 10 and 23.)

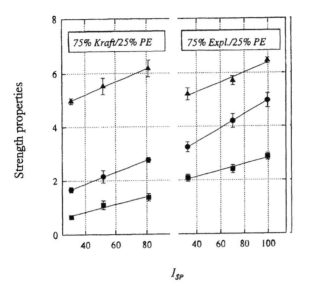

I_{SP}

FIG. 6   Strength properties of LG/PE pulp paper, for 75:25% blend, as a function of the specific donor–acceptor interaction parameter, as changed through surface oxyfluoration of PE pulp fiber. ▲, Tear index (mN.m^2/g); ●, B. length (km); ■, Burot index (kPa.m^2/g). (From Refs. 10 and 23.)

a proportion of 25% untreated PE fiber, properties increase anywere from 30% to 50% with treatment 2 on PE fiber, and this correlates well with concomitant increases in I_{sp} values obtained from IGC. In composite papers such as these, any increase in mechanical properties must come from an increase in interfiber adhesion, in this case related to donor–acceptor interactions.

VI. CONCLUSION

IGC results correlate well with experimental mechanical properties of composite papers. Although oxyfluoration treatments increase donor–acceptor interactions to a large extent, the dispersive characteristics of the treated fibers did not change much. The contact angle with water shows decreasing values with enhanced oxyfluorine treatment of fiber, which is consistent with enhanced surface polarity.

The dispersive component is not very sensitive or influenced by chemical treatments, but it can be put to good use, e.g., measuring the depth of the interphase, because of this relative insensitivity.

The donor–acceptor surface characteristics can be somewhat more difficult to obtain but can be highly sensitive to chemical surface modification. This is closely related to adhesion properties generated across the interface. These characteristics can be further broken down in acid–base or base–acid contribution due to both phases, remembering that many substrates, especially natural ones, have an amphoteric nature.

ACKNOWLEDGMENT

The authors gratefully acknowledge the financial help of the Conseil de récherches en sciences naturelles et en génie du Canada, of the Faculté de foresterie et de géomatique de l'Université Laval and of the Fonds pour la formation de chercheurs et l'aide à la recherche du Québec.

SYMBOLS AND ACRONYMS

A	surface area of stationary phase
AN	acceptor number of probe
θ	angle of contact of fiber paper with water
DN	donor number of the probe
F	force exerted on sample
γ_w	surface tension of water
$-\Delta G_A^0$	standard free energy of adsorption of probe
$-\Delta G_A^0(-CH_2-)$	free energy of adsorption of one methylene group
$-\Delta H_A^0$	standard enthalpy of adsorption of the probe
$-\Delta H_A^{SP}$	enthalpy of adsorption component of the probe due to specific interactions
I_{sp}	interaction (donor–acceptor) parameter
IGC	inverse gas chromatography
j	James-Martin compression correction term
K_a	acid (acceptor) number of nonmobile phase
K_b	basic (donor) number of nonmobile phase
K_A^{lig}	acid (acceptor) number of paper fiber
K_B^{lig}	basic (donor) number of paper fiber
K_A^{PE}	acid (acceptor) number of modifed polyethylene
K_B^{PE}	basic (donor) number or modifed polyethylene
K_s	partition coefficient
L	perimeter of sample
LG	lignocellulosic
P_a	atmospheric pressure
$P_{s,g}$	absorbate (probe) vapor pressure in gaseous standard state
P_w	saturated vapor pressure of water at ambient temperature
P_1	pressure drop in the column $+P_a$
PE	polyethylene

PMMA	polymethyl methacrylate
PS	polystyrene
Π_s	spreading pressure of absorbed film
Q	corrected flow rate of carrier gas
Q_0	measured flow rate
$-\Delta S_A^0$	standard entropy of adsorption of probe
t	time of retention of the gas probe in column
T_a	ambient temperature
T_c	temperature of column
t_m	time of retention of methane in column
V_N	volume of retention
$V_N(C_{n+1}H_{2n+4})$	volume of retention of an incremental methylene unit
$V_N(C_nH_{n+2})$	volume of retention of an alkane
V_G^0	specific net retention volume
W	weight of nonmobile phase
XPS	X-ray photoelectron spectroscopy

REFERENCES

1. J. E. Guillet, in *New developments in Gas Chromatography* (J. H. Purnell, ed.), Wiley-Interscience, New York, 1973, p. 187.
2. D. R. Lloyd, T. C. Ward, H. P. Schreiber, and C. C. Pizaña, *Inverse Gas Cromatography Characterization of Polymers and Other Materials*, ACS Symposium Series No. 391, American Chemical Society, Washington, D.C., 1989.
3. D. J. Gray, in *Progress in Polymer Science*, Vol. 5 (A. D. Jenkins, ed.), Pergamon Press, Oxford, 1977, p. 1.
4. J. M. Braun and J. E. Guillet, Adv. Polym. Sci. *21*:108 (1976).
5. O. Olabisi, Macromolecules *8*:316 (1975).
6. D. Patterson, Y. B. Tewari, H. P. Schreiber, and J. E. Guillet, Macromolecules *4*: 356 (1971).
7. Z. Y. Al-Saigh and P. Munk, Macromolecules *17*:803 (1984).
8. F. M. Fowkes, J. Adhesion Sci. Technol. *1*:7 (1987)
9. H. Chtourou, B. Riedl, and B. V. Kokta, J. Colloid Interface Sci. *158*:96 (1993).
10. H. Chtourou, Ph.D. thesis, Université Laval, Québec, Canada, 1994.
11. R. S. Drago, G. C. Vogel, and T. E. Needham. J. Am. Chem. Soc. *93*:6014 (1971).
12. V. Gutmann, *The Donor–Acceptor Approach to Molecular Interactions*, Plenum Press, New York, 1983.
13. H. K. Lee and P. Luner, Nordic Pulp Paper Res. J. *2*:164 (1989)
14. H. K. Lee and P. Luner, J. Colloid Inter. Sci. *146*:195 (1991).
15. H. K. Lee and P. Luner, J. Wood Chem. Technol. *13*:127 (1993).
16. Å. Lundqvist, L. Ödberg, and J. C. Berg, Tappi J. *78*:139 (1995).
17. D. P. Kamdem, S. K. Bose, and P. Luner, Langmuir *9*:3039 (1993).
18. D. G. Gray and J. E. Guillet, Macromolecules *3*:316 (1972).

19. J. H. De Boer, *The Dynamic Character of Adsorption*, Oxford University Press (Clarendon), Oxford, 1953.

20. G. M. Dorris, Ph.D. thesis, McGill University, Montréal, Québec, Canada, 1979.

21. Saint-Flour and C. Papirer, Ind. Eng. Chem. Prod. Res. Dev. *21*:666 (1982).

22. J. Schultz and L. Lavielle, in *Inverse Gas Cromatography Characterization of Polymers and Other Materials* (D. R. Lloyd, T. C. Ward, H. P. Schreiber, and C. C. Pizaña, eds.), ACS Symposium Series No. 391, American Chemical Society, Washington, D.C., 1989, pp. 185–202.

23. H. Chtourou, B. Riedl, and B. V. Kokta, Tappi J. *80*:141 (1997).

24. H. Chtourou, B. Riedl, and B. V. Kokta, J. Adhesion Sci. Technol. *9*:551 (1995); also appeared as chapter in *Polymer Surface Modification: Relevance to Adhesion* (K. L. Mittal, ed.) VSP, Utrecht, Netherlands, 1996, pp. 455–478.

25. G. M. Dorris and D. G. Gray, Chem. Soc. Faraday Trans. I *77*:725 (1981).

26. G. M. Dorris, and D. G. Gray. J. Colloid Interface Sci. *71*:93 (1979).

27. J. Felix and P. Gatenholm, Nord. Pulp Paper Res. J. *8*:200 (1993).

28. B. Riedl and P. D. Kamdem, J. Adhesion Sci. Technol. *6*:1053 (1992); also published in *Contact Angle, Wettability and Adhesion* (K. L. Mittal, ed.), VSP, Utrecht, Netherlands, 1993, pp. 625–638.

29. P. D. Kamdem and B. Riedl, J. Wood Chem. Technol. *11*:57 (1991).

30. J. Simonsen, Z. Hong, and T. G. Rials, Wood Fiber Sci. *29*:75 (1997).

31. W. B. Jensen, in *Surface and Colloid Science in Computer Technology* (K. L. Mittal, ed.), Plenum Press, New York, 1987.

32. J. Borch, J. Adhesion Sci. Technol. *5*: 523 (1991).

33. Q. Shi, D. G. Gardner, and J. Z. Wang, in *Fourth International Conference on Woodfiber-Plastic Composites*, Forest Products Society, Madison, Wisconsin, 1997, pp. 245–256.

34. D. T. Quillin, D. F. Caulfield, and J. A. Koutsky, Mat. Res. Soc. Symp. Proc. *266*: 113 (1992).

35. H. Haidara, M. -F. Pinhas and J. Schultz, J. Adhesion *54*:155 (1995).

36. J. C. Berg, in *Wettability* (J. C. Berg, ed.), Surfactant Science Series No. 49, Marcel Dekker, New York, 1993, pp. 75–148.

37. A. E. Bolvari and T. C. Ward, in *Inverse Gas Cromatography Characterization of Polymers and Other Materials* (D. R. Lloyd, T. C. Ward, H. P. Schreiber, and C. C. Pizaña, eds.), ACS Symposium Series No. 391, American Chemical Society, Washington, D.C., 1989, pp. 12–19.

4

Inverse Gas Chromatography: A Method for the Evaluation of the Interaction Potential of Solid Surfaces

EUGÈNE PAPIRER and HENRI BALARD   Institute of Chemistry of Surfaces and Interfaces, French National Center of Scientific Research, ICSI–CNRS, Mulhouse, France

I.	Introduction	146
II.	Inverse Gas Chromatography	147
	A. Definition	147
	B. Practice of IGC	148
III.	IGC at Infinite Dilution (IGC-ID)	149
	A. Fundamental equation of IGC at infinite dilution conditions	149
	B. London interaction potential	150
	C. Specific interaction potential	152
	D. Importance of the surface nanomorphology	157
IV.	IGC at Finite Concentration Conditions	160
	A. Acquisition of the adsorption isotherms using IGC-FC	161
	B. Thermodynamic quantities calculated from the isotherms	162
	C. Estimation of the solid's surface heterogeneity using adsorption energy distribution functions	163
	D. Examples of distribution functions of the adsorption energies: influence of the nature of the probe on the distribution function shape	164
V.	Advantages and Limitations of IGC	168
VI.	Conclusion	168
	References	169

I. INTRODUCTION

The role of solid–solid, solid–liquid, and solid–gas interfaces is presently well recognized in numerous phenomena, especially in chromatographic processes, and in numerous applications of divided solid surfaces such as fine-powdered mineral oxides used as fillers in elastomers or paints, catalysts supports, and thixotropic agents. For instance, the rheological behavior of powder mixes, either in the dry state or in suspension in liquids, clearly depends on the particle morphology, size, and size distribution, but also on the physicochemical interactions between particles themselves or with their environment. Similarly, the stability of colloidal dispersions is influenced by a complex balance between various forces (attractive and repulsive van der Waals, electrostatic forces, hydrophobic forces, acid–base interaction forces) that act in the dispersion. Efficient gas adsorption phenomena that allow high selectivity in gas chromatography, and hence fine separation, are essentially due to specific interactions between adsorbent and adsorbate. Major efforts have consequently been spent to reach a better insight into and description of those forces. However, the experimental evaluation of the interaction potential of a solid surface involving those forces is far from being evident, especially for powders. The development of the surface force apparatus [1] and, more recently, of near field electron microscopes [2], as well as new theories [3], has given new impetus in this research domain. These methods apply fairly well to "model" surfaces such as cleaved muscovite in the case of both the surface force apparatus and the atomic force microscope, but so far divided solids cannot be simply examined with these methods. The most familiar ways of determining the surface energy (potential for reversible or physical interactions) are wetting or contact angle measurements. Here, drops of liquids of known characteristics are deposited on a flat and nonporous solid surface and the contact angles are evaluated using well-accepted procedures leading to the determination of the surface energy value [4]. Such a procedure hardly applies to powders, even when they are compressed in a platelet form on which liquid drops may be deposited, because it is quite impossible to prepare surfaces of adequate smoothness. In all cases, careful and time-consuming verifications of the surface smoothness are required [5] before reaching meaningful conclusions.

For the study of powders, which are the preferred supports for chromatography purposes, one generally calls on either gas adsorption or calorimetric methods for the evaluation of their interaction potential, which in turn allows the prediction of interaction energy. Immersion calorimetry, combined with a gas adsorption method [6] monitoring the degree of solid surface coverage, delivers enthalpies and entropies of adsorption. But again, these methods require special equipment and skill. Inverse gas chromatography (IGC) is, at least at first sight, a straightforward technique for the characterization of a variety of solid surfaces independently of the extent of their specific surface area or of the particle or fiber morphology.

In this chapter, first we shall recall the physicochemical quantities that are accessible through IGC measurements and then we shall illustrate the possibilities of IGC in terms of examples of studies performed essentially in our laboratory. Finally, we shall state the advantages and limitations of that method.

II. INVERSE GAS CHROMATOGRAPHY

A. Definition

The term *inverse* relates to the fact that chromatography is performed in an unusual way. Commonly, chromatography is applied for the separation and possible identification of the components of a given mix. Generally, limited attention is paid to the chromatographic phase that fills the column as long as it operates properly. In IGC, the situation is exactly the "inverse" because one becomes interested in the chromatographic support and in order to characterize it one injects solutes of known properties. For instance, when injecting alkanes, only London-type or dispersive interactions are susceptible to intervene whatever the nature of the GC column filling. Intuitively already, it seems obvious that a long retention time (duration necessary for the alkane to cross the column) is indicative of intense London interactions between the alkane probe molecule and the solid's surface. Solutes other than alkanes may be used as molecular probes to test the solid surface's other characteristics such as its specific interaction (polar, acid-base, etc.) capacity.

IGC has been used for years for the measurement of solution thermodynamics of polymers that are coated on an inert chromatographic support; for the determination of their physical characteristics such as glass transition temperature, crystallinity, and melting point; but also for the evaluation of adsorption isotherms on various oxides or carbons, etc. Several books describe these possibilities [7,8] and in recent years a symposium was devoted to this peculiar type of chromatography [9].

IGC may be practiced in two different ways: The first method is IGC at *infinite dilution conditions* (IGC-ID) when injecting minor amounts of gaseous solutes, at the limit of detection of the most sensitive detectors. Hence, we are operating under conditions that give us the unique possibility of performing measurements at true infinite dilution condition, for which the theories are usually established. For the physicochemical surface characterization of a solid support, we see also advantages because the very few adsorbed probe molecules will be well separated from each other, avoiding the complication of interpretation due to lateral interactions between adsorbed neighboring molecules.

The second method is IGC at *finite concentration conditions* (IGC-FC). Here we inject a known quantity of liquid solute, sufficient to entirely cover the solid's surface. This mode will allow the determination of adsorption isotherms providing complementary information.

B. Practice of IGC

A commercial gas chromatographic apparatus, equipped with a highly sensitive flame ionization detector, is generally used for IGC studies, with a precision manometer for monitoring the input pressure of the chromatographic column. Helium is the most used carrier gas, at a flow rate depending on the diameter of the chromatographic column (about 20 mL/min for a 1/8-in, column and 60 mL/min for a 1/4-in, column). Depending on the size of the stationary particles in the column, which governs the pressure drop through the column, the length of the column varies from 5 cm (particles having a mean diameter around 5 μm) to 1 m (for coarse particles having a mean diameter of 500 μm). However, for a given chromatographic support, the most suitable flow rate, corresponding to the highest GC efficiency, needs to be determined in the usual way from the van Dempter curve, which links the theoretical plate number of the chromatographic column to the flow rate of the carrier gas.

It is important to point out a major difficulty in the IGC practice originating mainly from the elaboration of solid particles having adequate size for the column filling. If commercial chromatographic silicas can be used as purchased, often particles of the solid of interest are too small (diameter <2 μm) for being directly employed as chromatographic supports. In that case, pellets may be prepared, either by solvent agglomeration of a particle slurry or by powder compression under a pressure of about 100 Pa, that are then carefully hand-crushed and sieved to select the fraction of particles having diameters between 250 and 400 μm. For the former technique, the solvent chosen has to be as inert as possible toward the solid; otherwise, the physicochemical surface properties may be modified during the agglomeration process. For the latter, good mechanical properties of the elementary particles are required to prevent grinding or sintering of the solid during the compression process. This process could not be applied to a very soft mineral-like talc but was successfully used in the case of pyrogenic silicas [10] or clay minerals [11]. When neither compression process nor agglomeration can be used, then one can fill short columns of 10 cm length and a 4-mm inner diameter. This technique was successfully employed for micas [12] and talc [13].

The probes used for checking the surface properties of the powder are generally common molecules such as alkanes (for the evaluation of the dispersive interactions) and their branched isomers (for the determination of the solid's surface nanomorphology), or polar probes (determination of specific interactions) like chloroform, ether, terahydrofuran, alcohols, short oligomers [10], and so on.

The choice is limited by the necessity of having volatile solutes, showing no irreversible adsorption on the solid of interest and allowing the probe to be eluted from the chromatographic column in a reasonable time. This is a serious limitation, especially in the case of solids exhibiting a high surface energy, particularly when looking at the surface properties using polar probes. It strongly restricts the choice of polar probes and consequently the IGC ability to provide entire information on the capacity of the studied solid to exchange specific interactions.

For chromatographic experiments at infinite dilution conditions, a very small amount (about 10 μL of vapor) of the probe is injected in the column, so that intermolecular probe interactions can be neglected. On the contrary, for IGC at finite concentration conditions, some μL of liquid solutes, from 0.2 μL to about 20 μL, depending on the extent of the solid's surface area contained in the column, is injected.

III. IGC AT INFINITE DILUTION (IGC-ID)

A. Fundamental Equation of IGC at Infinite Dilution Conditions

The net retention time t_r' of the probe is the fundamental physical quantity measured by IGC-ID. It is equal to:

$$t_r' = t_r - t_0 \tag{1}$$

where t_r is the retention time of the probe and t_0 the retention time of a nonretained molecule, generally methane (measurement of the dead volume of the chromatographic column). The retention volumes V_n (volume of carrier gas necessary to push the probe through the column) is related to the net retention time by:

$$V_n = JDt_r \tag{2}$$

where D is the flow rate at the column output at the temperature of measurement, J is the James-Martin coefficient given by $J = (1 + P/P_a)^2/(1 + P/P_a)^3$, where P is the input and P_a is the output pressure (atmospheric pressure). This coefficient takes into account the gas compressibilty. Finally, the retention volume V_n is related, at equilibrium, to the surface partition coefficient K between the gaseous and adsorbed phases and to the area of the sorbent actually accessible to the probe by the equation:

$$V_n = KA \tag{3}$$

where A is the total surface area of the powder in the chromatographic column.

Intuitively, it is already understandable that a larger V_n will correspond to a higher affinity of the chromatographic support for the alkanes. Thermodynamically, affinity indicates a standard variation in the free energy of adsorption (ΔG_a^0). ΔG_a^0 and V_n are related by:

$$\Delta G_a^0 = -RT \ln(V_n/C) \tag{4}$$

where C is a constant depending on A and on the choice of a reference state for the adsorbed alkane molecule. When the chosen reference state is that of Kemball and Rideal [14], ΔG_a^0, expressed in kJ/mol, and $V_n(cm^3)$ are related by:

$$\Delta G_a^0 = -RT \ln \{8.186.10^4[V_n/(Tm S\rho)]\} \tag{5}$$

In this expression, T is the temperature of measurement, m the amount of powder in the column, S its specific surface area, and ρ the density of the powder. Because a perfect surface smoothness of the solid is assumed in that theory, it is not obvious that this bidimensional reference state will be valid for all of the studied solid surfaces, especially in the case of grafted solid surfaces or solid surfaces that are rough at a molecular level.

The use of thermodynamic values derived from the study of the variations of the free adsorption energy with an experimental parameter (temperature or molecular probe parameters) is the simplest way to overcome the difficulty encountered with the choice of the bidimensional reference state.

The temperature of measurement can easily be changed in IGC experiments. This constitutes one of the main advantages of the IGC technique. So, from the variation of ΔG_a^0 with temperature, the enthalpy of adsorption can readily be computed according to:

$$\Delta H_a = \partial(\Delta G_a^0/T)/\partial T = -R\partial(\ln(V_n))/\partial T \tag{6}$$

The study of the variation of the adsorption enthalpies of probes is very useful to understand the evolution of the surface properties of solids submitted to thermal or grafting treatments [15,16]. Transition phenomena such as a glassy temperature transition or melting point can be detected using IGC measurements of adsorption enthalpies [17,18]. Moreover, looking at the influence of some molecular parameters on the free energy values, such as polarizability, acid–base properties, or stereochemistry of the probe, is another way to get additional information on the surface interaction capacity according to the nature of the interaction forces that take place between the probe and the surface: London forces, specific interaction, or size exclusion effects.

B. London Interaction Potential

The London interactions, also called dispersive, nonspecific, or universal interactions, originate from the existence of instantaneous dipoles due to instantaneous distributions of electrons with respect to the positively charged nucleus. The generated electrical field polarizes any nearby atom or molecule inducing in this way a dipole moment in it. Hence, even nonpolar molecules such as alkanes are capable of undergoing London-type interactions. In order to evaluate the London interaction capacity that might also be expressed in terms of the London component of the surface energy (γ_s^d in mJ/m^2), we use the method originally proposed by Dorris and Gray [18]. For this purpose, a homologous series of n-alkanes is injected in very small amounts, at the limit of detection of the FID detector, in the column containing the solid of interest and the ΔG_a^0 is calculated from the retention times [Eq. (4)]. Experimentally, it is commonly observed that ΔG_a^0 of alkanes varies linearly with the number of carbon atoms of the alkane probes as shown in Fig. 1. As seen, the slopes of these straight lines vary in a large range. The higher the slope, the higher is the strength of interaction between the solid surface and a methylene group. Moreover, the ordinate at the origin of the lines in Fig. 1 is mainly related to the extent of surface area contained in the chromatographic column, and this can be used to estimate the specific surface area of the solid [19]. From the slope of the straight line, it is possible to compute an incremental value, ΔG_{CH_2}: the free adsorption energy variation corresponding to a CH_2 group.

$$-\Delta G_{CH_2} = RT \ln(V_{n+1}/V_n) \tag{7}$$

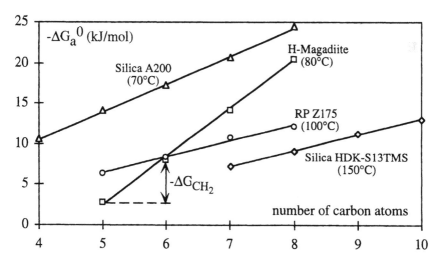

FIG. 1 Variation of ΔG_a^0 with the number of carbon atoms of the injected n-alkane probes for a crystalline silica (H-magadiite), a precipitated silica (Rhône-Poulenc Z175), a pyrogenic silica (Degussa A200), and a sylylated pyrogenic silica (Wacker Chemie HDK-S13). (Between brackets: temperature of measurement.)

This in turn is related to γ_s^d according to the following equation:

$$\gamma_s^d = \frac{1}{\gamma_{CH_2}}\left(\frac{\Delta G_{CH_2}}{2Na_{CH_2}}\right)^2 \tag{8}$$

In this equation all is known (a_{CH_2}, the area of an adsorbed CH_2 group; N Avogadro's number; γ_{CH_2}, the surface energy of a solid made entirely of CH_2 groups, i.e., polyethylene) or measurable (ΔG_{CH_2}), except γ_s^d. The γ_s^d value found for a pyrogenic silica sample amounts to 70 mJ/m². This value compares fairly well with the one obtained by Kessaissia et al [22] who applied a totally different method (contact angle or wettability measurements). Since a macroscopic method (wettability) and a molecular method (IGC) deliver the same value for γ_s^d, one may conclude that this silica sample exhibits a rather molecular smooth surface at least at the scale of the probing molecule, thus corresponding to the applicability requirements of the Dorris and Gray method.

The dispersive component of surface energy was determined for a series of solids and the results are collected in Table 1. γ_s^d values vary in a very large range from 20 mJ/m² for a silica silylated with a perfluorinated silane, up to more than 400 mJ/m² for carbon blacks. The variations of γ_s^d upon treatment of a given sample are caused by modification of the polarizability of surface atoms or surface atoms groups. This is based on the supposition that there are no other influencing factors, such as a change in accessibility of the surface or steric factors, pertaining to the molecular probe itself.

C. Specific Interaction Potential

IGC should also allow the estimation of the specific interaction potential of a polar solid surface just by injecting polar probes of known characteristics. In fact, this possibility exists but its application is far from evident for the following reasons: A polar probe will exchange, with a polar surface, London and specific interactions that contribute to its retention on the solid. Yet we collect just one chromatographic peak that contains the information from both types of interactions. The problem is to find a way to separate and evaluate both contributions. Several proposals may be found in the literature, but before entering in this field it might be useful to clarify the term *specific interactions*. It includes all types of interactions except London interactions. "Specific" also means that the types of interactions susceptible to be exchanged relate only to the two partners in contact.

A nonpolar surface will be able to undergo only London-type interactions even when the partner is polar. But with a polar surface, numerous possibilities appear, i.e., dipolar, H-bond-type, acid–base, metallic, magnetic, and hydrophobic interactions. Usually, for IGC, one operates in dry atmosphere that means

TABLE 1 Dispersive Component of the Surface Energy (γ_s^d) Measured on Various Solid Surfaces

Solids	S_{spe} (m^2/g)	γ_s^d (mJ/m^2)	Ref.
CaCO$_3$ (Socal Solvay)	9	52	
CaCO$_3$ (Rhône Poulenc)	35	60	13
Talc	3–25	110–200	21
MgO	7	95	22
ZnO	10	83	22
Natural muscovite	<1	38	12
Grinded muscovite samples	10–100	50–130	12
Silica (Aerosil A130)	130	80	23
H-magadiite	48	178	24
H-kenyaite	18	175	24
Al$_2$O$_3$ (γ)	100	92	25
Chrysotile	10	52	26
α-Fe$_2$O$_3$	15	55	27
Goethite (FeOOH) at 250°C	15	320	27
Zircone (ZrO$_2$)	53	220	27
Carbon black N110	130	400	28
Carbon black N220	110	450	28
Carbon black N330	78	240	28
Carbon black N550	41	195	28
Carbon black N773	30	95	

that the role of adsorbed water that plays such an important role in applications cannot be readily analyzed. In fact, it is possible to employ prewetted carrier gas to overcome this limitation. Hence, the solid surface will be partially or entirely precovered by water molecules leading to new adsorption conditions.

In recent years, and in the absence of electrostatic, magnetic, or metallic interactions, it has been shown that whenever possible acid–base–type interactions become prevalent [29] over dispersive interactions. However, to apply acid–base concepts, one needs first to establish acid–base scales. So far, theory does not allow to directly perform calculations of acid–base characteristics for the probes currently used for IGC. Therefore, it is necessary to call on semiempirical acid–base scales.

1. Acid–Base Scales

Several approaches were made to define, in the Lewis sense, the acidity and basicity of organic molecules:

The hard–soft acid base scale of Pearson [30]
The E-C relation of Drago [31]
The donor–acceptor numbers of Gutmann [32]
The solvatochromic parameters [33]

It would be out of scope to describe in detail those scales. A recent review by Mukhopadhyay and Schreiber [34] discusses all of these points. Only limited information will be given in this instance.

The *hard–soft acid–base scale of Pearson* [30] is based on quantum mechanical considerations or molecular orbital theories where the absolute hardness of a molecule is defined as equal to half the negative rate at which its electronegativity changes with a change of its electron population, at constant potential. Pearson-derived laws allowing one to predict the orientation of organic reactions between electron donor and electron acceptor molecules or groups. So far his concepts have not been used for IGC given the rather unknown and complex nature of solid surfaces.

The *four-parameter semiempirical equation of Drago* [31], on the contrary, is most valuable for surface characteristic determinations. The equation predicts acid–base reaction enthalpies in the gas phase or in poorly solvating solvents:

$$-\Delta H_{AB} = E_A E_B + C_A C_B \qquad (9)$$

The acid A and the base B are characterized by two parameters: an E value, which represents the ability of A and B to participate in electrostatic bonding (hardness), and a C value (softness), which indicates their tendency to form covalent links. Taking iodine as a reference substance for which $C_A = E_A$ and making a series of calorimetric measurements leads to C and E values of tens of molecules. The Drago approach has been used with success, starting from wettability measurements, for the evaluation of surface properties of oxides [35], yet not through IGC. But there is no peculiar reason for that. One major drawback of the Drago scale comes from the fact that a molecule is allowed to be either an acid or a base. In reality, most molecules exhibit amphoteric properties: they simultaneously possess partial acid and base behaviors. In that respect, the scale of Gutmann [32], which is also semiempirical, shows advantages.

The *Gutmann electron acceptor (AN) and donor numbers (DN)* that stand for acid and base properties of a molecule or substance, respectively, are determined experimentally. *AN* is defined as the enthalpy of formation of a 1:1 molecular adduct of a given molecule with a reference Lewis acid ($SbCl_5$). *DN* is measured by the nuclear magnetic resonance (NMR) shift of P when the given molecule is mixed with a solution of oxotriethylphosphine (Et_3PO), with the shift being normalized by taking the value 0 for the solvent (1,2-dichloroethane) and 100 for $SbCl_5$. Knowing *AN* of a pure acid and *DN* of a pure base, one may simply calculate the enthalpy of interaction using the following relation:

$$\Delta H_{AB} = AN \cdot ND/10 \tag{10}$$

For amphoteric molecules having respectively $(AN)_1$ and $(AN)_2$ as acceptor numbers and $(DN)_1$ and $(DN)_2$ as donor numbers, we [36] proposed use of the following equation:

$$-\Delta H_{sp} = (AN)_1 \cdot (DN)_2/100 + (DN)_1 \cdot (AN)_2/100 \tag{11}$$

This relation has since been often used in the literature for the evaluation of the acid–base properties of numerous supports; supplementary examples will be given later on.

The method of Kamlet and Taft [33] is called the solvatochromic method because the measurements were initially performed by spectrometry. Based on linear solvation energy relationships, the chromatographic stationary phase hydrogen bond acidity and basicity as well as its polarizability (London interactions) can be evaluated. Such analysis was performed by IGC, e.g., on modified silica surfaces [37].

But in all cases, neither of these methods will give complete information on the mechanism of interaction, in particular on the way the selected solute or molecular IGC probe contacts the solid surface. In other terms, factors limiting the access of the chosen molecular probe such as surface roughness or nanomorphology should be taken into account.

2. Acid–Base Measurements on Solid Surfaces

Most acid–base evaluations on chromatographic supports are based on the application of the relation:

$$-\Delta H_{sp} = (AN)_p(DN)_s/100 + (DN)_p(AN)_s/100 \tag{12}$$

In this relationship, ΔH_{sp} is the specific enthalpy of interaction: the one that is due to the sole specific interactions between the probe (p) and the surface (s). Of course, this raises the problem of the identification of ΔH_{sp}. In practice, one injects a polar probe in the column containing the polar support of interest. As mentioned before, this probe will exchange with the support both London and specific interactions. Yet only a single peak will be recorded that contains the contributions of both London and specific interactions. For example, various attempts were made to solve the problem by comparing the behaviors of linear alkanes and polar probes of the same molecular weights. We [38] suggested to use the vapor pressure as a molecular descriptor of the properties of the solutes. The first reason of that choice is obviously the fact that volatility is a perquisite for a solute in an IGC experiment and that vapor pressure values are readily found in the literature. However, the vapor pressure at saturation reflects only the molecular interactions that similar molecules are able to undergo and this underlines the limits of our method. Various improvements have been proposed:

1. The use [39] of a corrected vaporization enthalpy: the ΔH_v being corrected for London-type interactions
2. The use [40,41] of the molar deformation polarizability of probes. This approach is based on van der Waals theory that states that London interactions are proportional to the polarizabilities of the partners in interaction.
3. The use [42] of a quantity $a(\gamma_L)^{1/2}$ where a is the area of the adsorbed probe and γ_L the London component of the surface energy of the liquid probe, with the difficulty being to evaluate for each new support the true value of the adsorbate's section
4. The use [43] of topology indices of the probes. This procedure that looks most promising is under current investigation in our laboratory.

All of these different descriptors of the probe molecule characteristics are applied as illustrated in Fig. 2. From the ΔG_{sp} values determined, from the departure from the alkane line of the representative point of the polar probe, at several temperatures, one readily calculates the ΔH_{sp} by application of Eq. (6), and the donor and acceptor numbers of the chromatographic support using Eq. (12).

The determination of specific interaction parameter is based on the comparison of the retention properties of polar probes and *n*-alkane probes. One supposes that both types of molecules will have access to the same area of adsorption, i.e., that no size exclusion effects occur during the chromatographic process due to a particular nanomorphology of the surface. We shall see that this condition is not always filled, thus leading to erroneous values.

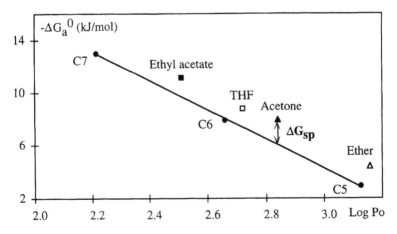

FIG. 2 Determination of the specific interaction parameter (I_{sp}) on natural graphite at 63.5°C.

D. Importance of the Surface Nanomorphology

As pointed out by Papirer et al. [20,44], the common assumption of surface planeity cannot be assumed for solids having a lamellar structure such as crystalline silicas or talc. Partial insert on of the flexible alkanes between silica layer defects takes place on the crystal edges and very efficient adsorption of alkanes will take place. From this insertion phenomena, it results that unexpected high values of γ_s^d were determined on those materials and correlatively some negative I_{sp}, especially in the case of bulky molecules such as chloroform or α,ω-dienes that do not have access to the same adsorption sites [44]. Similar size exclusion phenomena were observed on other lamellar solids like muscovite [12], talc [45], or clays [11], but also on carbon blacks. Each time, "abnormal" values of γ_s^d were calculated (see values in italic in Table 1) that cannot be accounted for by simple physical considerations. Rapidly, a particular surface structure and the existence of very efficient adsorption sites for alkanes were suspected to be at the origin of the unexpected γ_s^d values determined on those materials [46]. In the case of carbon blacks, which also exhibit very high values of γ_s^d, scanning tunneling microscopy [47] demonstrated that the surface of carbon black is made of scales of polyaromatic nature, on the boarder of which n-alkanes strongly and preferentially adsorb. Molecular modeling studies confirm the preferential adsorption possibilities of linear hydrocarbons at the edges of those scales.

Hence, since in infinite dilution conditions only very few probe molecules are used to estimate γ_s^d, the information, extracted from their retention volume (V_n), will only concern mainly the more active adsorption sites. Taking into account that the residence time of a molecule on an adsorption site is proportional to the exponential of its energy of interaction [following Eq. (13)], they will contribute mainly to the retention time of the probe:

$$\tau = \tau_0 \exp(\varepsilon/RT) \tag{13}$$

where τ_0 is a constant and ε the interaction energy of the molecule. The retention time is mainly dependent on the number of sites having the highest energy; the thermodynamic values calculated from the former are nonrepresentative of the interaction capacity of the whole solid surface.

For lamellar solids, partial insertion of n-alkanes between crystal layers is possible. Those inserted probes will be submitted to a much stronger force field, in comparison with molecules adsorbed on a flat surface, and will thus be more efficiently retained; following, the calculated γ_s^d will take abnormal values. Therefore, it is necessary to define a morphology index that would relate the accessibility of the solid surface on the one hand and the shape or morphology of the molecular probe on the other hand [44]. Theoretical considerations do not allow solution of the problem because little is known about the surface morphology of divided amorphous solids. A semiempirical method was therefore proposed based, on the one hand, on the definition and choice of a reference silica surface and, on the other hand, on the use of molecular probes of known morphologies, with the values (retention volumes) determined on this model silica surface being taken as references. By model silica surface, one understands a silica surface that is smooth and chemically homogeneous considering the dimensions of the molecular probe used to test it. Pyrogenic silicas, such as Aerosil from Degussa or HDK from Wacker (Germany), apparently meet these requirements. Concerning Aerosil, several experimental facts plead in its favor. For instance, the behavior of grafted alcohol and diols is only explainable if the silica surface is indeed flat at the molecular level [48].

Moreover, Barthel et al. [49] determined a fractal dimension of pyrogenic silicas using a series of analytical methods like gas adsorption and small-angle x-ray scattering. Their results point to a fractal dimension of 2.0. Comparatively, the fractal dimension of a precipitated silica is close to 3.0.

Coming back to the fundamental equation [Eq. (3)] of IGC at infinite dilution, one may write the ratio:

$$\frac{V_N^B}{V_N^L} = \frac{K^B A^B}{K^L A^L} \tag{14}$$

where K^B and K^L are the equilibrium constants for branched and linear hydrocarbons, respectively; A^B and A^L being the corresponding actual areas of adsorbed molecules. By definition, the morphology index is given by:

$$I_m = 100 \frac{V_N^B}{V_N^L} = \frac{V_N^L \text{ref}}{V_N^B \text{ref}} \tag{15}$$

where $V_N^L \text{ref}$ and $V_N^B \text{ref}$ are the retention volumes measured on the reference surface, i.e., pyrogenic silica. The branched alkanes commonly used are tetramethylbutane (Te MB), 2,2,4-trimethylpentane (224 TMP), 2,2,4-trimethylpentane (234 TMP), and 2,5-dimethylhexane (25 DMH). Some values of morphology indexes are reported in Table 2. The lower the nanomorphology index, the higher is the surface roughness of the solid. It is seen that the precipitated silica is more rugged than the pyrogenic silica. However, the most significant differences are noted with lamellar minerals, i.e., surfaces that also exhibit very high values of IGC-measured γ_s^d. To highlight a possible relationship between γ_s^d values and the nanomorphology index, we plotted (Fig. 3) the variation of γ_s^d versus the variation of the nanomorphology index for three families of solids.

Indeed, a clear correlation between the γ_s^d values and the index of morphology is observed for carbon blacks, for grinded muscovite samples, and for clays, demonstrating that a high value of γ_s^d, when measured using IGC-DI, is related to insertion phenomena of n-alkane probes. Nonetheless, it was observed recently in the laboratory that abnormally high γ_s^d values, determined on iron oxides, are

TABLE 2 Morphology Index for Some Powders

	Te MB	25 DMH	Ref.
Aerosil (A130)[a]	100	100	
Precipitated silica	72.9	90	20
H-kenyaite	0.6	2.1	20
H-magadiite	0.02	0.09	19
Talc	13.7	40.9	44
Muscovite	—	50–60	12
Illite	11 to 23	26 to 58	11
Kaolinite	7 to 19	39 to 53	11

[a] By definition.

not systematically related to such insertion phenomena. Ionic solids having semi-conductive superficial properties also lead to such abnormal γ_s^d values. These last considerations highlight the importance of the knowledge of the surface characteristics of the solid itself on the interpretation of thermodynamic values calculated from IGC-DI results, keeping in mind that the fundamental equation of IGC-DI totally ignores surface heterogeneity. Therefore, the problem to be solved is to estimate the level of heterogeneity of a solid: IGC at finite concentration conditions (IGC-FC) may held promise of a solution.

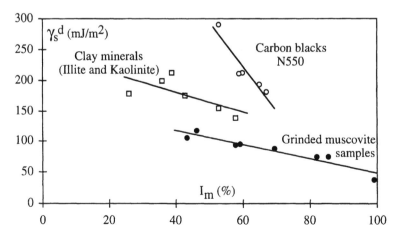

FIG. 3 Evolution of the γ_s^d values with the index of nanomorphology (I_m) for three families of solids: carbon blacks, grinded muscovite samples, and clays.

IV. IGC AT FINITE CONCENTRATION CONDITIONS

Atoms present on a solid surface are rigidly connected and cannot rearrange themselves to create a surface that is statistically homogeneous as is the case for liquid surfaces. This rigidity, combined to the fact that real solid surfaces are never "perfect," is responsible for the apparition of surface defects like cracks, corners, steps, porous structures, and presence of impurities besides the nonregular distribution of chemical groups. Consequently, the surface energy concept, easily defined and measurable in the case of pure liquids and used in the frame of IGC-ID conditions, does not hold for a heterogeneous real solid surface when considered at a molecular level.

The surface heterogeneity of a solid was evidenced years ago by looking at the variation, with the surface coverage degree (θ), of the heats of adsorption (ΔH_a) of chosen solutes. For an energetically homogeneous surface, ΔH_a would obviously be independent of the surface coverage. For actual solids, ΔH_a generally decreases most steeply as θ increases, indicating the presence of active adsorption sites that are rapidly occupied by the first adsorbing molecules. Very often, deviations are observed, at low relative pressures, between the BET model and the measured isotherm; this is also a supplementary proof of solid surface heterogeneity. Both observations suggest that information on the surface heterogeneity may be extracted from the analysis of the isotherm shape. Rudzinski and Everett [50] and Jaroniec and Madey [51] published extensive reviews on the determination of surface heterogeneity from adsorption measurements. Information that can be extracted from the isotherm analysis is recalled in Fig. 4. Three

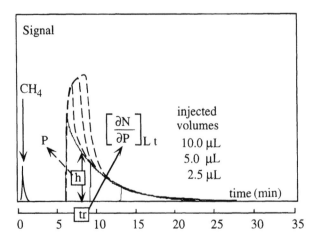

FIG. 4 Acquisition of the adsorption isotherm from the chromatographic peak.

main techniques can be used for the experimental acquisition of an adsorption isotherm in the case of a powder: volumetric measurement, the most widely used method; gravimetric measurement; and IGC-FC. We prefer IGC-FC.

A. Acquisition of the Adsorption Isotherms Using IGC-FC

Different IGC exploitation possibilities for the determination of adsorption isotherms have been reviewed by Conder [7]. The simplest and the most efficient one, from the point of view of the analysis duration, is the elution characteristic point (ECP) method, which allows the acquisition of a part of the desorption isotherm from a unique chromatographic peak. Using this method, the first derivative of the adsorption isotherm can be readily calculated starting from the retention times and the signal height of characteristic points taken on the diffuse descending front of the chromatogram as shown in Fig. 5. The first derivative of the adsorption isotherm is obtained according to the following equation:

$$\left(\frac{\partial N}{\partial P}\right) = \frac{JDt'_r}{mRT} \tag{16}$$

where N is the number of absorbed molecules, P the pressure of the probe at the output of the column, t'_r the net retention time of a characteristic point on the

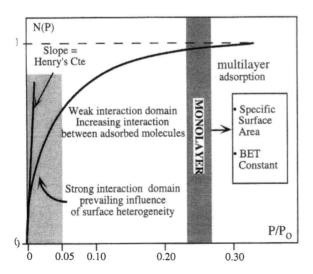

FIG. 5 Exploitation of chromatographic peaks using the elution characteristic point method.

rear diffuse profile of the chromatogram, J the James-Martin coefficient that allows correction of data for gas compression in the column, D the output flow rate, and m the mass of adsorbent in the column.

An alternative method that calls on the same equation consists of measurement of the retention times of the chromatographic peaks corresponding to increasing injected solute amounts. This method is, of course, highly time consuming because a chromatographic experiment has to be done for each point of the wanted isotherm. Furthermore, it is not necessarily more accurate because of the unavoidable slight variations of experimental conditions from one analysis to another over a long period.

Whatever the chosen method, the experimental conditions have to be carefully controlled in order to obtain a physically meaningful first derivative of the isotherm. In particular, one must pay special attention to the analysis temperature, which should be chosen so that the contribution of the probe to the total flow rate can be considered negligible, i.e., less than 5% at the maximum chromatographic peak. On the other hand, the injected amount has to be sufficient in order to reach coverage ratios corresponding to the monolayer domains—a required condition for the calculation of the main isotherm characteristics using BET theory. Finally, the shape analysis of the calculated isotherm will also lead to adsorption energy distribution functions that may be considered as fingerprints of the surface heterogeneity and that may be used, for the sake of comparison, for the study of the influence of physical or chemical treatments on the solid's surface heterogeneity as will be seen below.

B. Thermodynamic Quantities Calculated from the Isotherms

From the adsorption isotherm, it is possible to compute the specific surface area and the BET constant for an organic probe and, by linearization of the initial part of the isotherm, the Henry constant becomes available. Table 3 gathers the values of the Henry and BET constants and of the specific surface areas, determined from the IGC desorption isotherms, using four different probes, on a pyrogenic silica HDK-S13 (Wacker):

n-Heptane, an apolar molecule
Benzene, a weak basic molecule
Hexamethyldisiloxane (HMDS), a weak basic and sterically hindered molecule
Propanol-2, which can exchange strong hydrogen bonds with the silanol groups

From this table, it is worth pointing out that

1. The specific surface area, which gives essentially texture information, is quite independent from the nature of the probe, taking into account that the exact

TABLE 3 Molecular Section (A), Specific Surface
Area (S), BET, and Henry's Constants (C)

Probes	A_{mol} $(\text{Å}^2)^a$	S_{spe} (m^2/g)	C_{BET}	C_{Henry} $(\mu mol/Pa)$
n-Heptane	54	119	5	0.85
Benzene	42	110	7	0.99
HMDS	84[b]	130	29	0.29
Propanol-2	37	132	180	3.10

[a] Average molecular area from Ref. 52.
[b] Calculated from the nitrogen BET surface.
HMDS, hexamethyldisiloxane.

value of the molecular areas of the adsorbed probes are not well known and, moreover, that they vary with the physicochemical properties of the solid surface itself [52].

2. The BET constant, which leads to information about the interaction capacity of the solid surface at high surface coverage ratio, increases with the specific interaction capacity and with the molar mass of the probe. Therefore, a low value is observed for n-heptane and a high one for propanol-2.

3. Henry's constant, which is related to the interaction capacity at very low coverage ratios, follows the increasing order of the polarity of the probes.

Knowing two adsorption isotherms at two close temperatures [53], the isosteric heat of adsorption of a probe can be calculated. Its variation with the coverage ratio gives a first evaluation of the surface heterogeneity, as pointed out earlier, but the determination of the distribution function of the adsorption energies brings more information on this important characteristic of the solid surface.

C. Estimation of the Solid's Surface Heterogeneity Using Adsorption Energy Distribution Functions

All approaches described in the literature for the estimation of the distribution functions of the adsorption energies are based on a physical model that supposes that an energetically heterogeneous surface, with a continuous distribution of adsorption energies, may be described in simple terms as a series of homogeneous adsorption patches. Hence, the amount of adsorbed molecules (probes) is given by the following integral equation:

$$N(P_m, T_m) = N_0 \int_{\varepsilon_{min}}^{\varepsilon_{max}} \theta(\varepsilon, P_m, T_m) \, \chi(\varepsilon) d\varepsilon \qquad (17)$$

where: $N(P_m, T_m)$ is the number of molecules adsorbed at pressure P_m and temperature T_m of measurement, N_0 is the number of molecules needed for the formation of a monolayer, $\theta(\varepsilon, P_m, T_m)$ is the local isotherm (generally the Langmuir isotherm), ε the adsorption energy of a site, and $\chi(\varepsilon)$ is the distribution function (DF) of the sites seen by the probe. The range of adsorption energies is included between minimal (ε_{min}) and maximal (ε_{max}) values.

From a mathematical point of view, solving the former integral equation is not a trivial task because there is no general solution. The simplest way to solve this equation is to admit the condensation approximation χ^{CA} (DFCA) instead of the Langmuir isotherm as kernel of the integral equation. The condensation approximation supposes that the sites of adsorption of given energy are unoccupied below a characteristic pressure and entirely occupied above it. DFCA is directly related to the first derivative of the isotherm, which can be computed easily from the desorption profile of the chromatographic peak.

This approximation is all the better as the temperature of measurement approaches absolute zero. But at the usual chromatographic measurement temperature, we are far away from those conditions. Then the actual distribution function can be approached using various calculation techniques [50,51,54–56]. One of them, the extended approximation of Rudzinski-Jagiello [57], supposes the knowledge of the even derivatives of the DFAC, which may be determined using Fourier transforms. This procedure provides an efficient way to separate the signal from the experimental noise [58].

D. Examples of Distribution Functions of the Adsorption Energies: Influence of the Nature of the Probe on the Distribution Function Shape

The reduced adsorption isotherms of n-heptane, benzene, HMDS, and propanol-2 on a pyrogenic silica HDK-S13 from Wacker are plotted in Fig. 6. It is seen that the more polar probe induces the highest curvature of the recorded isotherms. From the first derivative of these adsorption isotherms, the adsorption distribution functions were calculated according to the extended Rudzinski-Jagiello approximation.

Figure 7 displays the distribution functions of the adsorption energies calculated for the n-heptane, benzene, the HMDS, and propanol-2 chromatographic peaks measured on silica HDK-S13. Only the distribution function (FD) of propanol-2 is clearly bimodal. This bimodality suggests that the HDK-S13 silica surface is not energetically homogeneous but exhibits two types of domains. The first type is rich in siloxane bridges and correlatively poor in silanol groups. It is possibly related to the peak centered around 18 kJ/mol, which is relatively narrow because of the weak variation of dispersive interaction with the chemical nature of the solid surface. The second type is rich in silanol groups that interact

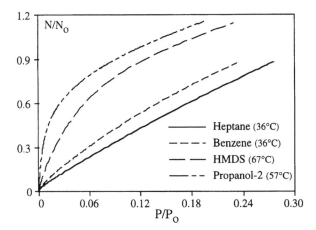

FIG. 6 Adsorption isotherms of heptane, benzene, hexamethyldisiloxane, and propanol-2 on a pyrogenic silica (HDK-S13 from Wacker Chemie). (Between brackets: temperature of measurement.)

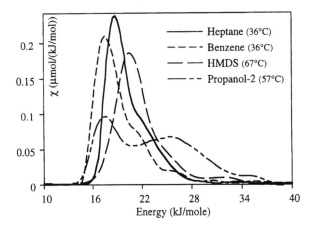

FIG. 7 Adsorption energy distribution functions of heptane, benzene, hexamethyldisiloxane, and propanol-2 on a pyrogenic silica (HDK-S13 from Wacker Chemie). (Between brackets: temperature of measurement.)

mainly through hydrogen bonds leading to higher energies of interaction. Therefore, the broad peak centered around 22 kJ/mol is attributed to this type of adsorption site. Its width is obviously related to the large range of silanol group environments that can take place on a silica surface.

Less interactive probes, like n-heptane, HMDS, or benzene, which interact with the surface mainly through dispersives forces, always lead to almost monomodal distribution functions (see silica DF in Ref. 59), sometimes showing a shoulder in the high-energy part of the FD. This lack of selectivity is again to be attributed to the lack of sensitivity of the dispersive forces, comparatively to the polar forces toward the solid surface. One may conclude that such probes do not give much information on the surface heterogeneity of a solid.

This is true when the surface heterogeneity is mainly related to the local variation of the chemical composition and structure that is in a short range from the point of view of the molecular size. However, n-heptane, HMDS, and benzene may become interesting probes for the study of a solid exhibiting surface heterogeneity on a larger scale. Such long-range heterogeneity may, for instance, stem from the crystalline structure and are encountered in the case of lamellar products such as talc [59] or ground mica samples [60]. Here we may speak about ''macroheterogeneity,'' which is due to the presence of both basal and lateral surfaces corresponding to much higher dimensions in comparison with the probe size. The fact that lateral surfaces are generally largely more interactive than basal ones leads to bimodal distribution functions. This long-range heterogeneity does not exclude a concomitant short-range heterogeneity on both basal and especially on the lateral surfaces that is not as well defined as the former. An example of such a distribution function that permits the estimation of the lateral/basal surface area ratio is shown in Fig. 8 for a grinded muscovite sample.

Another way to increase the surface heterogeneity is to substitute part of the surface chemical functions by noninteractive groups such as aliphatic groups. In the case of silica, this modification can be easily performed by a partial and controlled surface silylation of the silanol groups with trimethylchlorosilane. The distribution function of such trimethylsilylated silica is displayed in Fig. 9 and compared with that of the initial unmodified silica. Comparison with the distribution function recorded on the initial silica shows that the partial silylation leads to the emergence of a second peak in the lower energy domain. By partial silylation, some silylated domains become much larger than the probe itself and this leads to surface structures with a significantly lower interaction potential with the HMDS probe. Additional examples that highlight the great interest of this method for the evaluation of clays [61] or carbon blacks can be found in the literature.

IGC-FC, combined with IGC-ID measurements, seems to be a very promising method to improve our level of knowledge of solid surface interaction capacities. Nevertheless, this method has certain limitations.

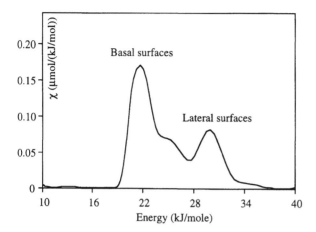

FIG. 8 Adsorption energy distribution functions of *n*-octane on a grinded muscovite sample.

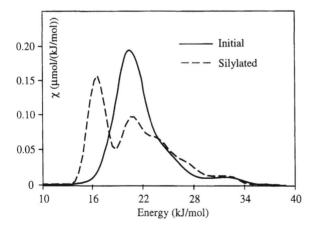

FIG. 9 Comparison of adsorption energy distribution functions of hexamethyldisiloxane, at 57°C, on an initial (HDK-S13) silica and on a partially trimethylsilylated sample (1.0 TMS/nm²).

V. ADVANTAGES AND LIMITATIONS OF IGC

IGC presents advantages. When the column is prepared, i.e., when the support of interest has been adequately conditioned and introduced in the column (an operation that is not always evident), then the measurements are rapid, usually reproducible, and accurate. The conditioning of the column under inert gas flow allows one to prepare surfaces freed from adsorbed gases and volatile pollutants. Moreover, solids may be modified in situ with adequate chemicals just before being characterized again. In addition, carrier gases with known amounts of humidity may be prepared, permitting examination of the alteration of the adsorption process of chosen probes. This is most important because divided solids often operate in the presence of humidity in actual applications. Since the experiments are rapid, series of probes may be selected to explore, at the molecular level, the surface peculiarities of a solid surface. Furthermore, the amounts of solute injected may be precisely controlled. In other terms, when injecting minor amounts, at the limit of detection, we are operating at infinite dilution conditions. This is a rather unique possibility of performing physicochemical measurements at true infinite dilution conditions, for which the theories are usually established. For the surface characterization of a solid support we see also advantages because the very few adsorbed probe molecules will be separated from each other, thus avoiding complications of interpretation due to lateral interactions between adsorbed neighboring solutes. On the other hand, we may inject known quantities of solutes sufficient to entirely cover the solid's surface providing complementary information such as adsorption energy distribution functions. Finally, as previously mentioned, the solid acting as the chromatographic support may be chemically modified inside the column or heat-treated, without return to the atmosphere before renewed control. Last but not least, the IGC equipment needed to perform those evaluations is available in most laboratories. Indeed, the number of situations that may be analyzed using IGC is impressive depending on the skill and imagination of the person applying it.

The limitations of IGC are twofold: the necessity to select volatile probes and nonvolatile stationary phases withstanding the GC temperature conditions. In addition, difficulties of interpretation of complex IGC peaks (tailing due to nonideality of the chromatographic process) may also show up and need special consideration. Even so, for difficult situations (solids of low surface area, solids of high morphological or energetic heterogeneities) it is worth trying IGC in order to learn more about the interaction potential of those solids.

VI. CONCLUSION

The application of IGC for the characterization of solid surfaces has been explored since the early 1960s. IGC at finite concentration was then used to determine principally adsorption isotherms that were exploited for the calculation of the specific surface area of the solid and adsorption energies. In recent years, adsorption energy distribution calculations complement previous information. This became possible following the remarkable development of computer science. IGC was also used since the start to detect characteristics of phases deposited on inert chromatographic supports. This allowed, for example, the determination of infinite dilution behavior and thermodynamics of test solutes, especially when the deposited phase was a polymer. In the late 1970s, IGC was employed to evaluate surface properties in terms of surface energy. This chapter underlines the difficulties encountered when trying to adopt this concept for actual solid surfaces that are highly heterogeneous. But IGC at infinite dilution conditions can be safely used to detect surface events (chemical rearrangements, structural reorganization) that appear upon heating a solid, for instance. This is due to the remarkable sensitivity of that method, a major advantage of IGC that is one of the few surface analysis techniques, since adsorption phenomena are truly surface phenomena.

This chapter demonstrates that IGC is still in its infancy and that progress is necessary so as to take full advantage of this unique method. It is our opinion that IGC is not a straightforward method providing the whole answer to a given problem, but often it indicates peculiarities that are unexpected and that may come to light when IGC is associated with other modern and rather sophisticated techniques such as NMR, electron spectroscopy for chemical analysis, or computer simulation methods.

REFERENCES

1. J. N. Israelachvili and G. E. Adams, Nature *40*:77 (1976).
2. G. Binning and H. Rohrer, Physica *127B*:37 (1984).
3. J. N. Israelachvili, in *Intermolecular and Surface Forces*, Academic Press, London, 1997.
4. F. M. Fowkes, Ind. Eng. Chem. *56*:40 (1964).
5. Z. Kessaissia, E. Papirer, and J. B. Donnet, J. Colloid Interface Sci. *82*(5):26 (1981).
6. H. Malendrini, R. Sarraf, B. Faucompré, S. Partyka, and J. Douillard, Langmuir *13*: 1337 (1997).
7. J. R. Conder, in *Physico-Chemical Measurements by Gas Chromatography*, John Wiley and Sons, New York, 1979.
8. R. J. Lamb and R. L. Pecsok, in *Physico-Chemical Applications of Gas Chromatography*, John Wiley and Sons, New York, 1987.
9. D. R. Lloyd, T. C. Ward, and H. P. Schreiber, in *Inverse Gas Chromatography: Characterization of Polymers and Other Materials*, ACS Symposium Series No. 391, 1989.
10. A. Khalfi, E. Papirer, H. Balard, H. Barthel, and M. Heinemann, J. Colloid Interface Sci. *184*:586 (1996).
11. A. Saada, E. Papirer, H. Balard, and B. Siffert, J. Colloid Interface Sci. *175*:212 (1995).
12. H. Balard, O. Aouadj, and E. Papirer, Langmuir *13*:1251 (1997).

13. H. Balard, A. Saada, J. Hartmann, O. Aouadj, and E. Papirer, Makromol. Chem., Makromol Symposia, *108*:63 (1996).
14. C. Kemball and E. K. Rideal, Proc. Roy. Soc. A *187*:53 (1948).
15. G. Ligner, A. Vidal, H. Balard and E. Papirer, J. Colloid Interface Sci. *134*:486 (1990).
16. E. Papirer, H. Balard, A. P. Legrand, L. Facchini, and H. Hommel, Chromatographia *23*:639 (1987).
17. E. Papirer, H. Balard, and M. Sidqi, J. Colloid Interface Sci. *169*:238 (1993).
18. G. M. Dorris and D. G. Gray, J. Colloid Interface Sci. *77*:353 (1980).
19. H. Hadjar, H. Balard, and E. Papirer, Colloids Surf. *99*:45 (1995).
20. G. Ligner, A. Vidal, H. Balard, and E. Papirer, J. Colloid Interface Sci. *133*:200 (1989).
21. J. Jagiello and E. Papirer, J. Colloid Interface Sci. *142*:232 (1991).
22. Z. Kessaissia and E. Papirer, J. Chim. Phys. *75*:709 (1974).
23. E. Papirer, H. Balard, J. Jagiello, R. Baeza, and F. Clauss, in *Chemically Modified Surfaces*, A. Mottola and J. R. Steimetz (eds.), Elsevier, Amsterdam, 1992, p.334.
24. J. Kuczinski and E. Papirer, Eur. Polym. J., *27*:653 (1991).
25. E. Papirer, J. M. Perrin, B. Siffert, and G. Philipponneau, J. Phys. III:697 (1991).
26. E. Papirer, A. Eckhardt, F. Muller, and J. Yvon, J. Mat. Sci., *25*:5109 (1990).
27. E. Brendlé and E. Papirer J. Colloid Interface Sci. (in preparation).
28. C. Lansinger, Ph.D. thesis, Mulhouse, 1990.
29. F. M. Fowkes and M. A. Mostafa, Ind Eng. Chem. *17*:3 (1978).
30. R. G. Pearson, J. Am. Chem. Soc. *110*:7684 (1988).
31. R. S. Drago and B. Wayland, J. Amer. Chem. Soc. *87*:3571 (1965).
32. V. Gutmann, in *The Donor–Acceptor Approach to Molecular Interactions*, Plenum Press, New York, 1978.
33. M. J. Kamlet and R. W. Taft, J. Am. Chem. Soc. *98*:377 (1976).
34. P. Mukhopadhyay and H. P. Schreiber, Colloids Surf. *100*:47 (1995).
35. M. J. Marmo, M. Mostafa, H. Jinnal, F. M. Fowkes, and J. A. Manson, Ind. Eng. Chem. Prod. Res. Dev. *15*:206 (1976).
36. E. Papirer, in *Composite Interfaces*, Elsevier, New York, to appear in 1998.
37. J. H. Park and Y. C. Weon, Anal. Sci. *12*:733 (1996).
38. C. Saint Flour and E. Papirer, J. Colloid Interface Sci. *91*:69 (1983).
39. M. Chehimi and E. Pigois-Landureau, J. Mat. Sci. *47*:41 (1994).
40. J. B. Donnet, S. J. Park, and H. Balard, Chromatographia *31*:434 (1991).
41. S. Dong, M. Brendlé, and J. B. Donnet, Chromatographia *28*:85 (1989).
42. J. Schultz, L. Lavielle, and C. Martin, J. Adhesion *23*:45 (1987).
43. E. Brendlé and E. Papirer, J. Colloid Interface Sci. *194*:217 (1997).
44. H. Balard and E. Papirer, Progr. Org. Coat. *22*:1 (1993).
45. E. Papirer and H. Balard, in *Adsorption and Chemisorption on Inorganic Sorbents* (A. Dabrovski and T. Tertykh, eds.), Elsevier, Amsterdam, 1995, pp. 479–502.
46. G. Ligner, M. Sidqi, J. Jagiello, H. Balard, and E. Papirer, Chromatographia *29*:35 (1990).
47. J. B. Donnet and E. Custodéro, Carbon *30*(5):813 (1992).
48. H. Balard, M. Sidqi, E. Papirer, J. B. Donnet, A. Tuel, H. Hommel, and A. P. Legrand, Chromatographia *25*:707 (1988).

49. H. Barthel, M. Heinemann, L. Rösch, and J. Weiss, *Proc. Eurofillers 95*, Mulhouse, France, 1995.
50. W. Rudzinski and D. H. Everett, *Adsorption of Gases on Heterogeneous Surfaces*, Academic Press, London, 1992.
51. M. Jaroniec and R. Madey, *Physical Adsorption on Heterogeneous Solids*, Elsevier, Amsterdam, 1988.
52. A. L. McClellan and H. F. Harnsberger, J. Colloid Interface Sci. *23*:577 (1967).
53. Z. Kessaissia, E. Papirer, and J. B. Donnet, J. Colloid Interface Sci. *82*(2):526 (1981).
54. M. V. Szombathely, P. Brauer, and M. Jaroniec, J. Comput. Chem. *13*(1):17 (1992).
55. J. Jagiello, Langmuir *10*:2778 (1994).
56. N. M. Nederlof, W. H. Riemsdjik, and K. Koopal, J. Colloid Interface Sci. *135*(2): 410 (1991).
57. J. Jagiello and J. A. Schwarz, J. Colloid Interface Sci. *146*(2):415 (1991).
58. H. Balard, Langmuir *13*:1260 (1997).
59. H. Balard, A. Saada, J. Hartmann, O. Aouadj, and E. Papirer, Makromol. Chem, Makromol. Symp. *108*:63 (1996).
60. H. Balard, O. Aouadj, and E. Papirer, Langmuir *13*:1251 (1997).
61. H. Balard, A. Saada, B. Siffert, and E. Papirer, Langmuir *13*:1256 (1997).

5
Chromatography of Colloidal Inorganic Nanoparticles

CHRISTIAN-HERBERT FISCHER   Department of Solid State Physics,
Hahn-Meitner-Institut Berlin, Berlin, Germany

I.   Introduction   174

II. Size Exclusion Chromatography   174
     A. Size determination   175
     B. Comparing transmission electron microscopy and size
         exclusion chromatography   190
     C. Concentration effects in size exclusion chromatography of
         colloidal inorganic nm particles   195
     D. Memory effect   195
     E. Investigation of the electrical double layer on nm particles by
         size exclusion chromatography   198
     F. Kinetics   202
     G. On-line optical spectroscopy   205
     H. Preparative size exclusion chromatography for monodisperse
         nm particles   212
     I. General considerations for establishing size exclusion
         chromatography conditions for colloidal inorganic nm particles 215

III. Hydrodynamic Chromatography   216
     A. Classical hydrodynamic chromatography   216
     B. Wide-bore hydrodynamic chromatography   218

IV. Optical Chromatography   219

V.   Comparison of Different Methods for Size Analysis of nm Particles 222

     References   224

I. INTRODUCTION

Dispersed particles can exist as colloids because repulsion by surface charges or steric effects prevent coagulation and flocculation. Inorganic particles in the nanometer (nm) size regime are interesting for several reasons. Colloidal silica is commercially applied as a protection colloid, inorganic binder, and for flame protection. Iron oxide colloids can be used in medicine as contrast medium for nuclear spin tomography. Semiconductor particles have a potential in solar energy conversion, and applications in electronics are being discussed. Finally, over the last 15 years so-called Q particles of semiconductors and metals in the lower nm size regime have become a rapidly growing topic of physicochemical research in their own right because of their extraordinary properties, such as size-dependent optical absorption or fluorescence (size quantization or Q effect; see below Sec. II.G.1). These investigations are closely connected with the names of Henglein, Brus, and others [1].

For all applications, control of particle size is crucial. Several methods exist for its determination. The classical method for size determination is transmission electron microscopy (TEM). It provides a direct image that can be measured. However, there are also disadvantages, such as lengthy sample preparation and measurement, tedious particle measuring and counting (if no image processing is available), possible radiation damage to the objects in the TEM itself [2], and rather bad statistics. There is a need for fast methods, especially when unstable (i.e., rapidly growing) colloidal particles are under investigation.

II. SIZE EXCLUSION CHROMATOGRAPHY

Since Porath and Flodin [3] and Moore [4] introduced crosslinked dextran (1959) and polystyrene gels (1964), respectively, size exclusion chromatography (SEC) has become a standard method for molar weight determination of dissolved organic polymers. Also, size separation of organic particles, e.g., latex, has been widely used. Though aqueous SEC is well studied [5] and polymer ferric hydroxide nitrate was already investigated by SEC in 1966 [6], inorganic compounds [7] have been characterized much less frequently than organic ones and especially inorganic particles have been rarely analyzed until the research on Q particles intensified. Reasons might be lack of such materials and problems with adsorption at the stationary phase.

SEC works with porous column packings. The sample particles are passed with the mobile phase along the packing surface. If they are not excluded from the pores, they are able to enter them (Fig. 1). The smaller they are, the deeper they can diffuse and the longer they stay in the stagnant liquid without forward motion. As a consequence, smaller particles elute later than larger ones. All analytes leave the column between the elution of the interstitial volume (excluded

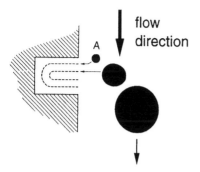

flow
direction

FIG. 1 Mechanism of size exclusion chromatography (SEC). Separation within a single pore. (Modified from Ref. 8.)

species) and the elution of the total permeation volume (smallest species). Theoretical treatment of SEC can be found, e.g., in Tijssen and Bos [8].

A. Size Determination

1. Silica, Polysilicic Acid, Siloxanes

In 1970, Tarutani described the application of SEC to the study of the polymerization of monosilicic acid [9]. The separation was carried out on Sephadex G25 or G100, respectively, with 0.1 mol/L NaCl/H_2O as eluent adjusted with HCl to pH 2. Fractions were taken and the content of silicic acid determined colorimetrically with the molybdate reagent after conversion of polysilicic acid to the monomer. Hamielec and Singh [10] published a general study of axial dispersion phenomena in chromatography of suspensions. Colloidal organic and silica particles were investigated; material and particle size of the column packing as well as flow rates were varied. The eluent consisted of Aerosol OT as anionic surfactant and potassium nitrate, both at a concentration of 1 g/L. Figure 2 shows a calibration plot of log diameter versus elution volume, which resembles those of dissolved organic polymers concerning the linear relationship in a certain range. Remarkably, in this universal calibration, the values for materials as different as silica, polystyrene (PS), styrene-methacrylic acid copolymer (SMA), and butadiene-acrylonitrile (BD/AN) copolymer lay more or less on the same curve. This was due to a rather low ionic strength and the fact that the stationary phase saw the same surface of adsorbed surfactant molecules for all kinds of colloidal particles. (Here "universal calibration" is not used in the same sense Benoit used it. Dispersed rigid inorganic particles do not affect viscosity.) The chromatographic peak was considered as Gaussian and its variance was taken as a direct measure of axial dispersion. The variance increased with flow rate and with particle size.

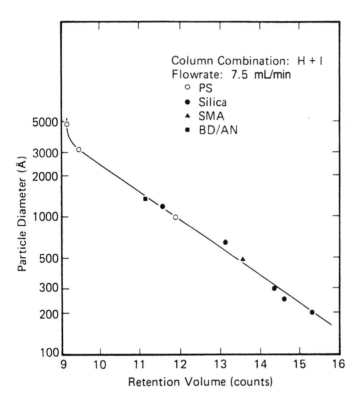

FIG. 2 Universal particle diameter–retention time calibration curve for SEC. Column: CPG 2500 and CPG 1500 (200–400 mesh, 4 ft × 3/8 in. each), eluent: 1 g/L Aerosol OT and 1 g/L potassium nitrate. (From Ref. 10.)

After a theoretical treatment of axial dispersion correction factors were given. Kirkland [11] also reported SEC separation of colloidal silica by high-performance liquid chromatography (HPLC) on various porous silicas. The upper size limit for elution was found to be 200 nm, and larger particles were retained on the column. However, in some cases the limit was much lower, depending strongly on composition (ionic strength, pH, etc.) of the mobile phase and the stationary phase. The results summarized in Table 1 were explained in terms of density of surface charges and van der Waals forces. Column plate height H as a function of flow rate was determined for different pore sizes and particles sizes of column packings as well as various sizes of colloidal particles. The resulting Van Deemter plots were very different from those for small molecules: H increased with flow rate, especially when small particles were separated on small porous particles with large pores, and the plots had no minimum (Fig. 3). The

TABLE 1 Elution of Silica from Various Systems

Mobile Phases, A: 0.02 M Na_2HPO_4-NaH_2PO_4; pH 7.2; B: 0.02 M Triethanolamine, pH 8.5 with HNO_3; C: 0.025 M NH_4NO_3, pH 8 with NH_4OH; D: 0.001 M NH_4OH; E: Methanol with 0.5% of 1 M HNO_3; F: 0.01 M KNO_3 with 0.05% Aerosol OT (Anionic Surfactant), pH 8.0 with NH_4OH; G: 0.01 M KNO_3 with 0.05% Aerosol OT, pH 3.5 with HNO_3; H: 0.25% Ludox-RM in 0.001 M NH_4OH; I: 0.1 M Na_2HPO_4-NaH_2PO_4, pH 8.0; J: 0.002 M Triethanolamine, pH 8.6 with HNO_3

Packing designation	Av. particle size (μm)	Av. pore size (nm)	Mobile phase	Range of silica sols eluted[a] (nm)	Silica sols retained (nm)
PSM-500	7.7	22	A	—	100–140
			D	5–200	
PSM-800	6.0	30	A	24–80	100–200
			B	5–80	
			C	—	50–200
			D	5–80	
			I	5–24	
PSM-1500	8.9	75	B	5–24	50–140
			C	24	100–140
			D	5–80	100–140
			J	24	100–500
Porasil C	41	50	A	6–140	
			B	100–200	
			C	≤200	
Spherosil XOB-030	38	30	B	60–200	380–500
			D	50–140	
Spherosil XOC-005	38	300	A	—	100–200
			D	—	100–200
			E	—	50–140
			F	24–80	200
			G	24–50	100–140
			H	24–1400[b]	

[a] Samples, 25 μL of 0.5–2%.
[b] The 100- to 140-nm sols retained about half again beyond total permeation volume.

various contributions to the plate height were discussed on the base of the experimental findings. However, one difficulty for the interpretation was the unknown contribution of size distribution of colloidal particles. The paper also gives an overview on the theoretical background of SEC of colloidal samples.

In 1994 Kirkland [12] improved conditions for SEC of silica. Porous spherical silica coated with "diol"-silane groups was recommended as stationary phase, the eluent being water brought to pH 7.0–7.5 by triethanolamine. Strong basic

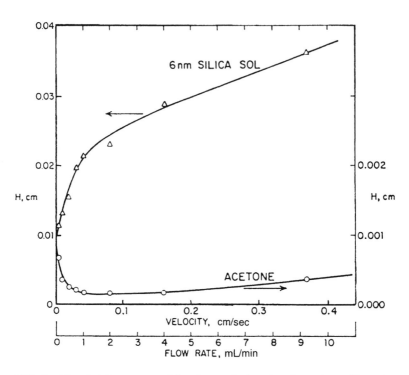

FIG. 3   Plate height versus mobile phase velocity plots for porous silica microsphere column with 30-nm pores. Column: PSM-800 (6 μm), 10 × 0.85 cm, mobile phase: 0.001 mol/L NH$_4$OH; sample 6.9 μL 2% 6 nm silica sol (Ludox-RB) and 5 mg/mL acetone. (From Ref. 11.)

or acidic mobile phases should be avoided, as negatively charged surfaces result in ion exclusion effects, and positively charged ones in irreversible retention of the negatively charged colloids. Under these conditions silica particles with diameters larger than 60 nm were irreversibly retained. The R_f method was suggested to improve the precision of the measurement, i.e., an internal standard such as potassium dichromate is added and the ratio of its retention time to that of the colloid is used.

Bürgy and Calzaferri [13] separated spherical silsesquioxanes (HSiO$_{3/2}$)$_n$(n = 8–18) abbreviated as HT8-HT18 with diameters between 0.96 and 1.64 nm by SEC on PLGel (5-μm, 5-nm pore size) with toluene or hexane (Fig. 4). In addition to the preparative fractionation, these species were used as test cases for the hard-sphere solute SEC retention theory [14]. The radii of the various silsesquioxanes were calculated as follows:

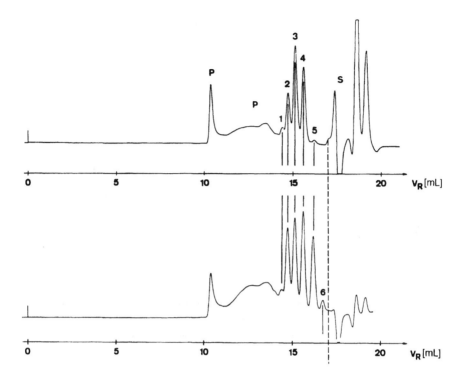

FIG. 4 Comparison of chromatograms of silsesquioxanes $(HSiO_{3/2})n$ obtained on PLGel, (pore size 5 nm) with hexane. Product synthesized in benzene (top), in toluene (bottom). Abbreviations: numbers $n = HT_n = (HSiO_{3/2})n$, P = Polymer, S = solvent. Peaks on and to the right of the dashed line belong to the solvent. (From Ref. 13.)

$$V_R = V_0 + K_{SEC}V_i \qquad (1)$$

where V_R is retention volume; V_0 is moving volume of the mobile phase; K_{SEC} is equilibrium constant for the SEC process, representing the ratio of the average solute concentration in the pores to that outside the pores, and V_i is stagnant volume of the mobile phase.

The following distribution coefficients K_e for pores of different geometrical shape were used:

$$K_e = (1 - r/a_c)^2 \qquad \text{for cylindrical pores} \qquad (2)$$

$$K_e = (1 - 2r/3a_s)^3 \qquad \text{for spherical pores} \qquad (3)$$

$$K_e = \exp(-2r/a_e) \qquad \text{for random plane pores} \qquad (4)$$

TABLE 2 Values for the Equilibrium Constants and
Calculated and Structural Molecular Radii (A)

Molecule	K_{SEC}	r	r_{struct}
HT8	0.7237	3.7	4.8
HT10	0.6614	4.7	5.5
HT12	0.5968	5.7	6.1
HT14	0.5425	6.6	6.8
HT16	0.4952	7.4	7.5
HT18	0.4559	8.1	8.2

where a_c, a_s, a_e are the radii of the pore types, and r_f is the radius of a solute molecule. Neglecting surface effects K_e equals K_{SEC}. The resulting radii assuming cylindrical pores with $a_c = 2.5$ nm were in good agreement with the radii r_{struct} from geometrical construction (Table 2). The reverse procedure, calculation of the pore sizes from r_{struct} and K_{SEC} also for other pore shapes [Eqs. (3) and (4)] showed best fit with the expected 2.5 nm for cylindrical pores (Table 3). Finally, it should be mentioned that SEC can be used for studies of sol-gel processes, e.g., controlled hydrolysis of tetraethyl orthosilicate [15].

2. Cadmium Sulfide

Size exclusion chromatography of nanoparticles has been used more frequently with the developments in the field of Q particles. Many of the latter are relatively stable only in the presence of complex forming stabilizers, e.g., polyphosphates, polyvinyl alcohol, alkanethiols, etc., which cover the surface and protect it against direct contact with other particles. This mostly prevents coagulation and precipitation. However, these bonds to the surface are usually not very strong, so that complexes are in equilibrium with free ligands. In a chromatographic process the ligands would be separated from the particles resulting in a destabilization of the

TABLE 3 Values of the Pore Radii (A), Calculated
from r_{struct} and K_{SEC}

Molecule	$\bar{a}_c$	$\bar{a}_s$	$\bar{a}_e$
HT8	32.1	31.3	29.7
HT10	29.4	28.5	26.6
HT12	26.8	25.7	23.6
HT14	25.8	24.6	22.2
HT16	25.3	23.9	21.3
HT18	25.2	23.7	20.9

colloid. Particle combination would lead to alteration; in the worst case, grown particles would be filtered off by the column. Furthermore, the free particle surface is often very active, so that reversible or even irreversible adsorption at the column material can occur, disturbing the pure SEC mechanism. In contrast to water-soluble organic polymers, this problem can be solved only partially by adding electrolytes to the eluent because colloids are quite sensitive to high ionic strength (their protective electrical double layer would collapse, leading to precipitation). Therefore other approaches were suggested, such as the addition of stabilizer to the mobile phase and nonpolar stationary phases. The first separations of Q particles were carried out by Fischer et al. [16,17] with CdS colloids by low-pressure chromatography on Sephacryl gel S200+S300+S500, the mobile phase being 6×10^{-3} mol/L sodium polyphosphate (referring to the phosphate units) and 1×10^{-3} mol/L cadmium perchlorate in water. Optical detection operated at 250 nm, where CdS particles have the same extinction coefficient regardless of their size.

For colloid chemists the diameter is more conspicuous and useful than the molar weight because the particles are rigid and do not swell or shrink as polymer coils do. Therefore, analogous expressions as in polymer chemistry were used: weight average diameter d_w, number average diameter d_n, and diameter polydispersity D_d [Eq. (5)–(9)]:

$$d_w = \sum n_i d_i w_i / \sum n_i w_i \qquad (5)$$

$$w_i \sim d_i^3 \qquad (6)$$

$$d_w = \sum n_i d_i^4 / \sum n_i d_i^3 \qquad (7)$$

$$d_n = \sum n_i d_i / \sum n_i \qquad (8)$$

$$D_d = d_w / d_n \qquad (9)$$

where n_i = number of particles in class interval i with diameter d_i and weight w_i.

In order to calibrate the system several colloid samples of different particle size and narrow size distribution were injected onto the SEC column. At the same time they were investigated by TEM to determine the weight average diameters d_w. In Fig. 5, top, the overlay chromatograms of seven CdS colloids are presented. From retention times and TEM diameters, a typical semilogarithmic calibration plot log $d = f(t)$ was obtained as shown in Fig. 5, bottom. Afterward the column can be used for the size analysis of unknown colloids of the same material.

The time scale of the method only allows analysis of quite stable or slowly growing samples [17]. Moreover, the Sephacryl gel bed becomes compressed with use, leading to changing calibrations. These arguments encouraged tests with HPLC, especially with wide-pore silica, because of its good mechanical

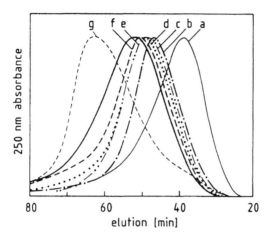

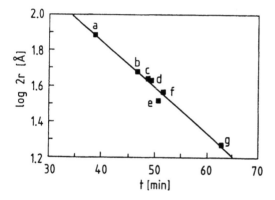

FIG. 5 (top) Low-pressure chromatograms of seven CdS colloids used for calibration. Conditions: column, Sephacryl S200, S300, and S500; eluent, 1 mol/L Cd(ClO$_4$)$_2$, 6 mol/L sodium polyphosphate. (bottom): Calibration of the system. Mean particle diameter was determined by TEM. (From Ref. 17.)

stability. The surface activity of bare silica resulting in peak tailing was strongly reduced by alkyl-modified silica such as Nucleosil C$_4$ or C$_{18}$. For the same size regime of CdS particles, packings with larger pores were recommended (50 + 100 nm) [27]. From the normalized chromatograms in Fig. 6 the influence of the eluent composition can be seen [18]: The same CdS colloid was injected four times onto the same column combination (Nucleosil C$_4$ 500 + Nucleosil C$_4$ 1000, 7 μm) using the following mobile phases:

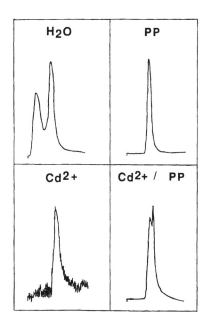

FIG. 6   Optimization of the eluent. Normalized chromatograms of the same aqueous CdS sol (stabilizer polyphosphate) on Nucleosil 500C$_4$ + Nucleosil 1000C$_4$ (7 μm) with four different eluent compositions: water, water with 1 mmol/L polyphosphate (PP), water with 1 mmol/L Cd(ClO$_4$)$_2$ (Cd^{2+}), water with 1 mmol/L Cd(ClO$_4$)$_2$ plus 6 mmol/L polyphosphate (Cd^{2+}/PP). Only the time range between 2.5 and 4.7 min is shown. Elution from left to right. (From Ref. 18.)

1. H$_2$O
2. Sodium polyphosphate (6 × 10^{-3} mol/L) /H$_2$O("PP")
3. Cd(ClO$_4$)$_2$ (1 × 10^{-3} mol/L) /H$_2$O ("Cd^{2+}")
4. Sodium polyphosphate (6 × 10^{-3} mol/L), Cd(ClO$_4$)$_2$ (1 × 10^{-3} mol/L) / H$_2$O ("Cd^{2+}/PP")

With water many CdS particles coagulated during the passage due to the loss of stabilizer, indicated by a second peak at earlier retention time. With Cd(ClO$_4$)$_2$/H$_2$O the situation was even worse: Most of the particles were either irreversibly adsorbed at the stationary phase or were coagulated and filtered off. The bad recovery was evident from the low signal-to-noise ratio. Ions adsorbed at the surface of the colloidal particles decreased net surface charge and lowered stabilizing repulsion. In the presence of the stabilizer polyphosphate one large late peak eluted. It is known from the preparation of CdS colloids that smaller species are formed with an excess of cadmium ions over sulfide ions. Also in

SEC the addition of some $Cd(ClO_4)_2$ to the polyphosphate let a second peak appear at longer retention time and improved the resolution. Therefore this eluent composition became the standard and for other colloidal sulfides the perchlorate of the relevant cation was used.

Calibration of the system was carried out by means of TEM as described for low-pressure SEC. Figure 7 shows chromatograms and calibration plot. The following advantages are obvious as compared to the Sephacryl separations (Fig.

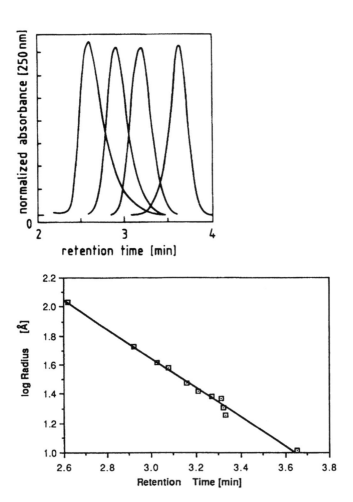

FIG. 7 (top) HPLC chromatograms of four CdS samples with narrow size distribution. Conditions: column, Nucleosil 500C$_4$ + Nucleosil 1000C$_4$ (7 µm), eluent: water with 6 mmol/L sodium polyphosphate and 1 mmol/L Cd(ClO$_4$)$_2$. (bottom). Calibration plot from 11 CdS samples, the histograms of which were determined by TEM. (From Ref. 27.)

5): The peaks are more symmetrical and narrower, resulting in a higher resolution, and the reduction of the analysis time from 80 min to 4 min allows the investigation of very unstable colloids. It was checked that particles leave the column unchanged. Only very small CdS particles of about 1.3 nm grew during the passage through the C_4 column, whereas the same material outside the column remained unchanged [18]. In the diode array detector the eluted colloid showed an absorption spectrum shifted by about 30 nm to longer wavelengths, indicating a particle growth (size quantization effect; see below). The growth seemed to be catalyzed by nonmodified silanol groups on the packing surface. This effect did not occur on a C_{18} stationary phase, where the silica is better shielded. Inorganic colloidal particles can also be stabilized by organic compounds (e.g., alkanethiols) and then behave like organics with respect to solubility [19]. In these cases, bare silica columns were recommended in order to avoid nonpolar interactions [18]. For SEC of CdS stabilized in such a way the eluent consisted of the particular thiol and $Cd(ClO_4)_2$, both in a concentration of 1 mmol/L in tetrahydrofuran (THF) (Fig. 8). It should be mentioned that negative peaks can occur at the retention time of the alkanethiol if a sample contains less thiol than the mobile phase [19].

3. Zinc Sulfide

Aqueous ZnS stabilized with polyphosphate was separated under analogous conditions to CdS on Nucleosil C_4 (7 µm), pore size 50 nm and 100 nm [18]. The

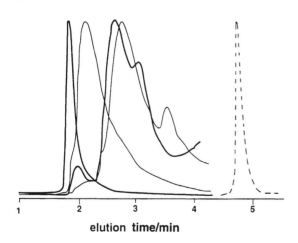

FIG. 8 Chromatograms of four CdS sols (solid lines) of various particle sizes in tetrahydrofuran with dodecanethiol as stabilizer. Columns: Nucleosil 500 + Nucleosil 1000; eluent 1 mM $CdClO_4$/1 mmol/L dodecanethiol/THF. Dotted line: dodecanethiol. (From Ref. 1.)

eluent consisted of an aqueous solution of sodium polyphosphate (6 mmol/L) and $Zn(ClO_4)_2$ (1 mmol/L). Figure 9 gives the calibration plot for ZnS and CdS. The scattering of the ZnS data is due to the occurrence of two crystal shapes. For a better fit, form factors have to be used. One can clearly see that calibration depends strongly on the material probably due to different surface charges.

4. Ferric Hydroxides Nitrates and Magnetic Fluids

In 1966 particles of polymer ferric hydroxide nitrate between 6 and 12 nm were separated from low molecular compounds by SEC [7]. Nunes et al. reported the fractionation of commercial magnetic fluids by SEC on Sepharose C1-4B 200. They found a change of magnetic properties as a function of particle size [20,21].

5. Gallium Arsenide

Gallium arsenide is a semiconductor with very good transport properties and is therefore of great industrial interest. Colloidal nm particles of GaAs prepared by Kher and Wells [22] were fractionated by SEC on silica with 10- and 30-nm pore size, the mobile phase being methanol with 1 mmol/L LiCl. The colored

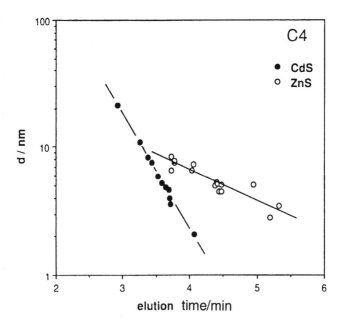

FIG. 9   SEC calibration for ZnS and CdS colloids on Nucleosil $500C_4$ + Nucleosil $1000C_4$. Eluent: water with 6 mmol/L sodium polyphosphate and 1 mmol/L $Zn(ClO_4)_2$ or $Cd(ClO_4)_2$, respectively. (From Ref. 18.)

larger particles remained on the column. However, seven peaks were obtained and attributed to very small GaAs particles based on the absorption spectra in the ultraviolet [23].

6. Gold

It is not difficult to prepare gold colloids with narrow size distributions that are very stable and consist of hard-sphere particles. Therefore, Holtzhauer and Rudolph suggested their use as a standard for characterization of column packings used in SEC [24]. Sepharose 4B or beaded Divicell cellulose were used as stationary phase with aqueous 0.05% PEG 20,000 plus 0.02% NaN_3 as mobile phase. Figure 10 shows the chromatograms of Au colloids with different average diameter on both columns and the plot particle diameter versus the coefficient of the distribution between the liquid in and outside the pores, here called K_{AV}. In this special case the hydrodynamic diameter of bovine serum albumin (BSA) fits well with the gold calibration. However, this cannot be generalized and it is expected that correlation coefficients must be used for other materials. Gold sols allow the determination of pore sizes and exclusion limits of hydrophilic gels. Interactions can be neglected and only limitations due to salt sensitivity of the colloid must be considered. Furthermore the mechanism of SEC was visualized by TEM micrographs showing the penetration of the packing by small particles and the exclusion of larger ones (Fig. 11).

Rapid size analysis of gold particles by SEC was demonstrated in Ref. 25. Bare silica (50- and 100-nm pore size) was chosen as stationary phase, since the organic stabilizer (1 mmol sodium citrate in water) was assumed to cause interaction with nonpolar silica. Interestingly, separation was successful only on rather coarse silica material (15–25 μm) With finer material most of the gold particles remained adsorbed on the column. In Fig. 12 five gold samples are presented with their TEM micrographs and size histograms and with the corresponding SEC chromatograms. In some cases SEC showed a bimodal distribution in contrast to TEM, where obviously the smaller particles were hidden under the larger ones. After SEC fractionation smaller Au species were indeed found in the second peak by TEM. The calibration for gold is given in Fig. 13. Later it was discovered that Nucleosil C_{18} (50- and 100-nm pore size, 7 μm) worked much better and with higher resolution, i.e., citrate stabilizer does not cause disturbing interaction at such a stationary phase as anticipated [23].

7. Silicon

Littau et al. [26] employed HPLC-SEC for luminescent colloidal silicon particles prepared by high-temperature aerosol reaction. Concentration and size distribution was analyzed and particular size fractions were collected. The separation was carried out on Zorbax 60-S and Zorbax 300-S columns with a mobile phase of methanol/ethylene glycol 60:40 mixture containing sodium methylate (1.5

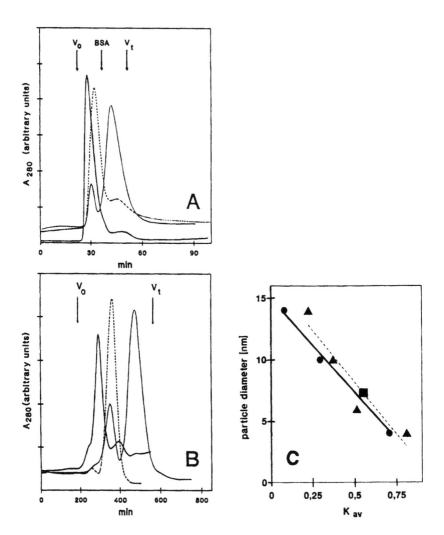

FIG. 10 SEC of colloidal gold on (A) Divicell (beaded cellulose) and (B) Sepharose 4B (beaded agarose); eluent: water containing 0.05% PEG 20,000 and 0.02% NaN_3. Elution profiles for the respective gold sols: thick solid lines, Au (14 nm); dashed lines, Au (10 nm); thin dotted lines, Au (4 nm). V_o, elution volume for blue dextran 2000 (void volume); BSA, elution volume for bovine serum albumin; V_t, elution volume for DNP-Ala (total volume). (C) Plot of K_{av} versus particle diameter for colloidal gold on Divicell (circles, solid line) and Sepharose 4B (triangles, dotted line). Square = respective hydrodynamic diameter and K_{av} values for BSA, run on the same column. (From Ref. 24.)

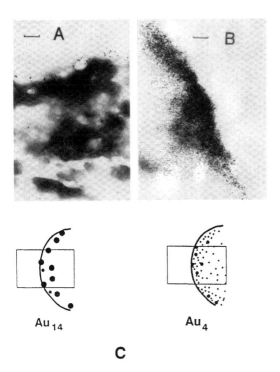

FIG. 11 TEM (original magnification ×58,000) of partial exclusion of larger particles (mean diameter 14 nm) from Divicell beads (A) and complete penetration of colloidal gold (mean diameter 4 nm) in the matrix (B). (C) schematic representation of A and B, respectively. The rectangles illustrate the cut of the surface area of cellulose beads as shown in A and B. Bars represent 200 nm. (From Ref. 24.)

mmol/L) and tetrabutylammonium bromide (0.1 mol/L). It was assumed that the base ionizes the surface hydroxyl groups on both the colloidal particles and the packing, preventing adsorption. When ionic strength was lower, poor resolution and low pore volume was observed because of a thick electrical double layer (see below) and ionic repulsion between particles and stationary phase. Sodium polystyrene sulfonate standards of different molar weight were injected under the same conditions and the equivalent hard-sphere diameter of each polymer was used for calibration [14]. In Fig. 14 the separation of a Si colloid with three size populations is shown.

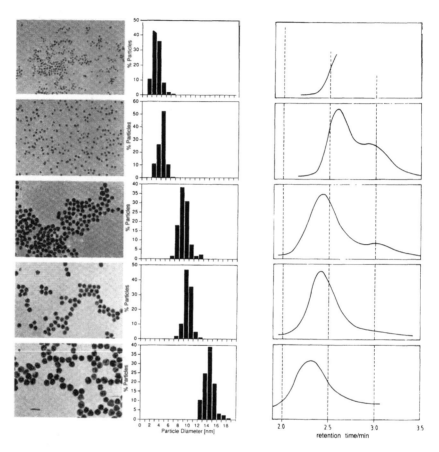

FIG. 12   Electron micrographs, histograms of size distribution obtained by TEM, and chromatograms obtained by monitoring the absorption of different gold sols at 520 nm. Conditions: column, Nucleosil 500 and Nucleosil 1000 (15–25 µm); eluent: 1 mmol/L sodium citrate in water. Scale bar 20 nm. (From Ref. 25.)

B.   Comparing Transmission Electron Microscopy and Size Exclusion Chromatography

1.   Average Size and Size Distribution

In order to compare the results of SEC with those of TEM, colloids were measured by both methods [27]. Figure 15 shows the electron micrographs and the SEC size distributions as well as the distribution histograms of two CdS samples.

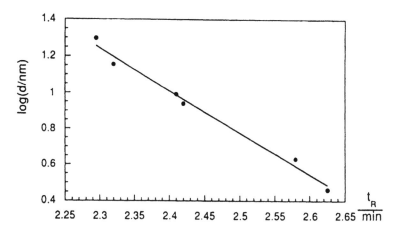

FIG. 13   SEC calibration for gold sols, conditions as in Fig 12. Semilogarithmic plot of the diameter as a function of the retention time. (From Ref. 25.)

Good agreement was found not only with respect to average size, but also to size distribution. However, one should keep in mind that the SEC result is obtained within a few minutes instead of several hours.

2.   Resolution

Fischer et al. [18] compared the resolution in SEC and TEM quantitatively. The lateral resolution is used for characterization of an electron microscope. This parameter is independent of the absolute size. In the case described it was 0.18 nm. The corresponding parameter in SEC is the standard deviation σ_d for a measured diameter d_i. It can be calculated from the standard deviation of the retention time σ_t by using the equation of the calibration line. Because of the semilogarithmic relation the lateral resolution is a function of absolute size. The calculation revealed values between 0.07 nm (1.9%) and 0.24 nm (1.3%) for CdS particles in a diameter range between 2 and 20 nm (analysis on two Nucleosil C_4 columns, 125 × 4 mm ID). From the upper part of Fig. 16 it can be seen that the lateral resolution in TEM is better for larger particles but more than three times worse for smaller particles.

In addition to the lateral resolution, the separation resolution, i.e., resolution in the chromatographic sense, must be discussed, though such a characterization is not used in TEM. It describes the resolution of particle populations different in size and could be estimated from the half peak width $b_{0.5}$. Due to the lack of strongly monodisperse inorganic colloids the half-width of the peak of a potas-

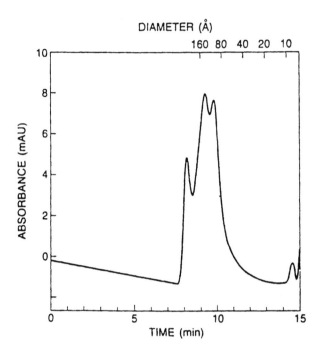

FIG. 14 SEC of colloidal nanocrystalline silicon particles. Bottom scale: elution time; top scale: size (calibrated with a set of sodium polystyrene sulfonate polymers and their hard-sphere diameters). Exclusion limit occurs at 8.0 min and the molecular fluid front occurs at 14.8 min. Conditions: column, Zorbax 60-S and 300-S; eluent, 60:40 methanol-ethylene glycol with 0.0015 mol/L sodium methylate and 0.1 mol/L tetrabutylammonium bromide at 50°C. (From Ref. 26.)

sium iodide solution was used. Such small ions have the longest possible retention time in SEC and the broadest possible peak width of all monodisperse species. Therefore it represents the worst case; for all larger species the peak width is smaller. $b_{0.5}$ was transformed into size values as before. From selected retention times t_i as well as from $t_{i \pm 0.5} = t_i \pm 0.5 b_{0.5}$ the corresponding diameters d_i and $d_{i \pm 0.5}$ were calculated by the equation of the calibration line. $d_{i \pm 0.5}$ is shown as a function of d_i in Fig. 16, lower part, dashed line; for comparison the error of the diameter is represented by the area between the full lines. Since the standard deviation of diameter of typical inorganic colloids is 15% or more the resolution of SEC is sufficient.

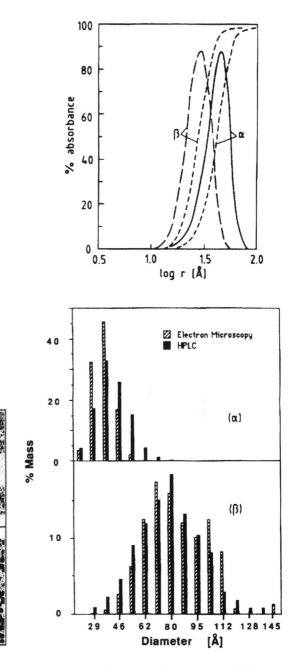

FIG. 15 TEM micrographs of two samples of colloidal CdS (left). Integral and differential size distributions from SEC (right, upper part). Mass distributions of the two samples, obtained by TEM and SEC, respectively. (From Ref. 27.)

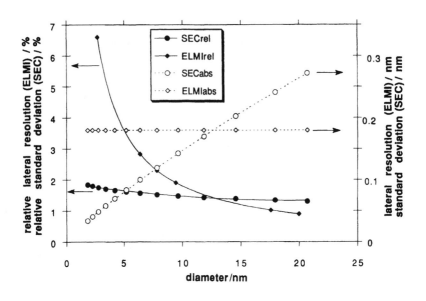

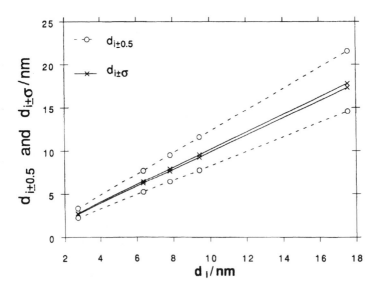

FIG. 16 Resolution of TEM and SEC as a function of particle diameter. Aqueous CdS sols (stabilizer polyphosphate) on Nucleosil $500C_4$ + Nucleosil $1000C_4$. (top) Absolute and relative lateral resolution in ELMI and absolute and relative standard deviation of the measured diameter in SEC as a function of the diameter. (bottom) Lateral and separation resolution in SEC as a function of d_i. The full lines enclose the area between $d_{i+\sigma}$ and $d_{i-\sigma}$. The area inbetween the dotted lines is limited by $t_{i+0.5}$ and $t_{i-0.5}$. (From Ref. 18.)

C. Concentration Effects in Size Exclusion Chromatography of Colloidal Inorganic nm Particles

In preparative SEC experiments with colloids it was found that recovery decreased with increasing sample concentration [28]. Fischer and Siebrands investigated the concentration effects in analytical scale SEC of CdS by dilution experiments [29]. They found that recovery was good up to CdS concentrations of 5 mmol/L but dropped dramatically above 10 mmol/L CdS (Fig. 17, top). This loss was primarily due to the higher concentration of the accompanying electrolytes present from the colloid preparation [Eq. (10)]. A higher ionic strength destabilizes colloids leading

$$Cd(ClO_4)_2 + H_2S \xrightarrow{\text{Na-polyphosphate}} CdS + 2HClO_4 \tag{10}$$

to irreversible and reversible adsorption at the stationary phase. The reversible adsorption became evident from increasing peak width and tailing. Fortunately, no preferential adsorption of certain particle sizes was found on reinjection. When instead of water a solution of all electrolytes present in the original sample (polyphosphate and perchloric acid) was used for dilution of a concentrated colloid, the CdS peak area was almost proportional to the CdS concentration (Fig. 17, bottom), but the recovery was generally lower than for samples diluted with water. In other words, *particle* concentration has practically no effect on recovery. These findings on adsorption of inorganic colloids in a flow system are in agreement with previous results obtained by Freundlich et al. [30] and Buzagh [31] on stationary experiments with iron trioxide and arsenic trisulfide on silica, respectively. They found adsorption to be independent of particle concentration as long as the ionic strength was constant. The effect of sample concentration on the retention time was not very strong. However, when water was used for dilution both particle and electrolyte concentrations were changed and retention *increased* with increasing concentration due to decreasing thickness of the electrical double layer (see below). When the electrolyte concentration was kept constant, retention *decreased* with increasing particle concentration, probably because of weak and reversible agglomeration. No comparable effect was observed on silica with small pores (Nucleosil 120 C_4), where adsorption should be predominant over any SEC mechanism.

D. Memory Effect

Closely connected to the adsorption described in the previous chapter is a memory effect of the column generally referred to as conditioning which is extraordinarily pronounced in this case [29]. When the same sample was injected five times on extensively purged Nucleosil C_4 columns, the CdS peak area increased by about

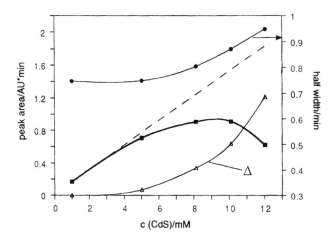

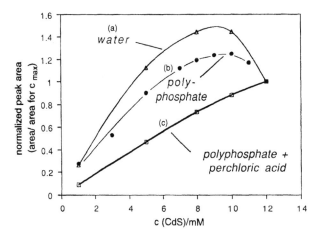

FIG. 17 SEC of a CdS sol at different concentrations (conditions: column, Nucleosil 500C$_4$ + Nucleosil 1000C$_4$ (7 μm); eluent, water with 6 mmol/L sodium polyphosphate and 1 mmol/L Cd(ClO$_4$)$_2$). (top) A 12 mM CdS sol stepwise diluted with water. Area (■) and half-width (●) of the CdS peak as a function of the CdS concentration. The difference area (△) is the loss of CdS calculated as the difference between the area curve and its extrapolated initial slope. (bottom) A 12 mM CdS sol stepwise diluted with (a) water (△), (b) 72 mM polyphosphate (●), or (c) 72 mM polyphosphate/24 mM perchloric acid (□) solutions. Normalized CdS peak area (referring to the peak area of the undiluted original sample). (From Ref. 29.)

10% to a plateau, i.e., the surface of the packing was more and more covered by a CdS layer, which had a lower surface activity than the original C_4-silica. This adsorption was proved by injections of methanol–water mixtures. They wetted the C_4 phase and enhanced desorption resulting in an eluting peak that shows the typical CdS absorption spectrum in the diode array detector.

The pronounced memory effect was also demonstrated in a dilution experiment of a CdS colloid where ionic strength was kept constant (Fig. 18). The series was injected twice, once in increasing order of CdS concentration and once in decreasing order. Between both series the column was purged extensively. Whenever the same sample saw a cleaner column packing, the peak area was smaller than in the case of a stationary phase covered by CdS. This effect was most dramatic for the most diluted samples, where the relative loss was up to 50%.

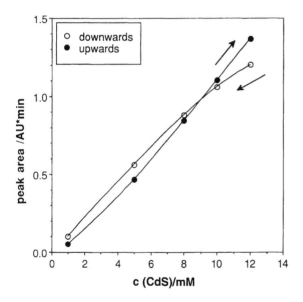

FIG. 18 Memory effect of the column. SEC of a CdS sol (12 mmol/L) stepwise diluted with a polyphosphate/perchloric acid solution 2 mmol/L and 24 mmol/L, respectively. (conditions: column, Nucleosil 500C$_4$ + Nucleosil 1000C$_4$ (7 μm); eluent, water with 6 mmol/L sodium polyphosphate and 1 mmol/L Cd(ClO$_4$)$_2$). Order of injection: increasing concentration (●); decreasing concentration (○). Before each series the column was purged extensively. Peak area as a function of CdS concentration. (From Ref. 29.)

E. Investigation of the Electrical Double Layer on nm Particles by Size Exclusion Chromatography

1. Theory

At all interfaces between a charged solid phase and a liquid phase containing electrolytes, a so-called electrical double layer is formed by the charge of the surface and that of dissolved counterions being concentrated toward the interface. SEC of particles is strongly affected by this phenomenon. On the other hand, SEC offers a unique tool for its investigation and visualization. In the field of organic polymers the influence of ionic strength on chromatographic behavior has been widely discussed [32–38]. Theoretical considerations about electrostatic double layer interactions for spherical colloids in cylindrical pores were published by Smith and Deen [39,40].

In analogy to the Debye-Hückel model, the Gouy-Chapman model describes the potential distribution by combining the Boltzmann distribution of energy and the Poisson equation, which describes the relation between potential and charge density [41]. For a flat surface the potential Ψ at a distance x from the interface is given by

$$\Psi = \Psi_0 e^{-\kappa x} \tag{11}$$

where Ψ_0 is the potential at $x = 0$. For a spherical surface (with radius a) the potential distribution is given as a function of the radial coordinate, r, by

$$\Psi = \Psi_a(a/r) \, e^{\kappa(a-r)} \tag{12}$$

In both equations, parameter κ is

$$\kappa = (8 \, \pi n z^2 e^2 / \, \varepsilon k T)^{0.5} \tag{13}$$

with number of ions n, their valency z, elemental charge e, dielectric constant ε, Boltzmann constant k, and temperature T. Equation (12) is valid as long as the product $a \, \kappa$ is not much larger than unity.

From Eqs. (12) and (13) the electrical potential as a function of the distance from the surface and its dependence on electrolyte valency and concentration can be calculated. The potential Ψ drops to Ψ_0/e at distance $x = 1/\kappa$, which is called the thickness of the diffuse double layer, or Debye length. $1/\kappa$ is dependent on the ionic strength J (which is $J = 0.5\Sigma \, z_i^2 m_i$, with m_i, the molarity of the ions):

$$1/\kappa \sim J^{-0.5} \tag{14}$$

For dilute electrolyte solutions the thickness of the double layer, $1/\kappa$, is in the range of 10^3–10^2 nm, for more concentrated ones about 1–10^{-1} nm.

One could say that colloidal particles wear a coat of ions that is thinner the higher the ionic strength. For the stability of colloids this double layer is essential as its repulsion avoids direct contact of surfaces with the same charge, which would lead to coagulation or adsorption. Therefore, high ionic strengths destabilize colloids. It has been shown in chromatography of polyelectrolytes that the double layer is quite rigid. Its influence on SEC is striking. An effective particle size $d_{p,eff}$ has to be considered, which is the sum of particle core diameter d_p and twice the double layer thickness d_{dl}. In other words, the colloidal peak elutes earlier than expected for the core size due to repulsion between particles plus surrounding ions and the packing.

2. Retention Time Dependence of Electrolyte Concentration in the Mobile Phase

When nanoparticles migrate through the SEC column, they are immediately separated from the small ions present in the sample and are surrounded by the eluent with given electrolyte content. Thus Fischer and Kenndler [42] brought colloidal CdS particles in media of well-defined ionic strengths and observed the resulting change of $d_{p,eff}$ in terms of retention time. Figure 19, top, shows chromatograms of a CdS sol (average size 5.7 nm) obtained on the same column combination, whereby NaCl was added at concentrations from 3 mmol/L to 30 mmol/L to the standard eluent (1 mmol/L $Cd(ClO_4)_2$, 6 mmol/L sodium polyphosphate). Increasing retention times were found with increasing ionic strength together with decreasing recovery due to irreversible particle adsorption at the stationary phase (the double layer also acts as a protection coat against adsorption at the packing). Unchanged size of eluting particles was checked by reinjection and several factors indicated that the *eluting* particles were not much affected by adsorption. When $MgSO_4$ was added to the mobile phase instead of NaCl, retention times increased much more quickly (Fig. 19, center). This was in accordance with theory [Eq. (14)] predicting an inverse proportionality between $1/\kappa$ and z^2 (valency).

3. Comparison of the Chromatographically Measured Double-Layer Thickness with the Debye Length

Since the SEC system was calibrated under standard conditions for CdS core diameters by TEM, a calibration for the effective particle size $d_{p,eff}$ was necessary. The retention times of the same CdS sol increased with increasing salt concentration in the eluent and reached saturation at about 20 to 25 mmol/L NaCl (Fig. 19, middle). The similarity to the dependence of the Debye length on electrolyte concentration can be seen from Fig. 19, bottom. The ionic strength of the standard eluent was determined to be $J = 8$ mmol/L. For ionic strengths about 30 mmol/L $1/\kappa$ is still about 1.8 nm (Fig. 19, bottom). Therefore the following approximation was made: When the retention time no longer increases with increasing ionic strength, the effective particle size $d_{p,eff}$ equals the core diameter d_p plus 1.8 nm for the double-layer thickness in this concentration range. Thus a calibration plot was constructed from d_p plus 1.8 nm and the particular maximum retention times corresponding to the plateau for several CdS samples (Fig. 20). The linear fit in

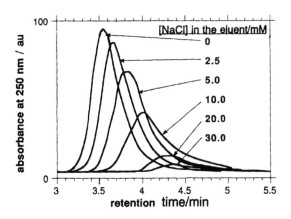

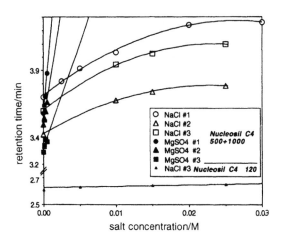

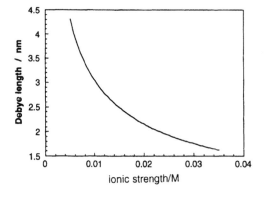

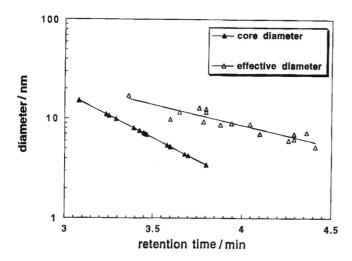

retention time / min

FIG. 20 SEC calibration for particle core diameters d_p and effective diameters $d_{p,\text{eff}}$ of CdS sols on Nucleosil 500C$_4$ plus Nucleosil 1000C$_4$. Full triangles: Standard calibration. Diameter of the particle core d_p, determined by TEM as a function of retention time with standard eluent composition (1 mM Cd(ClO$_4$)$_2$/6 mM sodium polyphosphate). Open triangles: effective diameters $d_{p,\text{eff}}$ as a function of the plateau retention time t_{max} in Fig. 19, center, obtained with NaCl added to the standard eluent in high concentration (typical ionic strength of the eluent $J = 0.03$ M). $d_{p,\text{eff}}$ was calculated from the diameter of the particle d_p, determined from SEC under standard conditions (standard calibration based on TEM) plus 1.8 nm, the theoretical Debye length $1/\kappa$ for an ionic strength $J = 0.03$ M. These samples were different from those for standard calibration. Both retention times were obtained under different conditions (with and without NaCl in the eluent). (From Ref. 42.)

FIG. 19 Effect of electric double layer on SEC of colloidal particles. SEC of a CdS sol (average size 5.7 nm) on Nucleosil 500C$_4$ + Nucleosil 1000C$_4$ with the standard eluent (1 mM Cd(ClO$_4$)$_2$/6 mM sodium polyphosphate) containing various salt concentrations. (top) Chromatograms obtained with standard eluent containing NaCl from 0 mM to 30 mM. (center) Retention time as a function of salt concentration for CdS sols of various particle size. NaCl (open symbols), MgSO$_4$ (full symbols). For comparison the retention times of colloid 3 on Nucleosil 120C$_4$ are given (small triangles). (bottom) Debye length $1/\kappa$ as a function of ionic strength for an 1:1 electrolyte. (From Ref. 42.)

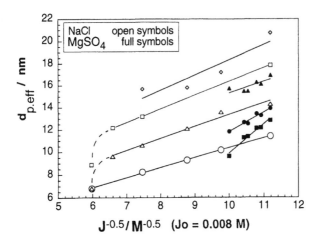

FIG. 21   Effective particle diameter $d_{p,eff}$ for CdS sols of different particle size, chromato-graphically determined in the standard eluent (1 mM $Cd(ClO_4)_2$/6 mM sodium polyphos-phate) with and without NaCl or $MgSO_4$ addition as a function of the reciprocal square root of ionic strength J according to Eq. (14). For the ionic strength J_0 of the standard eluent a value of 8 mM was taken. Details in the text. (From Ref. 42.)

the plot $d_{p,eff}$ versus $J^{-0.5}$ was good proof that the retention time shift was due to the changing electric double layer. Only the values at highest ionic strengths seemed to be influenced by adsorption (Fig. 21). Potschka described similar be-havior of polystyrene latex beads [36]. In Fig. 22 experimental values for d_{dl} at various particle core diameters were compared with the Debye length for selected ionic strengths. The agreement over a wide range revealed that at least in this case $1/\kappa$ can be used in the estimation of the effective particle size $d_{p,eff}$ in SEC.

F.   Kinetics

1.   Particle Growth

As already mentioned, small nm particles are thermodynamically unstable and have the tendency to grow. There are two mechanisms. In the case of Ostwald ripening [Eqs. (15) and (16)] smaller particles dissolve preferentially and the ions precipitate on larger particles, leading to gradual growth. Particle combination [Eq. (17)] results in a very sudden increase in diameter.

Ostwald ripening

$$(CdS)_m \rightarrow (CdS)_{m-1} + Cd^{2+} + S^{2-} \tag{15}$$

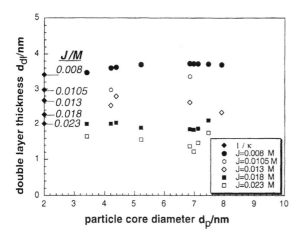

FIG. 22   Experimentally determined double-layer thickness d_{dl} of CdS particles as a function of the particle size d_p for various ionic strengths. (From Ref. 42.)

$$(CdS)_n + Cd^{2+} + S^{2-} \rightarrow (CdS)_{n+1} \tag{16}$$

typically $n > m$

Particle combination

$$(CdS)_m + (CdS)_n \rightarrow (CdS)_{m+n} \tag{17}$$

The short analysis times of HPLC-SEC meant that the rapid growth of 2-nm CdS particles could be investigated [27]. The chromatograms of a colloid just after preparation and at various stages of aging are shown in Fig. 23, top. From the diameters corresponding to the peaks, from peak heights and growth time a three-dimensional diagram was constructed (Fig. 23, bottom). One recognizes the decrease of the initial population without pronounced change in size. The new population, significantly larger, initially exhibited a sudden increase in size indicating particle combination. Since the final growth was more gradual, it was concluded that Ostwald ripening was then the predominant mechanism.

2. Photocorrosion

Size exclusion chromatography is very useful for photolysis studies of colloidal semiconductor particles under different conditions, e.g., under oxygen or inert gas [19] or on exposure to light of different wavelengths [27]. For example, CdS particles partially dissolve when they are illuminated under air due to oxidation to $CdSO_4$. In an irradiation experiment using light with wavelengths > 390 nm, the chromatograms of Fig. 24 were obtained. The original CdS sample had two

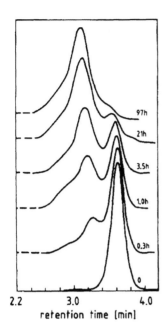

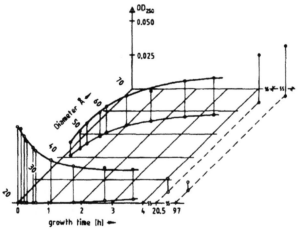

FIG. 23 Thermal aging of small CdS particles. (top) Chromatograms after different aging times. Conditions: column, Nucleosil 500C$_4$ + Nucleosil 1000C$_4$ (7 μm), eluent: water with 6 mmol/L sodium polyphosphate and 1 mmol/L Cd(ClO$_4$)$_2$. Dashed line: zero line of chromatogram. (bottom) Three-dimensional diagram showing the height of the peaks in the chromatograms and the corresponding diameters as a function of time. (From Ref. 27.)

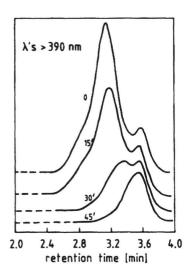

FIG. 24 Chromatograms of a CdS sample at various times of illumination with $\lambda >$ 390 nm. Conditions: column, Nucleosil 500C$_4$ + Nucleosil 1000C$_4$ (7 μm), eluent: water with 6 mmol/L sodium polyphosphate and 1 mmol/L Cd(ClO$_4$)$_2$. Detection, absorbance at 250 nm. Dashed line: zero line of chromatogram. (From Ref. 27.)

size populations, but only the larger particles were dissolved. The small particles of the second peak were not affected because they do not absorb light of such long wavelengths (see Sec. II.G.1).

G. On-Line Optical Spectroscopy

1. Q Effect

Particles of semiconductors in the low-nm size regime change their optical absorption spectra with diameter. This behavior, known as quantum size effect or Q effect [43], is explained in terms of the band gap between valence band and conductivity band. The bandwidth depends on the number of molecules the crystal is formed of, since it determines the number of electronic states overlapping to a band (Fig. 25). As a consequence, smaller crystallites have a larger band gap and therefore start absorbing light at shorter wavelengths than larger crystallites or macrocrystals. It should be stressed that these nanoparticles really change color with size, e.g., 2-nm CdS particles are almost colorless! This is no effect of scattering; species of that size practically do not scatter light anymore. Chromatographic separation according to size is extremely useful for basic research in the field of these new materials especially when coupled with optical spectroscopy.

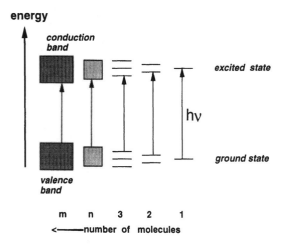

FIG. 25 Schematic explanation of the size quantization effect (Q effect).

The first experiments were done off-line by collecting fractions from low-pressure chromatography of CdS [16] followed by conventional measurement of absorption spectra. Later a modification of a conventional spectrophotometer was described, which allowed permanent up-and-down scanning with computer storage of the data [17]. From the resulting 3D plot (wavelength–absorbance–retention time) chromatograms at any recorded wavelength could be extracted. The results belonged to the early proofs of the Q effect.

Such a permanent scan technique was much too slow for the time scale of HPLC. Today commercially available diode array spectrometers from numerous companies allow measurement of entire UV/visible (UV-Vis) spectra every second or even faster [27]. Figure 26, top, represents the 3D plot of an HPLC separation of a CdS colloid with three size populations on Nucleosil C_4 (50- and 100-nm pore size, 7 μm). In the normalized chromatograms of the same run it is evident how the peak of small and later of medium-sized particles disappear when

FIG. 26 SEC separation of a CdS colloid containing three different size populations with on-line optical spectroscopy by diode array detection. Column: Nucleosil 500C_4 + Nucleosil 1000C_4; eluent: water with 6 mmol/L sodium polyphosphate and 1 mmol/L $Cd(ClO_4)_2$. (top) 3-D plot. (center) Chromatograms at different detection wavelengths. (bottom) Optical absorption spectra on-line measured at the retention times marked with 1–4 in the center plot. (From Ref. 44.)

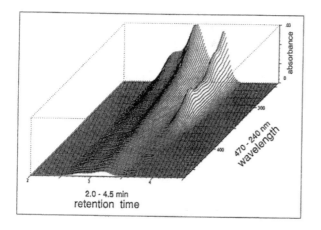

absorbance

470 - 240 nm
wavelength

300

400

2.0 - 4.5 min
retention time

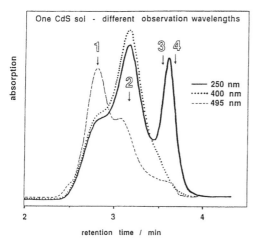

One CdS sol - different observation wavelengths

absorption

1

2

3 4

—— 250 nm
········ 400 nm
------ 495 nm

2   3   4

retention time / min

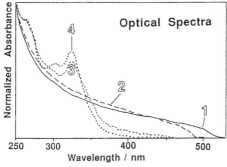

Normalized Absorbance

Optical Spectra

4

3

2

1

250   300   400   500

Wavelength / nm

the detection wavelength is increased (Fig. 26, center). The lower part of Fig. 26 shows the optical spectra at the three chromatographic peaks. With decreasing particle diameter (increasing retention time) the onset of adsorption shifts to shorter wavelengths due to the Q effect. It is interesting that spectra 3 and 4 taken in the third peak are still significantly different, indicating separation within the peak, i.e., the peak width still reflects size distribution [44].

2. "Magic" Agglomeration Numbers

Frequently, excitonic peaks are observed in the UV absorption spectra of very small semiconductor particles. They could be considered as electronic transitions to various excited states because these smallest species have a term scheme similar to that of a molecule rather than a conduction band with a high density of states [2,45,46]; or, on the other hand, such spectral maxima could occur when size distributions are structured, i.e., certain agglomeration numbers are more common than others in the neighborhood due to lower energy. These "magic" agglomeration numbers are explained by the shell model of Lippens and Lannoo [47]. This question could be answered by SEC with diode array detection [27]. A closer inspection of numerous CdS colloids revealed that typically two spectral maxima appeared and disappeared simultaneously during the SEC runs. Figure 27 gives two examples showing the spectra of CdS at two retention times marked in the insert chromatogram. It was concluded that a pair of spectral peaks belongs to the same magic agglomeration number because of the same retention time in SEC. The peak at longer wavelength is attributed to the transition to the first excited state, that at shorter wavelength to the transition to the second. The next blue-shifted pair of peaks originates from the next smaller magic number.

3. Plasmon Band in Gold Spectra

The optical absorption spectra of gold colloids also depend on particle size. However, in this case absorption of light generates oscillation of electrons in the conductivity band, called *plasmons*. These oscillations are damped by impacts of the electrons with the particle surface, which are more frequent in smaller particles. Therefore the intensity of the resulting plasmon band in the spectrum of gold colloids, situated at about 520 nm, decreases with the particle diameter (Fig. 28, top) [48,49].

By means of SEC with on-line spectroscopy the plasmon band and its size dependence can be investigated very conveniently [50]. Moreover, the plasmon band enables a quick test for monodispersity of particles within one SEC peak. As the extinction at 440 nm is independent of particle size, it is practical to normalize the extinction at 520 nm. The extinction ratio $R_1 = E_{520nm}/E_{440nm}$ is an intensity value for the plasmon band independent of concentration. The central part of Fig. 28 shows the chromatograms of two gold colloids (average diameters 14.3 nm and 4.2 nm). The ratio chromatograms R_1 calculated by the diode array

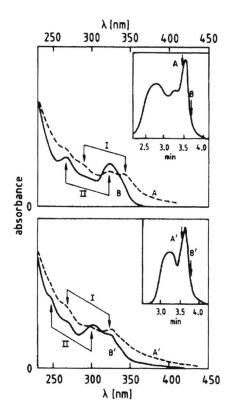

FIG. 27 Optical absorption spectra of two fractions of CdS samples measured with a diode array detector during SEC separation. The insets show the chromatograms of the samples. The times at which the spectra were recorded are marked with A, B and A', B', respectively. Details in the text. (From Ref. 27.)

system gave values of 1.7 and 1.2, respectively. A constant R_1 value indicates a monodisperse colloid. Constant values $R_2 = E_{440nm}/E_{400nm} = 0.95$ were found for all Au samples in accordance with previous findings that interband transitions in that area are independent of particle diameter [51]. In the final part of the second peak not only R_1 but also R_2 decreased, which could not be due to polydispersity of the Au sample. It is explained by beginning elution of tannin, a stabilizer of some gold colloids. Figure 28, bottom, comprises the results of seven samples.

4. Fluorescence Detection

Many cadmium sulfide colloids show intense fluorescence. The wavelength of the emitted light depends on the particle size in the same way as for absorption.

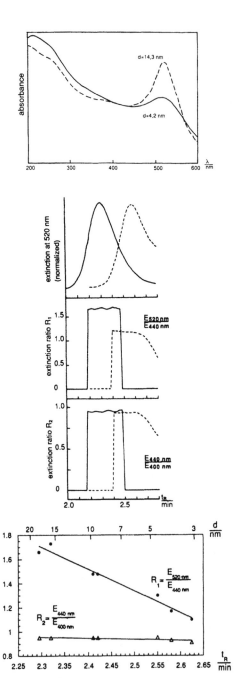

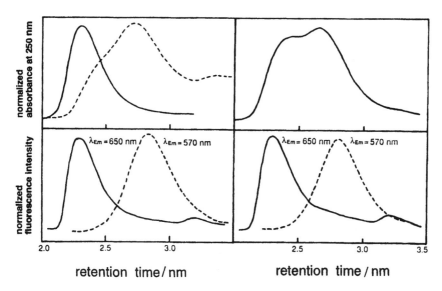

retention time / nm   retention time / nm

FIG. 29 Coupled diode array detection (top) and fluorescence detection (bottom) with emission wavelength 650 nm (solid line) and 570 nm (dashed line). SEC of two CdS colloids of different particle size. (left) Single samples. (right) 1:1 mixture of both samples. Conditions: columns, Nucleosil 500 and Nucleosil 1000, 5 μm; eluents: THF containing cadmium perchlorate (2×10^{-4} mol/L) and dodecanethiol (2×10^{-4} mol/L); fluorescence excitation, 325 nm. (From Ref. 50.)

Fluorescence of every color has been found. When a fluorescence detector is placed behind the diode array detector, not only can the Q effect of the fluorescence emission be demonstrated, but the selectivity of the detection is also enhanced [23]. In the experiment of Fig. 29 two CdS colloids of different particle size were injected in the SEC column separately (left side) and as a 1:1 mixture (right side). In the upper part the absorbance at 250 nm is given and in the lower

FIG. 28 Size dependence of the plasmon band (520 nm) in the UV-Vis spectra of gold nm particles. (top) Spectra of two gold sols with average particle sizes of 14.3 and 4.2 nm, respectively. (center) Normalized chromatograms of the two samples. Conditions: column, Nucleosil 500 and Nucleosil 1000 (15–25 μm, 120 $\times$ 4 mm each); eluent: 1 mmol/L sodium citrate in water). Ratiochromatogram $R_1 = E_{520}/E_{440}$. Ratio chromatogram $R_2 = E_{440}/E_{400}$. (bottom) Extinction ratios $R_1 = E_{520}/E_{440}$ and $R_2 = E_{440}/E_{400}$ as a function of retention time and particle diameter respectively. (From Refs. 25 and 50.)

part the emissions at 650 and 570 nm, respectively. The two peaks in the mixture chromatogram overlapped considerably. However, by setting the emission wavelength of the detector on the maximum of each sample the original chromatograms were obtained.

H. Preparative Size Exclusion Chromatography for Monodisperse nm Particles

Since the properties of nm particles depend strongly on their size, it is very important to possess colloids that are as monodisperse as possible. Preparative SEC is a way to narrow size distributions. However, one should keep in mind that SEC represents a very special situation, when fractions are cut out of one broad peak. In all other preparative chromatographic methods, where two or more peaks are to be fractionated, the column can be overloaded resulting in peak broadening. As long as the interesting peak is separated from its neighbors at the baseline, there is no loss of product purity. In SEC an entire broad peak has to be considered as a superposition of elemental chromatograms, each corresponding to species of a certain size. Any overloading leads automatically to broader elemental chromatograms with more overlap and consequently to a higher polydispersity of collected fractions. Thus a compromise between yield and quality of the product must be found unless the system is scaled up further. Fischer [28] investigated the influence of sample amount (concentration and volume), flow rate, and fraction volume on separation of aqueous CdS colloids on two preparative columns in series, each 125 mm long and 32 mm in diameter, filled with 7-μm Nucleosil C_4, pore size 50 nm and 100 nm. The fractions were characterized by diameter polydispersity D_d [Eq. (9)] obtained by injection in analytical SEC or by standard deviation of diameter from TEM.

1. Effect of Flow Rate

Since preparative SEC is typically carried out through many cycles in order to avoid larger (too expensive) columns, flow rate has a great influence on the yield per time. But when the flow rate is too high, particles can miss the entrance to the pores, which leads to poor efficiency. For the columns described above, a flow rate of 4.5 mL/min was found to be best. When central fractions of the peak were considered, efficiency of separation was not improved by further decrease of flow rate.

2. Effect of Sample Concentration and Volume

Loading can be increased either by concentration or by volume. With increasing concentration the quality of separation deteriorates much faster than with increasing volume. When a 10 mM CdS colloid was injected three times—500 μL nondiluted, 1000 μL diluted 1:1 with water, and 2000 μL diluted 1:3—the sample mass was the same in all three runs and the same CdS peak area would be expected. But the recovery dropped dramatically with increasing sample concentration (Fig. 30). Due to the higher electrolyte concentration (from colloid preparation) the protective electrical double layer shrank and the CdS particles were irreversibly adsorbed at the stationary phase. Such behavior has already been discussed in sections II.C and II.D. Interestingly, the diameter polydispersity D_d was the same in all three cases. The strong material loss in the concentrated samples took place at the beginning of the column and this improved the separation for the remaining particles so much that it equaled the dilute sample, where the situation was worsened by the larger sample volume. For preparative separations of organic polymers larger sample volumes with lower concentrations have been recommended, but the reason is better separation efficiency and not a recovery problem [52].

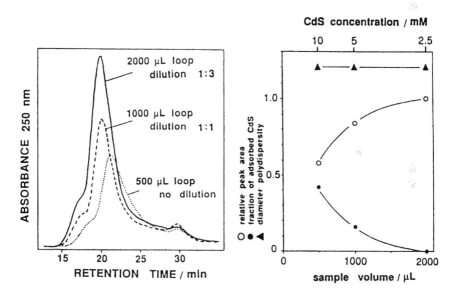

FIG. 30 Preparative separation of nanoparticles. Comparison of the combined effects of sample concentration and sample volume on the separation of colloidal cadmium sulfide. The sample volumes were 500, 1000, and 2000 μL, but the latter two were diluted 1:1 and 1:3, respectively, with water. Column: Nucleosil 500C$_4$ + Nucleosil 1000C$_4$ (7 μm, 125 × 32 mm each); eluent: water with 6 mmol/L sodium polyphosphate and 1 mmol/L Cd(ClO$_4$)$_2$ (4.5 mL/min). (left) Chromatograms of the three concentration/volume combinations. (right) Relative peak area (nondiluted = 1), fraction of adsorbed CdS (calculated from relative peak areas under the assumption that the most dilute sample has no loss by adsorption) and diameter polydispersity of the main fraction as a function of sample volume and concentration. (From Ref. 28.)

3. Effect of Fraction Volume

For studying the influence of fraction volume, 200 µL of a CdS colloid was separated. Original fractions of 0.9 mL were collected. Afterward in the peak maximum the fraction volume was expanded stepwise from 0.9 mL to 8.1 mL by alternate addition of the next four and previous four fractions. The diameter polydispersity D_d determined by analytical SEC is plotted as a function of fraction size in Fig. 31. It turned out that serious loss of product quality occurred only when the fraction volume exceeded the 15-fold sample volume.

4. Check of Fractionated Colloids by TEM

Two CdS colloids of different particle size were fractionated by preparative SEC. One main fraction of each run was reinjected in analytical SEC and measured by TEM. Diameter polydispersity D_d was reduced from 1.69 to 1.13/1.10 (SEC/TEM) and from 1.50 to 1.18/1.04 (SEC/TEM), respectively. Standard deviation of diameter σ_d could be improved down to 11%, a very good value for semiconductor colloids. Table 4 summarizes the results and shows again the good agreement between the two methods.

5. Preparation of Colloidal nm Particles in a Chromatographic Column

It should be mentioned that nm particles can also be generated in a chromatographic column (e.g., Lichrospher Alox T) by pumping the educt solutions separately and mixing the streams in a "zero volume" T piece or X piece immediately

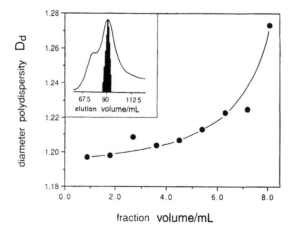

FIG. 31   Effect of fraction volume on the separation of 200 µL of a CdS sol. Diameter polydispersity of the main fraction as a function of fraction size. The inset shows the preparative chromatogram and the size of the fractions. (From Ref. 28.)

TABLE 4 Preparative Separation of Two CdS Colloids: Product Analysis by SEC and TEM

Sample	d_n[nm]		d_w[nm]		$D_d = d_w/d_n$	
	HPLC	ELMI	HPLC	ELMI	HPLC	ELMI
Original 1	2.32		3.91		1.69	
fraction	2.49	2.56	2.82	2.81	1.13	1.10
		$\sigma = 0.46$ (18%)				
Original 2	6.05		9.09		1.50	
fraction	5.66	5.77	6.73	5.97	1.18	1.04
		$\sigma = 0.65$ (11%)				

Source: Ref. 28.

before the column [53]. The advantages are ease of control of educt ratios, immediate product check in the diode array spectrometer, and—the most important feature—products of higher stability. The latter was obvious when the optical spectra of the particles after different aging times were compared with those of colloids obtained in the same way but without use of the column. The effect was explained by elimination of very small particles with high surface activity that behave as a glue for particle combination. The smallest crystalline CdS particles to date with a diameter of only 1.3 nm were prepared by this technique. They contain only 23 unit cells yet still show lattice planes in high-resolution TEM.

I. General Considerations for Establishing Size Exclusion Chromatography Conditions for Colloidal Inorganic nm Particles

When colloidal particles of a new inorganic material are to be separated by SEC, nonpolar stationary phases such as C_4 or C_{18} silica for HPLC or Sephacryl for low-pressure chromatography should be tried first, unless the colloid is stabilized by organic groups. In cases where alkanethiols or detergents are used, bare silica is a good packing material. Addition of stabilizers or detergents to the mobile phase is usually necessary to avoid adsorption. Those used for the preparation of the colloid should be tried first. In the case of colloidal sulfides it is best to add the particular cation in the form of a soluble salt in a modest concentration (about 1 mmol/L) to improve resolution, but high ionic strengths should be avoided for recovery reasons.

Whenever a method is elaborated, two checks are obligatory: (1) Compare injected and eluted sample amount. For comparison use $E_{s,\lambda}V_s$ ($E_{s,\lambda}$, optical absorption of the sample solution at wavelength λ; V_s, sample volume) and $E_{p,\lambda}V_p$ $= A_\lambda F$ ($E_{p,\lambda}$, integral optical absorption of all collected particles at the same wavelength λ and V_p, their volume; A_λ, total area of all colloid peaks at wavelength λ; F, flow rate). If both values are of comparable size, recovery is satisfactory. (2) Collect and reinject all eluted colloidal material. Compare the normalized chromatograms as a test for preferential loss of particles of a certain size. If the first check is negative but the second positive, i.e., a certain percentage of the whole distribution is lost without any preference, the method fails for quantitative analysis but could still be used for size determination under the condition that further careful tests by TEM give comparable results.

Since there is a lack of stable standards of inorganic colloids, calibration samples have to be measured once by TEM. Later, e.g., for calibration of a new column of the same type, another standard can be used. This can be an organic polymer that elutes under the same condition as the nm particles or even a substance with low molar weight as internal standard. This is also very helpful in compensating for retention variations due to changes in flow rate, temperature, etc. A specific calibration for each type of colloid seems to be the safer way, though a universal calibration for particles at low ionic strengths is reported [10,54]; the hydrodynamic volume of an organic polymer could also be used under the same conditions. When Q particles are to be separated, a diode array detector should be used, if available, as it provides useful additional information, e.g., about particle size, monodispersity of a peak, distinction between colloid and stabilizer peaks. For quantitative evaluation a detection wavelength must be chosen, where particles of all interesting sizes absorb light (i.e., this should be quite short) and, if possible, where the extinction coefficient is the same for all particle sizes. Otherwise a density detector is recommended that gives a signal proportional to the mass; however, this has the disadvantage of low sensitivity and large cell volume.

III. HYDRODYNAMIC CHROMATOGRAPHY

A. Classical Hydrodynamic Chromatography

Hydrodynamic chromatography (HDC) is mentioned but not in great detail because the method is rarely applied to inorganic colloids. However, HDC has a great general potential in particle separation that does not seem to be exhausted for inorganic nm particles. The method was first published by Small [55] and works with column packings of nonporous, spherical particles. Colloidal particles or dissolved macromolecules are separated according to size just by passing the void volume through the packing. In the capillary model of HDC the liquid flow in the interstitial void volume is considered as a Poiseuille flow in a capillary

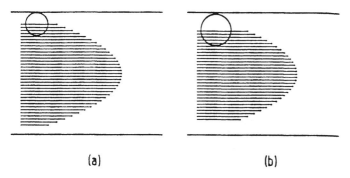

(a) (b)

FIG. 32 The capillary model of HDC. Colloidal particles are sterically prevented from enjoying the slowest velocities near the fluid wall interface. Larger particles are more excluded from this region (b) than are the smaller particles (a) and consequently move through the capillary with a higher mean velocity. (From Ref. 56.)

(Fig. 32) [56]. There are regions of slower flow velocities (close to the wall) and faster ones (in the center) across which a selective distribution of solutes according to their size is found. Brownian motion lets colloidal particles move radially through zones of different velocities, but only the smallest particles can reach the slowest regions. The larger the particles the more they are excluded from the "wall" zone and the earlier they elute. Here the ionic strength plays an important role, too, because of its influence on the electrical double layers on colloidal particles and "walls," causing repulsion due to similar charge. This effect is stronger at lower ionic strengths, resulting in a faster elution. Studies of experimental parameters, e.g., size of packing material and theoretical treatment of HDC in packed columns, can be found, e.g., in the paper of Stegemann et al. from the Amsterdam group [57].

In practice samples should contain small solutes, e.g., potassium dichromate, as a marker substance for compensation of flow rate changes, etc. Typically, surfactants are added to the eluent to diminish adsorption effects. Nevertheless, recovery can be a problem. That might be one reason why examples of HDC separation of inorganic colloids are rare, a second being that chromatographers might be less familiar with HDC. Kirkland separated silica particles in a range of 6–600 nm on 20-μm glass beads using the R_f method with a marker. Figure 33 shows the calibration [12].

HDC can also be carried out in microcapillaries a few μm in diameter [58–62]. The theoretical background of this special technique is discussed, e.g., in the papers of Tijssen et al. [63] and Silebi et al. [64]. An example with inorganic particles, namely silica, was given by Noel et al. [58]. However, the size was 10 μm. The calibration revealed that silica fits well with the values for latex, *Cereus*

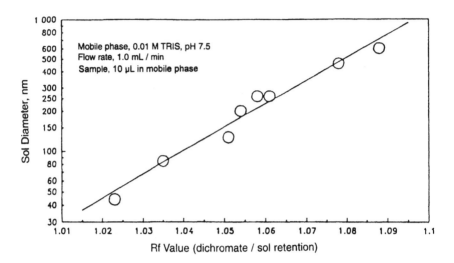

FIG. 33 HDC calibration for silica sols. Conditions: two 25 × 0.94 cm columns of 20-μm glass beads, 1 mmol/L potassium dichromate marker, samples 0.40% in mobile phase, UV detector 260 nm. (From Ref. 8.)

spores, and pollen. The results of HDC should encourage analysts to test the method more frequently for inorganic nanoparticles.

B. Wide-Bore Hydrodynamic Chromatography

As described above, classical HDC in narrow capillaries with typical diameters of 10–25 μm has successfully been used for separation of organic particles such as latex. However, these particles were not much smaller than 100 nm. For particles in the range of a few nanometers the diameter of the separation capillary has to be narrower, which might cause experimental difficulties with respect to reproducible, stable flow and detection sensitivity. When HDC is performed in packed columns problems with adsorption may occur [65] because of high specific surface.

In 1978 Mullins et al. [59] and in 1979 Noel et al. [58] reported fractionating of organic μm particles by pumping them through a relatively wide capillary (250 μm diameter). Kelleher and Trumbore [66] described the molar weight determination of biopolymers by the same principle. Beside UV detectors, special infrared (IR) detectors were also used, allowing the measurement of radial concentration gradients [67–69]. In the past this method was classified as flow injection analysis (FIA). Nevertheless it has typical elements of chromatography so that the name wide-bore hydrodynamic chromatography (wbHDC) was suggested [71].

1. Mechanism

To some extent the method of wbHDC resembles the classical HDC. A plug of sample solution is injected in a laminar flow through a capillary. Dissolved or dispersed species are transported forward by convection, but they can also move in other directions by diffusion. This diffusion depends on the size of the solute and can be neglected in comparison with the forward transport in the case of large colloidal particles that just follow the laminar flow. In contrast, small particles with a high diffusion coefficient undergo a rapid and effective exchange between fast-and slowly flowing regions of the laminar flow profile. Therefore the concentration over the cross-section is more uniform and the average migration velocity is slower than in the first case. Vanderslice et al. [70] calculated radial and axial concentration distributions in such capillaries as shown in Fig. 34, top. With the simple setup shown in Fig. 34, bottom, consisting of a pump, injection valve with sample loop, capillary (20 m long, 0.7 mm I D) and optical detector the flow profiles resulting from the inhomogeneous distributions are obtained (Fig. 34, center). Large species with low diffusion coefficient give an early, asymmetrical convection peak; small analytes with high diffusion coefficient have a late, rather symmetrical diffusion peak, and analytes in between show elements of both extremes.

2. Applications of wbHDC to Colloidal Systems

Fischer [71] applied wbHDC to CdS and gold colloids. SEC chromatograms and correspoonding wbHDC elution profiles are presented in Fig. 35, top, for CdS particles of different size. The samples are ordered according to increasing size. One can see the transition from diffusion peak to convection peak. For the evaluation of weight average diameter d_w the method of height ratios $R = h_1/h_2$ suggested by Trumbore et al. for molar weight determination of organic polymers was applied [72]. h_1 and h_2 are the heights at the position of convection peak and diffusion peak, respectively. A calibration plot $R = f(d_w)$ was constructed (Fig. 36, bottom), from which the diameters of unknown samples could be obtained. Based on studies of concentration dependence a 1 mmol/L colloidal solution is recommended for all wbHDC analyses. However, a 5-fold higher or a 10-fold lower concentration would cause only an error of 7% or 8%, respectively.

IV. OPTICAL CHROMATOGRAPHY

Recently, a new method of size separation of particles based on optical trapping [73] was published by Imasaka et al. [74]. A laser beam is focused in a capillary and the colloidal solution is pumped from the opposite end. The behavior of the

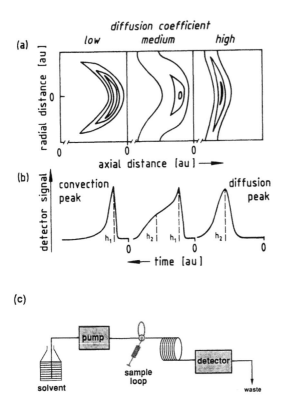

FIG. 34   Wide-bore HDC. Behavior of material with small, medium, and high diffusion coefficient during laminar flow through a "wide-bore" capillary. (a) Axial and radial concentration distribution expressed in equiconcentration lines, based on the theoretical calculations of Vanderslice et al. [70]. (b) Corresponding elution profiles. (c) Scheme of experimental setup in wide-bore HDC. (From Ref. 71.)

particles is recorded by a video camera. Figure 36 describes schematically the motion of colloidal particles in the capillary in several steps. After introduction of a particle (A) it is focused into the center line of the laser beam by the gradient force (B), turned around and accelerated by the scattering force (C) [75], and retarded with increasing distance from the beam waist (D). When the radiation pressure becomes identical with the force of the liquid flow, the particle drifts (D). The effect increases with increasing refractive index of the particle. Therefore larger particles are pushed further backward than smaller ones. The theoretical background is discussed in [76]. The advantages are easy control of separation by changing the beam focusing, collecting of particles in order of size by interrup-

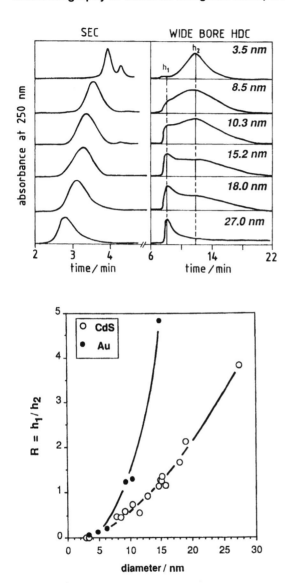

FIG. 35 (top) CdS sols of different particle sizes analyzed by SEC (left) and wide-bore HDC with marked heights h_1 and h_2 (right). Samples are sorted according to size. Weight average diameters d_w are given on the right-hand side. Conditions: HDC: 20 m long, 0.7 mm ID PEEK capillary; flow rate 0.8 mL/min, sample volume 100 μL. SEC: Nucleosil 500C$_4$ (7 μm) and Nucleosil 1000C$_4$ (7 μm). Mobile phase for both methods 10^{-3} mol/ L Cd(ClO$_4$)$_2$/6 × 10^{-3} mol/L sodium polyphosphate. (bottom) Calibration plot for the wide-bore HDC of gold and cadmium sulfide sols. It shows the ratio $R = h_1/h_2$ of the height of convection peak over diffusion peak as a function of weight average particle diameter d_w, determined by SEC or electron microscopy, respectively. (From Ref. 71.)

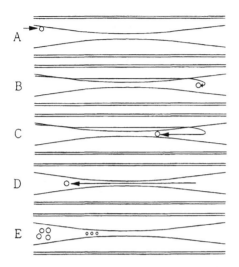

FIG. 36 Schematic motion of a particle in optical chromatography. The laser beam is introduced from the right-hand side and the liquid from the left-hand side. (A) Particle introduction. (B) Focus of particle into beam center by gradient force. (C) Acceleration of particle. (D) Deceleration of particle. (E) Particles drifting at equilibrium position. (From Ref. 74.)

tion of the laser pulse, observation of the whole system all of the time, and perfect recovery. In the publication, separation of 1-μm, 3-μm, and 6-μm particles was shown (material not defined), but it could be extended down to molecular dimensions by adapting laser power and wavelength. Though this method is limited to photochemically stable material, it seems potentially useful.

V. COMPARISON OF DIFFERENT METHODS FOR SIZE ANALYSIS OF nm PARTICLES

A good overview of the literature concerning all kinds of particle analysis can be found in the reviews of Barth et al. [77–82]. Provder considers the assessment of particle size distribution by various methods [89]. Kirkland compares SEC, sedimentation field flow fractionation (SdFFF), HDC, and capillary zone electrophoresis (CZE) of silica particles [12].

TEM provides images of particles that can be directly measured using a scale. This classic method is the basis for calibration of other methods. It can be coupled with x-ray spectroscopy for elemental analysis. However, typically not more than a few hundred particles or even much less can be counted for establishing a size distribution, resulting sometimes in rather bad statistics. Radiation damage has been observed during the analysis [2]. The drying process during sample preparation might be accompanied by aggregation or inhomogenization. Evaluation is time consuming unless an image processor is available.

SEC uses commonly available HPLC equipment (an isocratic pump is sufficient). Even with cheap peristaltic pumps low-pressure chromatography is possible. The method is fast enough (3–5 min for HPLC) for kinetic studies of unstable colloids. It gives a size distribution based on perfect statistics (if complete elution is proven). On-line coupling with UV-Vis absorption or fluorescence spectroscopy is possible, providing useful information in the case of Q particles. Standards are required for calibration, preferably of the same material as the analyte. Selection of stationary and mobile phase (pH, stabilizer, ionic strength) can be difficult. There is still a need for special software for colloids allowing corrections for axial dispersion. Easy scaling up or cyclical operation allows preparative separations.

HDC with packed columns requires normal HPLC pumps, etc. Due to small elution volumes, dead volumes have to be minimized carefully. The quite poor resolution can be improved by special software with band deconvolution and corrections for band broadening [83]. Standards and thorough optimization of eluent composition are required. Special equipment for HDC in narrow-bore capillaries is available.

The advantages of *wbHDC* are the almost complete absence of disturbing surfaces as compared to classic HDC and SEC and the simplicity of the experimental setup, where even peristaltic pumps can be used. The result is limited to the average size of the colloidal particles unless in the case of Q particles additional information is obtained via elution profiles at different wavelengths from a diode array detector.

Field flow fractionation (FFF) introduced by Giddings [84] comprises several flow techniques. Particles in the μm and nm range can be separated. The resulting fractograms reflect the size distribution. Sedimentation FFF (SdFFF), especially with force-field programming, enables characterization of particles in a wide-diameter range with excellent resolution. However, it seems that no separations have yet been carried out in the very low nm range. SdFFF does not require standards, if the density of the particles is known. The method is quite fast (about 30 min), but unfortunately rather expensive.

With *ultracentrifugation* the size of colloidal particles, e.g., gold [85], can be determined without standards, but the density of the analyte must be known. Modern apparatus even allows coupling with UV-Vis spectroscopy. Such equipment is quite rare.

Capillary zone electrophoresis (CZE) has successfully been applied to silica [86] and to gold particles [87]. Resolution is higher than in SdFFF, SEC, or HDC [12] and the size range for silica lies at least between 6 and 600 nm. Standards are necessary. At the moment, it seems that CZE is not the appropriate method for routine analyses.

The well-established method of *light scattering* allows determination of average particle size and particle shape. Numerous articles, reviews, and books are cited in Refs. 77–82. For proper operation and correct results the sample has to be free of any dust. In some instances problems might occur during purification by filtration, where colloidal particles could be retained by adsorption. Furthermore, the exact particle concentration is necessary, the determination of which is not always easy. Under certain conditions information about particle size of nm particles can be gained from *UV-Vis spectroscopy*. In the case of very small semiconductor nm particles the onset of absorption is a function of diameter [88] or for gold colloids the relative intensity of the plasmon band depends in a certain range on particle size [48–50]. Measurement of optical spectra is fast and simple. However, no exact values are obtained, especially about the size distribution.

REFERENCES

1. (a) A. Henglein, in *Modern Trends of Colloid Science in Chemistry and Biology* (H.-F. Eicke, ed.), Birkhäuser, Basel, 1985, p. 126. (b) A. Henglein, Ber. Bunsenges. Phys. Chem. *99*:903 (1995). (c) L. E. Brus, J. Chem. Phys. *79*:5566 (1983). (d) H. Weller, Adv. Mat. *5*:88 (1993).
2. E. Zeitler (ed.), *Cryoscopy and Radiation Damage*, North-Holland, Amsterdam, 1982.
3. J. Porath and P. Flodin, Nature *183*:1657 (1959).
4. J. C. Moore, J. Polym. Sci. A2835 (1964).
5. P. L. Dubin (ed.), *Aqueous Size Exclusion Chromatography*, J. Chromatography Library, Vol. 40, Elsevier, Amsterdam, 1988.
6. T. G. Spiro, S. E. Allerton, J. Renner, A. Terzis, R. Bils, and P. Saltman, J. Am. Chem. Soc. *88*:2721 (1966).
7. M. Shibukawa and N. Ohia, in Ref. 5, pp. 77–115.
8. R. Tijssen, J. Bos, in *Theoretical Advancement in Chromatography and Related Separation Techniques* (F. Dondi and G. Guiochon, eds.), NATO ADI Series C, Kluwer, Dordrecht, 1992, p. 397.
9. T. Tarutani, J. Chromatogr. *50*:523 (1970).
10. A. E. Hamielec and S. Singh, J. Liq. Chromatography *1*:187 (1978).
11. J. J. Kirkland, J. Chromatography *185*:273 (1979).
12. J. J. Kirkland, in *Colloid Chemistry of Silica*, Adv. Chem. Ser. *234*:287, (1994).
13. H. Bürgy and G. Calzaferri, J. Chromatography *507*:481 (1990).
14. W. W. Yau, J. J. Kirkland, and D. D. Bly, *Modern Size Exclusion Chromatography*, John Wiley and Sons, New York, 1979, pp. 31–36.
15. J. Retuert, Bol. Soc. Chil. Quim. *40*:415 (1995).
16. Ch.-H. Fischer, H. Weller, A. Fojtik, C. Lume-Pereira, E. Janata, and A. Henglein, Ber. Bunsenges. Phys. Chem. *90*:46 (1986).
17. Ch.-H. Fischer, J. Lilie, H. Weller, L. Katsikas, and A. Henglein, Ber. Bunsenges. Phys. Chem. *93*:61 (1989).
18. Ch.-H. Fischer, M. Giersig and T. Siebrands, J. Chromatography A, *670*:89 (1994).

19. Ch.-H. Fischer and A. Henglein, J. Phys. Chem. *93*:5578 (1989).
20. A. C. Nunes and Z.-C. Yu, J. Magnetism Magnetic Mat. *65*:265 (1987).
21. A. C. Nunes and Z.-C. Yu, J. Magnetism Magnetic Mat. *78*:241 (1989).
22. S. S. Kher and R. L. Wells, Chem. Mat. *6*:2056 (1994).
23. Ch.-H. Fischer, unpublished results.
24. M. Holtzhauer and M. Rudolph, J. Chromatography *605*:193 (1992).
25. T. Siebrands, M. Giersig, P. Mulvaney, and Ch.-H. Fischer, Langmuir *9*:2297 (1993).
26. K. A. Littau, P. J. Szajowski, A. J. Muller, A. R. Kortan, and L. E. Brus, J. Phys. Chem. *97*:1224 (1993).
27. Ch.-H. Fischer, H. Weller, L. Katsikas, and A. Henglein, Langmuir *5*:429 (1989).
28. Ch.-H. Fischer, J. Liq. Chromatography *17*:3593 (1994).
29. Ch.-H. Fischer and T. Siebrands, J. Chromatography A *707*:189 (1995).
30. H. Freundlich and A. Poser, Kolloid-Beihefte *6*:297 (1934).
31. A. v. Buzágh and E. Kneppó, Kolloid-Zeitschrift *82*:150 (1938).
32. P. L. Dubin and M. M. Tecklenburg, Anal. Chem. *57*:275 (1985).
33. C.-H. Cai, V. A. Romano, and P. L. Dubin, J. Chromatography A *693*:251 (1995).
34. P. L. Dubin, S. L. Edwards, and M. S. Mehta, J. Chromatography A 635:51 (1993).
35. M. G. Styring, C. J. Davison, C. Price, and C. Booth, J. Chem. Soc. Faraday Trans. *80*:3051 (1984).
36. M. Potschka, J. Chromatography *587*:276 (1991).
37. M. G. Styring, H. H. Teo, C. Price, and C. Booth. Eur. Polym. J. *24*:333 (1988).
38. M. Potschka, J. Chromatography *441*:239 (1988).
39. F. G. Smith and W. M. Deen, J. Colloid Interface Sci. *78*:444 (1980).
40. F. G. Smith and W. M. Deen, J. Colloid Interface Sci. *91*:571 (1983).
41. E. J. W. Verwey and J. Th. G. Overbeek, *Theory of the Stability of Lyophobic Colloids*, Elsevier, New York, 1948.
42. Ch.-H. Fischer and E. Kenndler, J. Chromatography, *773*:179 (1997).
43. R. Rosetti, J. L. Ellison, J. M. Gibson, and L. E. Brus, J. Chem. Phys. *80*:4464 (1984).
44. Ch.-H. Fischer, thesis.
45. A. Fojtik, H. Weller, U. Koch, and A. Henglein, Ber. Bunsen Ges. Phys. Chem. *88*: 969 (1984).
46. L. Brus, J. Phys. Chem. *90*:2555 (1986).
47. P. E. Lippens and M. Lannoo, Phys. Rev. B *39*:10935 (1989).
48. U. Kreibig, J. Phys. (Paris) *38*:C2-97 (1977).
49. U. Kreibig and L. Genzel, Surf. Sci. *156*:678 (1985).
50. T. Siebrands, master's thesis, Berlin, 1993.
51. R. H. Doremus, J. Chem. Phys. *40*:2389 (1964).
52. L. E. Ekmanis, in *Detection and Data Analysis in Size Exclusion Chromatography* (T. Provder, ed.), Am. Chem. Soc., Washington, DC, 1987, p.47.
53. Ch.-H. Fischer and M. Giersig, Langmuir *8*:1475 (1992).
54. M. G. Styring, J. A. J. Honig, and A. E. Hamielec. J. Liq. Chromatography *9*:3505 (1986).
55. H. J. Small, J. Colloid Interface Sci. *48*:147 (1974).
56. H. Small, Acc. Chem. Res. *25*:241 (1992).

57. G. Stegemann, R. Oostervink, J. C. Kraak, and H. Poppe, J. Chromatography *506*: 547 (1990).

58. R. J. Noel, K. M. Gooding, F. E. Regnier, D. M. Ball, C. Orr, and M. E. Mullins, J. Chromatography *166*:373 (1978).

59. M. E. Mullins and C. Orr, Int. J. Multiphase Flow *5*:79 (1979).

60. A. W. J. Brough, D. E. Hillman, and R. W. Perry, J. Chromatography *208*:175 (1981).

61. R. Tijssen, J. P. A. Bleumer, and M. E. van Krefeld, J. Chromatography *260*: 297 (1983).

62. J. G. DosRamos and C. A. Silebi, J. Colloid Interface Sci. *133*:302 (1989).

63. R. Tijssen, J. Bos, and M. E. van Krefeld, Anal. Chem. *58*:3036 (1986).

64. C. A. Silebi and J. G. DosRamos, AlChE J. *35*:1351 (1989).

65. A. W. Thornton, J. P. Olivier, C. G. Smart, and L. B. Gilman, in *Particle Size Distribution: Assessment and Characterization* (T. Provder, ed.), ACS Symp. Ser. No. 332, Am. Chem. Soc., Washington, DC, 1987, p. 256.

66. F. M. Kelleher and C. N. Trumbore, Anal. Biochem. *137*:20 (1984).

67. J. Pawliszyn, Anal. Chem. *58*:3207 (1986).

68. D. O. Hancock and R. E. Synovec, Anal. Chem. *60*:2812 (1988).

69. D. O. Hancock, C. N. Renn, and R. E. Synovec, Anal. Chem. *62*:2441 (1990).

70. J. T. Vanderslice, A. G. Rosenfeld, and G. R. Beecher, Anal. Chim. Acta *179*:119 (1986).

71. Ch.-H. Fischer and M. Giersig, J. Chromatography A, *688*:97 (1994).

72. C. N. Trumbore, M. Grehlinger, M. Stowe, and F. M. Kelleher, J. Chromatography *322*:443 (1985).

73. A. Ashkin, Phys. Rev. Lett. *24*:156 (1970).

74. T. Imasaka, Y. Kawabata, T. Kaneta, and Y. Ishidzu, Anal. Chem. *67*:1763 (1995).

75. A. Ashkin, Biophys. J. *61*:569 (1992).

76. T. Kaneta, Y. Ishidzu, N. Mishima, and T. Imasaka, Anal. Chem. *69*:2701 (1997).

77. H. G. Barth and S. T. Sun, Anal. Chem. *57*:151 (1985).

78. H. G. Barth, S. T. Sun, and R. M. Nikol, Anal. Chem. *59*:142 (1987).

79. H. G. Barth and S. T. Sun, Anal. Chem. *61*:143 (1989).

80. H. G. Barth and S. T. Sun, Anal. Chem. *53*:1 (1991).

81. H. G. Barth and S. T. Sun, Anal. Chem. *65*:55 (1993).

82. H. G. Barth and R. B. Flippen, Anal. Chem. *67*:257 (1995).

83. G. R. McGowan and M. A. Langhorst, J. Colloid Interface Sci. *89*:94 (1982).

84. J. C. Giddings, Science *260*:1456 (1993).

85. D. G. Duff, A. Bailker, I. Gameson, and P. P. Edwards, Langmuir *9*:3210 (1993).

86. R. M. McCormick, J. Liq. Chromatography *14*:939 (1991).

87. U. Schnabel, Ch.-H. Fischer, and E. Kenndler, J. Microsep., in press.

88. L. Spanhel, M. Haase, H. Weller, and A. Henglein, J. Am. Soc. *109*:5649 (1987).

89. (a) T. Provder (ed.), *Particle Size Distribution Assessment and Characterization*, ACS Symp. Ser. No. 332, Am. Chem. Soc., Washington, DC, 1987, p.256. (b) T. Provder (ed.), *Particle Size Distribution*, vol. 2, ACS Symposium Series No. 472, Am. Chem. Soc., Washington, DC, 1991.

6

Chromatographic Behavior and Retention Models of Polyaromatic Hydrocarbons in HPLC

YU. S. NIKITIN and S. N. LANIN   Department of Chemistry, M. V. Lomonosov State University of Moscow, Moscow, Russia

I.	Introduction	228
II.	General Approach to Description of PAH Retention in HPLC	230
III.	General Model of PAH Retention in HPLC	231
IV.	Reversed Phase HPLC of Polycyclic Aromatic Hydrocarbons on Nonpolar Stationary Phase MCH-10	233
V.	Reversed Phase HPLC of PAHs on ODS and Phenyl Reversed Phases. Influence of the Nature and Composition of the Mobile Phase	237
VI.	Influence of Various Parameters of a Chromatographic System on the Retention of Unsubstituted PAHs in HPLC	240
VII.	Influence of the Molecular Parameters of PAHs on Their Retention by Bonded Cyanoalkyl Phases of Different Types	246
VIII.	Retention of Polyphenyls and Substituted Polycyclic Aromatic Hydrocarbons in the System Hydroxylated Silica-*n*-Hexane	251
IX.	Optimization of a Mobile Phase Composition and Identification of Components of PAH Mixes	258
	References	260

I. INTRODUCTION

The study of chromatographic behavior of polycyclic aromatic hydrocarbons (PAHs) in high-performance liquid chromatography (HPLC) has the interest both for the chemist-analysts in the analysis of objects of an environment and for development of retention theory and separation selectivity in liquid chromatography. PAHs are some of the most widespread and toxic pollutants of the environment. Many are cancer producing. PAHs have been established as priority pollutants by the U.S. Environmental Protection Agency. All of this causes increased attention to the investigations of chromatographic behavior of this class of substances and to development of methods of its determination.

PAH determination in industrial sewage and natural waters is an important and complex analytical problem. No less important are the study of a retention mechanism of PAHs and selectivity of their separation in different variants of liquid chromatography, that of establishment of correlation dependence between chromatographic parameters and structure of PAH molecules, and that of properties of a chromatographic system.

In connection with the wide application of computers in chromatography, mathematical modeling of the chromatographic process for determining the correlation dependence between molecular parameters and physical properties of sorbents and their retention with the subsequent computerized experiment holds promise. It enables one to predict the properties of a chromatographic system and to optimize its composition and the conditions for the separation (the direct task), as well as to study the molecular interactions, molecular structure of the sorbates, their retention mechanisms and physicochemical properties, in order to identify the chromatographic zones corresponding to the components of complex mixtures of unknown composition (the reverse task) [1–20].

Numerous attempts were made to find the correlation dependencies between the structure of PAHs and their retention in liquid chromatography. Some physicochemical parameters of the PAH molecules used to determine the retention are quite varied, namely, the number of aromatic cycles in a PAH molecule [21], number of the carbon atoms (n_c) [22–24], van der Waals volumes (V_w) [25–27], length-to-breadth ratio of the molecule (L/B) [28–31], hydrophobic factor (lgP) [32–36], the molecular connectivity index (χ) [37–39], π-electron energies [40,41], combination of several parameters and an additive scheme [11,12,42–44].

The main defect of the similar approach is that the received correlation dependencies are empirical in nature and the area of their application is frequently limited by the conditions of the chromatographic experiment (type of chromatographic system, normal phase (NP) or reversed phase (RP) HPLC, type of adsorbent, class of aromatic substance).

It is difficult to imagine that only one parameter describing the PAH molecule could describe its retention in various conditions of chromatographic experiment. The retention mechanism in HPLC of such complex molecules as PAH molecules apparently is reasonably complicated and is determined by various molecular interactions in the chromatographic system.

By consideration of PAH retention in NP HPLC it is possible to show the complex character of retention dependence on a structure of PAH molecules even when the properties of sorbent and mobile phase during chromatographic separation remain practically constant (isocratic-isothermic variant of HPLC).

Based on the availability of delocalized π electrons, molecules of aromatic hydrocarbons (AHs) are able to undergo specific interactions with polar groups on the surface of silica [45–48], aluminum oxide [49–51], or silica with chemically bonded functional groups capable of forming charge-transfer complexes with AH molecules [21,52–57].

In all cases the retention of AH molecules increases with increasing number of aromatic cycles in a molecule. However, in some cases this dependence is not observed. It has been shown [58] that on some polar adsorbents (e.g., with surface-bonded ester groups) the PAH molecules containing six aromatic cycles but belonging to a number of pericondensed PAHs are retained more weakly than cata-condensed PAH molecules containing five aromatic cycles.

The retention regularities of various PAHs on silica gel at elution with n-pentane have been investigated [48]. It has been shown that some AHs are eluted from chromatographic column before partially hydrogenated derivatives. So, for example, 9,10-dihydrophenanthrene is retained more strongly than phenanthrene and 4,5-dihydropyrene is retained more strongly than pyrene. The authors consider that the hydrogenated cycle in partially hydrogenated PAH molecules increases the electronic density in remaining nonhydrogenated aromatic cycles of a PAH molecule. This increases specific interactions with a silica gel surface. However, there are data [48] that show that on polar adsorbents, such as silica gel with surface-bonded alkylamine groups [59], or on polymer adsorbent Porapak T (sorbent on the basis of copolymer of divinylbenzene and ethylvinylbenzene modified by ethylene glycol metacrylate [60]), and by n-hexane elution, AH retention is increased with reduction of hydrogenization degree of AH molecules, i.e., with an increase of the number of aromatic cycles in an AH molecule.

Thus PAH retention on polar adsorbents is determined not only by the number but by the arrangement of aromatic cycles in a molecule.

Data have confirmed the elution order of the isomeric PAHs from a chromatographic column. It is seen [48] that in NP HPLC on polar adsorbents, which are not capable of forming the charge-transfer complexes with PAH molecules, the linearly annellated cata-condensed PAHs are retained more weakly than isomeric angular annellated compounds. Similarly to cata-condensed PAHs the distinction in retention of isomeric pericondensed PAHs are also observed. So, for example, benz[a]pyrene is more weakly retained than the isomeric compounds benz[e]pyrene and perylene. Some authors believe that such retention dependencies are stipulated by steric effects [48].

II. GENERAL APPROACH TO DESCRIPTION OF PAH RETENTION IN HPLC

In our opinion, the most general and theoretically justified approach to the description of the retention dependence of PAH molecules on their structure is that based on additivity of the contribution of various groups or atoms of molecules in retention. This approach is analogous to the approach widely used in organic chemistry for calculation of some physicochemical parameters of organic substances.

Almost simultaneously with creation of the bases of classical theory of a chemical structure, attempts were made to present the molecule's properties as a sum on its structural elements or group of structural elements (i.e., on atoms, pair of connected atoms, chemical bonds, etc.). Furthermore, methods of calculation of many properties of molecules and substances were developed that correlated well with experimental data and have received wide application for practical calculation of physicochemical characteristics of molecules and substances [61]. However, these methods were considered as purely empirical, without a common theoretical foundation. Tatevskij [62–64] proposed some extensive property (L) of any system that can be determined by integration on system volume, as the sum of individual parts, pairs, triplets, etc., of a system:

$$L = \int_V d\tau_1 \cdots \int_V d\tau_N F(X_1, Y_1, Z_1, \ldots, X_N, Y_N, Z_N) \tag{1}$$

The results have been discussed [62–64] and generalized relating to energy of formation and other physicochemical characteristics of molecules and substances.

However, the application of this approach has difficulties because function F in Eq. (1) depends not only on space coordinates but also on physical conditions under which the substance exists, e.g., pressure, temperature.

Equation (1) can be expressed as

$$L = \sum_\alpha L_\alpha + \Sigma L_{\alpha,\beta} + \cdots \tag{2}$$

where L_α is partial value of property L, attributed to volume V, $L_{\alpha,\beta}$ is partial value of property L, attributed to pair of volumes V_α and V_β, etc. Representation of any extensive property of a system in a form of a sum by separate parts, pairs, triplets, etc., of any system (from space object and up to nucleuses of atoms and elementary particles) is a direct consequence of the definition of extensive property [61].

In chromatography the study of dependence between a structure of a molecule and its chromatographic retention began about 50 years ago. In his fundamental work, Martin [65] stated that substituted groups change the partial values of factors that depend on the nature of substituted groups and on the nature of both phases used (mobile and stationary), but not on remaining ("constant") part of a molecule. This phenomenon has generally been acknowledged in particular case of a linear free energy relationship (LFER) [66].

In his outstanding and pioneering work, Snyder [67] considered in detail the principle of additivity of the contribution of separate groups of molecules in retention and proved the applicability of this principle in liquid–solid chromatography.

III. GENERAL MODEL OF PAH RETENTION IN HPLC

Since the unsubstituted PAHs represent a limited set of solutes (sorbates) with similar molecular structure it is quite possible for the analysis of their chromatographic behavior to apply the retention structure model. In other words, in order to study the dependence of chromatographic retention on molecular structure it is expedient to use the additivity principle according to which the sorption free energy of a molecule ΔG (or proportional to it value $\lg k$, where k is the constant of adsorption equilibrium), can be expressed in a form of a linear combination of free energies (ΔG_i), relating to various fragments, or parameters of molecules, or to various physicochemical properties of substances.

$$\Delta G = \Sigma \Delta G_i \qquad (3)$$

However, for construction of an adequate and universal (for both NP and RP HPLC) mathematical model of PAH retention one should include in a correlation equation a number of members such that mathematical model would be transformed in physically (physicochemically) model.

In our opinion the physicochemical model of retention and separation of polycyclic aromatic hydrocarbons should be based, on the one hand, on the principle of additivity of the contributions of the various groups or atoms of PAH molecules to retention; on the other hand, it should take into account all kinds of molecular interactions in a chromatographic system. Such a model would allow a better understanding of the retention mechanism of PAHs of different structure (including the isomeric PAHs) and to describe from the uniform point of view their retention for different chromatographic systems in HPLC.

In considering a PAH retention mechanism in HPLC one should note the domination of different types of molecular interactions (Table 1) in NP and RP HPLC, i.e., on a different retention mechanism of PAHs in NP and RP HPLC. In normal phase HPLC the PAH retention from nonpolar mobile phase is defined in the first approximation by specific adsorbate–adsorbent interaction (Table 1). The introduction in a mobile phase of a polar modifier, which is sorbed specifically on active groups of a polar adsorbent surface, results in reduction of PAH reten-

TABLE 1   Types of Molecular Interactions (Specific and Nonspecific, SI and NI, Respectively) in NP and RP HPLC

NP HPLC			RP HPLC		
Component of chromatographic system	Sorbate	Sorbent (polar)	Component of chromatographic system	Sorbate	Sorbent (nonpolar)
Sorbent (polar)	SI (NI)[a]		Sorbent (nonpolar)	NI (SI)[b]	
Mobile phase (nonpolar)	NI	NI	Mobile phase (polar)	SI (NI)	NI (SI)[b]
Mobile phase (weakly polar, modified by a polar additive)	NI, SI	NI, SI	Mobile phase (nonpolar)	NI	NI

[a] In parentheses are given the interactions that are realized but do not determine retention of the sorbate (secondary)
[b] On residual silanol groups, especially in the absence of the "end-capping" operation.

tion. At the same time, specific interactions between adsorbate and solute molecules in a mobile phase influence the separation selectivity of PAHs of different structure, and this permits regulation of their elution sequence.

In RP HPLC, on the contrary, the nonspecific interactions of PAH molecules with hydrophobic adsorbent dominate over specific and nonspecific interactions of adsorbate molecules with components of mobile phase.

In RP HPLC of PAHs, nonspecific interactions are related to the number of carbon atoms in the PAH molecule. Thus the distinction in electronic state of carbon atoms and types of bond plays an insignificant role for nonspecific interaction. The specific interaction of PAH molecules with polar component of mobile phase defines the separation selectivity of PAHs in RP HPLC. In NP HPLC, the specific interactions of PAH molecules with polar adsorbent and components of a mobile phase are stipulated by interactions of π electrons of these molecules with polar groups of surface and molecules of mobile phase, and are connected with number of π electrons and with distribution of electronic density in PAH molecules. Hence the energy of specific interactions of PAH molecules with components of chromatographic system should be influenced by distinction of electron state of carbon atoms in a PAH molecule. It is also dependent on dimension of PAH molecules because the density of the π-electron cloud in the PAH molecule is related to its dimension. The higher the number of electrons in a PAH molecule and the more compact the cloud of electrons (i.e., the smaller the size of the molecule), the stronger the specific interaction. On the basis of the above-described representations we have attempted to find a simple dependence to account for PAH retention in various chromatographic systems.

We shall take a look at this dependence by an example of PAH retention in RP HPLC.

IV. REVERSED PHASE HPLC OF POLYCYCLIC AROMATIC HYDROCARBONS ON NONPOLAR STATIONARY PHASE MCH-10

A simple linear correlation equation that accounts for the carbon atom environment in the molecule was proposed to describe unsubstituted PAH retention in reversed phase liquid chromatography (RP-LC):

$$\lg k' = B_1 n_I + B_2 n_{II} + B_3 n_{III} + B_4(n_\pi/L) + B_0 \tag{4}$$

where k' is capacity factor; n_I, n_{II}, n_{III} are the number of carbon atoms, which are distinguished in their environment that is included in one, two, or three aromatic cycles of a PAH molecule accordingly (see Fig. 1, atoms of types I, II, and III); n_π is the number of π electrons in the molecule; L is the PAH molecule length along a main axis [68], B_1, B_2, B_3, B_4, and B_0 are the equation coefficients. The types of the carbon atoms different in their environment in molecules of unsubstituted PAHs, e.g., in a pyrene molecule, are indicated below (Fig. 1).

According to our model, the first three members of Eq. (4) should reflect the differences, if any, in the specific interaction of the carbon atoms (types I, II, and III of the PAH molecules, which are characterized by different distribution of electron density [11,12,69]) with the polar mobile phase in the chosen RP chromatographic system. The fourth member (n_π/L) takes into account the mean density of the π-electron cloud in the PAH molecule and enables the identification

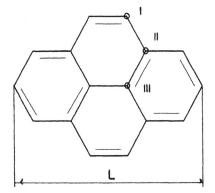

FIG. 1 Types of carbon atoms differing by the environment in the molecules of unsubstituted polyaromatic hydrocarbons.

of the differences in specific interactions of isomeric molecules with the polar mobile phases.

The significance of structural parameters and the number of π electrons of PAH molecules are given in Table 2. Table 3 gives the values of the coefficients in Eq. (4), the correlation coefficients and Fisher criteria for the calculation of the capacity factor logarithms. It follows from Table 3 that the proposed model describes the retention of the PAH molecule quite well (the coefficients of multiple correlation for all mobile phases exceed 0.999 and Fisher criteria have even greater values) in the chromatographic system (nonpolar sorbent–polar mobile phase). In all mobile phase compositions studied, the B_2 and B_3 coefficients have close values, which indicates a slight difference in the contribution of the carbon atoms of types II and III to the PAH retention and enables them to be combined into one member to simplify the model equation:

$$\lg k' = D_1 n_I + D_2(n_{II} + n_{III}) + D_3(n_\pi/L) + D_0 \qquad (5)$$

Table 4 presents the calculation results of capacity factor logarithms according to Eq. (5). As is seen in Table 4, the calculated values of the logarithms are

TABLE 2 Structural Parameters and Number of π Electrons of Unsubstituted Polycyclic Aromatic Hydrocarbons

	Substance	Chemical formula	n_Σ	n_{II}	n_{III}	n_π	L, Å
1	Benzene	C_6H_6	6	0	0	6	5,5
2	Naphthalene	$C_{10}H_8$	10	2	0	10	8,0
3	Anthracene	$C_{14}H_{10}$	14	4	0	14	10,5
4	Phenanthrene	$C_{14}H_{10}$	14	4	0	14	9,5
5	Pyrene	$C_{16}H_{10}$	16	4	2	16	9,5
6	Tetracene	$C_{18}H_{12}$	18	6	0	18	13,0
7	Tetraphene	$C_{18}H_{12}$	18	6	0	18	11,8
8	Chrysene	$C_{18}H_{12}$	18	6	0	18	11,8
9	Triphenylene	$C_{18}H_{12}$	18	6	0	18	9,5
10	Benz[a]pyrene	$C_{20}H_{12}$	20	6	2	20	11,5
11	Perylene	$C_{20}H_{12}$	20	4	2	20	10,5
12	1,12-Benzperylene	$C_{22}H_{12}$	18	6	4	22	10,5
13	Coronene	$C_{24}H_{12}$	24	6	6	24	10,5
14	1,2,7,8,-Dibenzanthracene	$C_{22}H_{14}$	22	8	0	22	13,5
15	1,2,5,6,-Dibenzanthracene	$C_{22}H_{14}$	22	8	0	22	13,6

TABLE 3 Values of Coefficients B_1, B_2, B_3, B_4, B_0, Correlation Coefficients (r)[a], and Fisher Coefficients (F) for the Dependency Eq. (4) at Various Mobile Phase Compositions (the Ethanol–Water Mixture); Column: MCH-10 "MicroPak," Temperature 39°C

Parameter	Water content in ethanol, vol%				
	0	10	20	30	40[b]
B_1	0.0219	0.0334	0.0392	0.0518	0.0664
r	0.8578	0.8845	0.8959	0.9099	0.9565
B_2	0.0890	0.0972	0.1057	0.1150	0.1243
r	0.9702	0.9686	0.9664	0.9600	0.9592
B_3	0.0925	0.0926	0.0956	0.1046	0.1134
r	0.0888	0.0657	0.0537	0.0506	0.1498
B_4	−0.0177	−0.0939	−0.1571	−0.2140	−0.2496
r	0.8920	0.8675	0.8521	0.8348	0.8250
B_0	−1.1028	−0.8265	−0.5565	−0.3545	−0.1672
Multiple correlation	0.9995	0.9996	0.9996	0.9995	0.9994
F	2244.6	2845.6	2383.0	2085.5	1443.7

[a] Coefficients of correlation of B_i with lg k'
[b] For 40 vol% the calculation was made with respect to 12 substances, in all other cases to 14 substances (see Ref. 11).

close to the experimental values. For the capacity factors themselves, the mean deviations of the calculated k' values from the experimental ones are 2–4%.

The values of the coefficients D_1, D_2, D_3, and D_0 of Eq. (5) are given in Table 5. The values of the coefficients D_1, D_2, D_3, and D_0 linearly depend on the water concentration in the mobile phase (Fig. 2). Coefficients D_1, D_2, and D_0 are directly proportional and D_3 is inversely proportional to the water concentration. This is in agreement with the increasing contribution of nonspecific interactions of carbon atoms of the PAH molecules with the nonpolar adsorbent surface (members n_1 and ($n_{II} + n_{III}$), and with specific (donor–acceptor) interaction of the PAH molecules (member n_π/L) with the polar mobile phase, upon increasing polarity of the latter. As shown in Fig. 2 and Table 5, the contribution to the retention of the carbon atoms of types II and III, characterized by less electron density as compared with the carbon atoms of type I, is greater ($D_2 > D_1$) than that of the carbon atoms of type I. The negative value of the coefficient D_3 is indicative of the predominant specific interaction of the π electrons of the PAH molecule with the mobile phase, which reduces the retention of PAHs.

TABLE 4   Logarithm Values of the PAH Capacity Factors Found
Experimentally [11] and Calculated According to the Dependency Eq. (5) for
Various Mobile Phase Compositions (the Ethanol–Water Mixture); Column:
MCH-10 "MicroPak," Temperature 39°C

	Water content in ethanol, vol%			
	0		30	
Substance	Exp.	Calc.	Exp.	Calc.
Benzene	−1.000	−0.990	−0.260	−0.284
Naphthalene	−0.770	−0.771	0.017	0.015
Phenanthrene	−0.553	−0.552	0.297	0.316
Anthracene	−0.538	−0.551	0.338	0.326
Pyrene	−0.387	−0.377	0.459	0.500
Triphenylene	−0.337	−0.340	0.570	0.565
Chrysene	−0.319	−0.332	0.618	0.630
Tetraphene	−0.337	−0.332	0.627	0.630
Perylene	−0.155	−0.161	0.760	0.778
Benz[a]pyrene	−0.137	−0.156	0.818	0.820
1,2,7,8-Dibenzanthracene	−0.125	−0.114	0.950	0.925
1,2,5,6-Dibenzanthracene	−0.125	−0.111	0.950	0.945
1,12-Benzperylene	0.009	0.014	0.958	0.963
Coronene	0.187	0.189	1.173	1.149

TABLE 5   Coefficients D_1, D_2, D_3, D_0, Correlation Coefficients (r),[a]
and Fisher Coefficients (F) for the Dependency Eq. (5) at Various
Mobile Phase Compositions (the Ethanol–Water Mixture); Column:
MCH-10 "MicroPak," Temperature 39°C

	Water content in ethanol, vol%			
Parameter	0	10	20	30
D_1	0.0213	0.0347	0.0419	0.0546
r	0.8578	0.8845	0.8959	0.9100
D_2	0.0894	0.0936	0.0988	0.1072
r	0.9952	0.9883	0.9832	0.9760
D_3	−0.0187	−0.0647	−0.1037	−0.1504
r	0.8638	0.8384	0.8229	0.8042
D_0	−1.0973	−0.8695	−0.6372	0.4477
Multiple correlation	0.9995	0.9994	0.9990	0.9989
F	3124.6	2959.1	1677.0	1478.5

[a] Coefficients of correlation of D_i with lg k'.

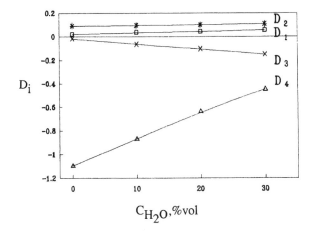

FIG. 2 Dependence of coefficients D_1, D_2, D_3, and D_0 of Eq. (5) on the water content in the mobile phase (ethanol–water system). Column: MCH-10 "MicroPak," temperature 39°C.

V. REVERSED PHASE HPLC OF PAHs ON ODS AND PHENYL REVERSED PHASES. INFLUENCE OF THE NATURE AND COMPOSITION OF THE MOBILE PHASE

In order to verify the applicability of Eq. (5) to other RP systems, we calculated the data (by computer) obtained by Hanai and Hubert [41]. Tables 6 and 7 contain the values of D_1, D_2, D_3, and D_0; and they show that the experimental data on PAH retention on the octadecyl (YMC ODS) and phenyl (YMC phenyl) reversed phases, with the use of acetonitrile or tetrahydrofuran-water solutions as eluents [41], are also well described by Eq. (5). Figure 3 shows that, in this particular case, a good linear dependence between coefficients D_1, D_2, D_3, and D_0 and water content in the eluent is observed. In this case the dependence is similar to that described for the ethanol–water mobile phase, i.e., $D_2 > D_1$, and $D_3 < 0$.

As shown in Refs. 41 and 70, a very low selectivity toward PAHs in RP HPLC is observed upon their elution with the tetrahydrofuran (THF)–water mixture. The variations in k' of PAHs upon transition from benzene to tetracene are very small, changing from 0.09% for 70% THF to 0.15% for 60% THF upon adsorption on the YMC ODS phase. This demonstrates the poor applicability of Eq. (5) to a system such as this. Nevertheless, also in this case the multiple correlation coefficient exceeded 0.99, and only for the 30% H_2O–70% THF system on the ODS phase did $r = 0.9344$.

TABLE 6   Coefficients D_1, D_2, D_3, D_0, Correlation Coefficients (r),[a] and Fisher Coefficients (F) for the Dependence Eq. (5) for Various Mobile Phase Compositions (the Acetonitrile–Water or Tetrahydrofuran–Water Mixture); Column: YMC ODS, Temperature 40°C

| | Water content in mobile phase, vol% | | | |
| | Acetonitrile | | Tetrahydrofuran | |
Parameter	10	30	30	40
D_1	0.0070	0.0302	0.0020	0.0173
r	0.9629	0.9732	0.7499	0.9679
D_2	0.1222	0.1368	0.0198	0.0124
r	0.9915	0.9860	0.6180	0.8733
D_3	−0.3382	−0.4250	−0.2014	−0.1317
r	0.8432	0.8252	0.2606	0.6102
D_0	0.0597	−0.4983	−0.3856	0.4243
Multiple correlation	0.9996	0.9996	0.9344	0.9904
F	1563.7	1701.5	9.20	68.4

[a] Coefficients of correlation of D_i with lg k'.

TABLE 7   Coefficients D_1, D_2, D_3, D_0, Correlation Coefficients (r),[a] and Fisher Coefficients (F) for the Dependency Eq. (5) for Various Mobile Phase Compositions (the Acetonitrile–Water or Tetrahydrofuran–Water Mixture); Column: YMC Phenyl, Temperature 40°C

| | Water content in mobile phase, vol% | | | | | |
| | Acetonitrile | | | | Tetrahydrofuran | |
Parameter	20	30	40	50	40	50
D_1	0.0250	0.0348	0.0493	0.0697	0.0261	0.0475
r	0.9784	0.9832	0.9869	0.9890	0.9887	0.9914
D_2	0.0468	0.0515	0.0580	0.0682	0.0040	0.0042
r	0.9875	0.9834	0.9784	0.9755	0.9153	0.9270
D_3	−0.0834	−0.1095	−0.1437	−0.1720	−0.0471	−0.0393
r	0.8398	0.8277	0.8153	0.8097	0.7051	0.7316
D_0	−0.4243	−0.2102	−0.0467	0.0766	0.0698	0.0663
Multiple correlation	0.9994	0.9996	0.9994	0.9992	0.9908	0.9918
F	1222.2	1733.6	1099.9	884.9	71.5	80.1

[a] Coefficients of correlation of D_i with lg k'.

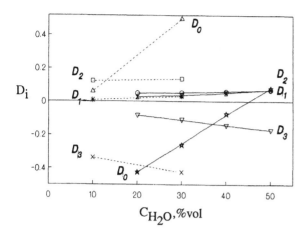

C_{H_2O},%vol

FIG. 3   Dependence of coefficients D_1, D_2, D_3, and D_0 of Eq. (5) on the water content in the mobile phase (acetonitrile–water) upon polyaromatic hydrocarbon adsorption on phenyl (unbroken line) and octadecyl (broken line) phase according to the data in Ref. 41.

It should be noted that, in comparison to correlation equations (using the connectiveness index), mostly distributed in RP-LC [38]:

$$\lg k' = H_1\chi + H_0 \tag{6}$$

equation (5) better describes the experimental data on the PAH retention [11]; the values of the correlation coefficient and Fisher criteria (cf. Tables 5 and 8) are higher. This seems to be due to the fact that the topological index of connectiveness slightly reflects the molecule geometry, which is most drastically seen in the case of isomers. For example, in the case of anthracene and phenanthrene, $\chi_a = 4.80940$ and $\chi_{ph} = 4.81538$ and differ by

$$\Delta\chi = \frac{2(\chi_{ph} - \chi_a)}{(\chi_{ph} + \chi_a)}100\% = 0.124\%$$

for chrysene and triphenylene ($\chi_{chr} = 6.22607$, $\chi_{thp} = 6.23205$; $\Delta\chi = 0.096\%$), whereas in the case of the same PAH pairs, the differences in the molecule length are equal to: $L_{ph} = 9.95000$ Å, $L_a = 10.36801$ Å, $\Delta L = 4.115\%$; $L_{chr} = 12.18336$ Å, $L_{tph} = 9.95000$ Å, $\Delta L = 20.181\%$).

We introduced one more parameter (L) into Eq. (6) to account for the linear dimensions of the PAH molecules.

$$\lg k' = H_1\chi + H_2(n_\pi/L) + H_0 \tag{7}$$

TABLE 8 Coefficients H_1, H_2, and H_0, Correlation Coefficients (r),[a] and Fisher Coefficients (F) for the Dependency Eqs. (6) and (7) for Various Mobile Phase Compositions (the Ethanol–Water Mixture); Column: MCH-10 "MicroPak," Temperature 39°C

		Water content in mobile phase, vol%			
Equation	Parameter	0	10	20	30
$\lg k' = H_1\chi + H_0$	H_1	0.1719	0.1869	0.1980	0.2192
	r	0.9888	0.9950	0.9962	0.9972
	H_0	−1.3670	−1.1291	−0.9135	−0.7342
	Multiple correlation	0.9888	0.9950	0.9962	0.9972
	F	525.8	1179.8	1566.3	2101.5
$\lg k' = H_1\chi + H_2(n_\pi/L) + H_0$	H_1	0.1429	0.1720	0.1920	0.2242
	r	0.9888	0.9950	0.9962	0.9972
	H_2	0.1908	0.0978	0.0390	−0.0330
	r	0.8920	0.8675	0.8521	0.8348
	H_0	−1.5051	−1.1999	−0.9424	−0.7102
	Multiple correlation	0.9944	0.9962	0.9964	0.9973
	F	487.7	722.0	755.0	1001.0

[a] Correlation coefficient H_1 with $\lg k'$.

However, this equation is also less suitable for describing the experimental data [41] than Eq. (5) (see Tables 5 and 8).

VI. INFLUENCE OF VARIOUS PARAMETERS OF A CHROMATOGRAPHIC SYSTEM ON THE RETENTION OF UNSUBSTITUTED PAHs IN HPLC

We have seen that the values of coefficients B_i of Eq. (4) are proportional to the increments of the distribution coefficients (between the mobile and stationary phases of the chromatographic system) belonging to different fragments of the PAH molecule. However, for a more complete description of PAH behavior in various chromatographic systems the member should be introduced in the correlation equation, reflecting the influence of the mobile phase composition of PAH retention. Thus we have:

$$\lg k' = E_1 n_\mathrm{I} + E_2(n_\mathrm{II} + n_\mathrm{III}) + E_3(n_\pi/L) + E_4 c + E_0 \qquad (8)$$

where c is content of the organic component (in the case of the RP HPLC) or the polar additive (in the case of the NP HPLC) in the binary mobile phase.

Previously, we have shown (see Fig. 2) that in the case of RP HPLC the Eq. (5) coefficients D_i are the linear functions of the content (vol%) of water or an organic component in the mobile phase:

$$D_1 = G_1c + G_{0,1} \tag{9}$$

$$D_2 = G_2c + G_{0,2} \tag{10}$$

$$D_3 = G_3c + G_{0,3} \tag{11}$$

$$D_0 = G_0c + G_{0,0} \tag{12}$$

where c is the content of the organic component in the polar mobile phase (RP HPLC), G_i and $G_{0,i}$ are the equation coefficients.

Introducing Eqs. (9)–(12) in Eq. (5), we get

$$\lg k' = c[G_1n_I + G_2(n_{II} + n_{III}) + G_3(n_\pi/L) + G_0] \tag{13}$$
$$+ [G_{0,1}n_I + G_{0,2}(n_{II} + n_{III}) + G_{0,3}(n_\pi/L) + G_{0,0}]$$

Comparing Eqs. (8) and (13), we get:

$$E_0 = G_{0,0} \tag{14}$$

$$E_1 = G_{0,1} \tag{15}$$

$$E_2 = G_{0,2} \tag{16}$$

$$E_3 = G_{0,3} \tag{17}$$

$$E_4 = [G_1n_I + G_2(n_{II} + n_{III}) + G_3(n_\pi/L) + G_0] \tag{18}$$

It is of interest to compare Eqs. (5) and (8) with the equation in Ref. 71:

$$\lg k' = \lg k_w - S\Phi \tag{19}$$

where Φ is the volume fraction of organic component in the mobile phase, S is the coefficient, k_w is the value of k' (found by extrapolation) for the water mobile phase (when $\Phi = 0$). The values of S are approximately constant, as Φ is varied, though each substance will have a characteristic value S; the values of S can also differ when the organic component of the mobile phase or chromatographic column is changed.

Having Eqs. (8) and (19), we see that

$$\Phi = c/100 \tag{20}$$

if c is presented in vol% and $\Phi = c$ if c is presented in volume fraction

$$S = E_4 \tag{21}$$

$$\lg k_w = E_1n_I + E_2(n_{II} + n_{III}) + E_3(n_\pi/L) + E_0 \tag{22}$$

Thus it follows that the coefficient S in Eq. (19) in the general case is not constant for a given chromatographic system, can depend on the structure of the adsorbate molecule, and is a sufficiently complex value.

For data on PAH retention [41] the mean values of coefficients E_1, E_2, E_3, and E_0 were calculated on the basis of the system of similar Eq. (8), then substituting them and the values of lg k' in Eq. (8), we found the values of $E_{4,i}$ for individual PAHs. Having solved the system of Eq. (19), we found the following values of coefficients G_i. $G_1 = 9.92 \times 10^{-5}$; $G_2 = -2.95 \times 10^{-5}$; $G_3 = -9.35 \times 10^{-5}$; $G_0 = -3.11 \times 10^{-2}$. These show that G_0 is three magnitudes of order greater than other G_i in terms of its absolute value and, consequently, is the determining value in calculating E_4. Thus, in our case G_0 does not appear to be function of the adsorbate structure and nature, i.e., it is equal for all unsubstituted PAHs, but coefficient $S = E_4$ is approximately constant and depends only on the nature of the organic component of the water–organic mobile phases (for the given sorbent and at the given temperature).

Tables 9 and 10 give the results of treatment of the experimental data on PAH retention [11,41] in RP HPLC systems by using Eq. (5) and the data on the retention of normal hydrocarbons (41) by using Eqs. (23) and (24):

$$\lg k' = H_1 n_c + H_2 c + H_0 \tag{23}$$

$$\lg k' = H_1 n_c + H_2 c + H_3 n_c c + H_0 \tag{24}$$

where n_c is the number of the carbon atoms in the n-alkane molecule; C is the concentration, in vol%; and H_1, H_2, H_3, and H_0 are equation coefficients.

Table 9 shows that the acetonitrile (ACN) mobile phase is more selective toward the separation of isomeric PAHs than THF in the case of both stationary phases of YMC (phenyl and octadecyl), the selectivity on C_{18} being higher than that on the phenyl phase. At the same time, the phenyl phase is characterized by greater values of E_1 as compared with C_{18} (i.e., it is characterized by greater affinity to the carbon atom of type I). It is also characterized by a higher selectivity toward the carbon atoms of various types (compare the values of coefficients E_1 and E_2) when the THF mobile phase is used. It should be noted that E_2 is less than E_1 only in this particular case (the YMC–phenyl–tetrahydrofuran system). Close values of E_4 (at THF and ACN) for all systems concerned are indicative of a relatively weak effect of the mobile phase organic component on retention in the case of both PAHs and n-alkanes (Table 10). The ethanol mobile phase is characterized by relatively high values of E_1, E_2, and E_3, but its selectivity is less than that of ACN mobile phase.

Tables 9 and 10 show that carbon atoms of type I PAH interact weakly with stationary phase as compared with the carbon atoms of n-alkanes (in the case of the ACN mobile phase it is more noticeable, though E_1 values for ACN phase are greater than for THF phase). This difference in the behavior of PAH and n-

TABLE 9 Retention of PAHs in RP HPLC. Coefficients E_1, E_2, E_3, E_4, E_0, Correlation Coefficients,[a] Multiple Correlation Coefficient (R), and Fisher Coefficients (F) for the Dependency Eq. (8)

Sorbent	Organic comp. of mobile phase	E_1	E_2	E_3	E_4	E_0	R	F
MCH-10[b] MicroPak	Ethanol	0.0442 (0.552)	0.1016 (0.590)	—	-0.0309 (-0.761)	1.8392	0.9959	1918
YMC[b] phenyl	Acetonitrile	0.0447 (0.496)	0.0561 (0.494)	0.1221 (0.514); -01272 (0.413)	-0.0304 (-0.856)	1.8242	0.9924	441
	Tetrahydrofuran	0.0368 (0.557)	0.0041 (0.519)	0.0432 (0.406)	-0.0223 (-0.811)	1.2930	0.9839	83
YMC[c] ODS	Acetonitrile	0.0186 (0.653)	0.1295 (0.667)	0.3816 (0.562)	-0.0306 (-0.734)	2.7299	0.9966	403
	Tetrahydrofuran	0.0096 (0.232)	0.0161 (0.203)	-01665 (0.126)	-0.0259 (-0.963)	2.0860	0.9941	203

[a] Correlation coefficients E_i with lg k' are given in parentheses.
[b] For calculations, the retention values for benzene, naphthalene, phenanthrene, anthracene, pyrene, triphenylene, chrysene, tetraphene, perylene, benz[a]pyrene, 1,2,7,8-dibenzanthracene, 1,2,5,6-dibenzanthracene, 1,2,-benzperylene, and coronene were used, the ethanol content in water from 60 to 100 vol%, column temperature 39°C
[c] For calculations, the retention values for benzene, naphthalene, phenanthrene, anthracene, pyrene, chrysene, tetraphene, and benz[a]pyrene on the YMC phenyl with the content of acetonitrile and tetrahydrofuran in water from 50 to 80 and 50 to 60 vol%, and on the YMC ODS from 70 to 90 and 60 to 70 vol%, respectively, were used.

TABLE 10   Retention of n-Alkanes in RP HPLC. Coefficients H_1, H_2, H_0,
Correlation Coefficients,[a] Multiple Correlation Coefficient (R), and Fisher Coefficients
(F) for the Dependency Eq. (23)

Sorbent	Organic comp. of mobile phase	H_1	H_2	H_3	H_0	R	F
	Acetonitrile	0.0854	−0.0336		1.9682	0.9912	564
YMC[b]		(0.3843)	(−0.8664)				
phenyl	Tetrahydrofuran	0.0540	−0.0288		1.7596	0.9863	161
		(0.6562)	(−0.7664)				
	Acetonitrile	0.2344	−0.0168	−0.0022	0.8393	0.9994	5567
YMC		(0.3842)	(−0.8664)	(−0.0888)			
ODS	Tetrahydrofuran	0.2008	−0.0161	−0.0027	0.5845	0.9996	3258
		(0.6562)	(−0.7364)	(0.3968)			

[a] Correlation coefficients H_i with lg k' are given in parentheses
[b] For calculations, the retention values for pentane, hexane, heptane, octane, decane, dodecane on
the YMC phenyl stationary phase were used (the content of acetonitrile and tetrahydrofuran in water
from 50 to 80 and from 50 to 60 vol%, respectively).

alkanes can be due to the fact that the distribution constant of saturated hydrocarbons between the hydrophobic sorbent and the polar mobile phase (as well as
the relative increments accounted for one carbon atom) is greater than that of
PAHs whose electrons are able to specifically interact with the mobile phase.

The results given in Table 11 show that the proposed model also can be used
in NP HPLC. As would be expected, inversion (in comparison with RP HPLC)
the phase polarity of the chromatographic system resulted in a change of the
sign of the coefficients E_3 and E_0. It accounts for the fact that the coefficient E_3
characterizing SI of π electrons with the polar groups of the components of the
mobile phases and sorbents will reflect an increase in PAH retention (in the case
of NP HPLC) and a decrease in their retention (in the case of RP HPLC). A
change in the sign of E_0 reflecting the contribution of NI to the PAH retention
corresponds to the character of the n-alkane distribution between the polar and
nonpolar (weakly polar) components of the chromatographic system.

The sign of the coefficients E_1 and E_2 does not depend on the type of the
chromatographic system (RP or NP). However, in NP HPLC $E_1 > E_2$, whereas
in the case of RP HPLC $E_1 < E_2$ since the atoms of type I having greater electron
density as compared with the atoms of type II and III interact more strongly with
a more polar mobile phase. An increased concentration [c_1 Eq. (8)] due to the
modifying additive to the mobile phase results in less retention of PAHs in all
cases $(E_4 < 0)$. It follows from Table 11 that the character of PAH retention at
small concentrations of the additive, which is indicated by smaller values of the

TABLE 11 Retention of PAHs in NP HPLC. Coefficients E_1, E_2, E_3, E_4, E_0, Correlation Coefficients,[a] Multiple Correlation Coefficient (R), and Fisher Coefficients (F) for the Dependency Eq. (8)

Sorbent	Polar comp. of the mobile phase	E_1	E_2	E_3	E_4	E_0	R	F
SiO$_2$[b]	Chloroform (0–5 vol%)	0.0702 (0.362)	0.0109 (0.347)	0.1360 (0.293)	−0.0620 (−0.848)	—	0.9330	
MicroPak	Chloroform (3–15 vol%)	0.0620 (0.548)	0.0106 (0.540)	0.1418 (0.463)	−0.408 (−0.787)	1.0737 1.2181	0.9873	454
NH$_2$-10[b]	Chloroform (0–25 vol%)	0.0706 (0.516)	0.0456 (0.588)	02051 (0.518)	−0.0360 (−0.755)	— 1.1886	0.9705	295
MicroPak	Chloroform (5–25 vol%)	0.0625 (0.625)	0.0420 (0.724)	0.2125 (0.645)	−0.0282 (−0.644)	—	0.9873	579
	Isopropanol (0–20 vol%)	0.0721 (0.474)	0.0406 (0.536)	0.2250 (0.475)	−0.0360 (−0.576)	1.2287	0.8015	33
	Isopropanol (3–20 vol%)	0.0643 (0.793)	0.0359 (0.912)	0.2363 (0.817)	−0.0108 (−0.285)	1.5443 1.8056	0.9885	643
CN-10[b]	Chloroform (0–20 vol%)	0.0730 (0.505)	0.0113 (0.476)	0.3816 (0.562)	−0.0331 (−0.823)	— 1.1618	0.9830	615
MicroPak	Chloroform (1–20 vol%)	0.0699 (0.522)	0.0106 (0.493)	−01665 (0.126)	−0.0314 (−0.811)	— 1.1450	0.9841	558

[a] Correlation coefficients E_i with lg k' are given in parentheses.

[b] For calculations, the retention values for naphthalene, anthracene, phenanthrene, pyrene, tetracene, tetraphene, chrysene, triphenylene, benz[a]pyrene, perylene, 1,2,-benzperylene, and 1,2,7,8,-dibenzanthracene were used at the column temperature 29°C.

multiple correlation coefficients and Fisher criteria when the chromatographic systems containing the modified and nonmodified mobile phases, are simultaneously described by Eq. (8). That may be due to the property of the surface, i.e., its energetic and geometric heterogeneity, which is practically imperceptible when 3–5 vol% of the modifying component is introduced. Nevertheless, the proposed model operates satisfactorily within a wide range of mobile phase compositions.

VII. INFLUENCE OF THE MOLECULAR PARAMETERS OF PAHs ON THEIR RETENTION BY BONDED CYANOALKYL PHASES OF DIFFERENT TYPES

It was shown earlier that using the lg k' versus carbon number relationship it is possible to identify the qualitative difference in the chromatographic properties of PAHs on cyanodecyl and commercial phases. In order to analyze the behavior of these phases, it is necessary to use a dependence that better reflects the effect of PAH structure on retention.

PAH retention on the studied polar stationary phases may be described as follows:

$$\lg k' = B_1 n_{\mathrm{I}} + B_2 n_{\mathrm{II}} + B_3 n_{\mathrm{III}} + B_4(n_\pi/L) + B_0 \tag{4}$$

Table 12 gives the values of the coefficients of Eq. (4) for the studied cyanoalkyl stationary phases (cyanopropyl C_3CN, cyanodecyl $C_{10}CN$, and cyanoisodecyl i-$C_{10}CN$). As can be seen, Eq. (4) describes the retention of PAHs on the analyzed stationary phases. This is confirmed by the high values of the multiple correlation coefficient $R \geq 0.9981$ and Fisher criteria $F \geq 578$. The values of coefficient B_1 are quite close for all stationary phases considered, whereas the value of coefficients B_2, B_3, and B_4 for silica gel and the C_3CN phase are quite different from those obtained for the $C_{10}CN$ and i-$C_{10}CN$ phases. This indicates a similarity in the adsorption properties of silica gel and silica gel modified by cyanopropyl groups, on the one hand, and the sorbents containing cyanodecyl groups on the other. The smaller (5–8 times less) values of the B_2 and B_3 coefficients for silica gel, as compared with the cyanodecyl phases, are indicative of considerably less specific interaction of the silanol groups with the "inner" carbon atoms of the PAH molecule (type II and III carbon atoms), as compared with the interaction of the nitrile groups in the cyanodecyl phases with these carbon atoms. The C_3CN phase represents an intermediate position. The B_2 and B_3 values of the cyanopropyl phase are approximately twice those for silica gel but are still much smaller than the corresponding values for the long-chain $C_{10}CN$ and i-$C_{10}CN$ phases.

TABLE 12 Coefficients B_1, B_2, B_3, B_4, B_0, the Correlation Coefficients (r),[a] and Fisher Coefficients (F) for Eq. (4). Stationary Phase: Silica Gel or Various Cyanoalkyl Phases. Mobile Phase: n-Hexane, Temperature 29°C

Parameter	SiO_2	C_3CN	$C_{10}CN$	$i\text{-}C_{10}CN$
B_1	0.1022	0.1118	0.1040	0.0894
r	0.9646	0.9557	0.8564	0.8705
B_2	0.0116	0.0182	0.1014	0.0818
r	0.8436	0.8523	0.9437	0.9360
B_3	0.0145	0.0289	0.0996	0.0831
r	−0.0969	−0.0670	−0.0315	−0.0308
B_4	0.1208	0.1265	0.1011	0.0455
r	0.6279	0.6462	0.7839	0.7648
B_0	−1.2154	−1.3846	−1.6040	−1.7125
Multiple correlation	0.9983	0.9981	0.9994	0.9993
F	672.39	577.97	1796.64	1607.58

This can be attributed either to poor modification of the C_3CN phase surface or to a high degree of interaction between the nitrile groups of the bonded layer and the surface silanol groups, which are weakly screened by the short hydrocarbon chains (propyl). In the case of cyanodecyl phases, the surface silanol groups are less available and thus the main adsorption centers appear to be the "free" nitrile groups in the normal phase chromatography of PAHs.

Analysis of the values of coefficients B_1, B_2, and B_3 has shown that in the case of adsorption on the cyanodecyl phases the electrostatic interaction of free nitrile groups and PAH molecules is weakly influenced by the difference in the distribution of the electron density of the peripheral (type I) and inner (type II and III) carbon atoms in the PAH molecules. In the case of the cyanopropyl phase and silica gel, the deficiency of the electron density of the inner carbon atoms in the PAH molecule results in less contribution to the retention of peri-condensed PAHs as compared to cata-condensed PAHs and in the corresponding changes of the elution sequence as compared with the cyanodecyl phases (see Table 13).

The difference in the retention of isomeric PAH molecules is determined to a great extent by the difference in their sizes (n_π/L). It is clear that the effect of this difference on retention will be more pronounced when the PAH molecule can simultaneously interact with several adsorption centers, i.e., at least two. This effect is more readily attained on silica gel and silica gel modified with cyanopropyl groups, as indicated by the maximal value of coefficient B_4 (0.1208 and

TABLE 13   Characteristics of the Support and Stationary Phases

Stationary phase	Symbol	Spec. surf. area of the initial silica gel, m²/g	Carbon conc.	Bonding density groups /nm²	mmole/g
Si-10 MicroPak	SiO_2	400–500	0	$(4.6)^a$	$(3.2–4.0)^a$
CN-10 MicroPak	C_3CN	400–500	n.a.[b]	—	—
CH-10 MicroPak	PC_{18}	400–500	26	—	1.2
MCH-10 MicroPak	MC_{18}	400–500	14	1.4–1.7	0.65
$C_{10}H_{22}CN$ Silasorb 600	$C_{10}CN$	527	13.4	1.5	1.01
i-$C_{10}H_{22}CN$ KCK-2	i-$C_{10}CN$	229	10.0	2.3	0.76

[a] Surface concentration of the silanol groups.
[b] No data are available.

0.1265, respectively). This type of interaction is less possible for the i-$C_{10}CN$ stationary phase ($B_4 = 0.0455$) due to the "dilution" of cyanoalkyl groups by hydrocarbon chains. An increase in the size of the PAH molecule makes the interaction of the PAH molecules with several nitrile groups more possible. We assume that this can explain the lower selectivity of the cyanodecyl phases toward the phenanthrene–anthracene pair and the higher selectivity for larger PAH molecules (Table 14).

When the weakly polar mobile phase, hexane containing 10% chloroform, is used, Eq. (4) also describes the retention of unsubstituted PAHs (Table 15). In this case we can observe a general reduction in the absolute values of coefficients B_1, B_2, and B_3 for all of the investigated stationary phases and an increase in the absolute values of coefficient B_4 that is indicative of the increasing difference in the retention of isomers. The highest selectivity toward the retention of PAH isomers under the given conditions could be observed on silica gel (maximal value $B_4 = 0.1996$), while the selectivity on the cyanodecyl phases becomes approximately similar ($B_4 \cong 0.11$).

It is interesting to compare the value of B_1 with A_1 in the equation:

$$\lg k' = A_1 n_c + A_0 \tag{25}$$

where n_c is the number of carbon atoms in the PAH molecule belonging to a series of linearly and angularly annellated cata-condensed PAHs, and A_1 and A_0 are constants. Equation (25) was used in the literature [43,70,72,73] for the evaluation of the chromatographic properties of various phases with respect to PAHs; the value of coefficient A_1 can be used as an integral characteristic of the specific interaction of a stationary phase with PAH molecules. Table 16 presents the values of coefficient A_1 in Eq. (25), and the values of the multiple correlation

TABLE 14 Selectivity (α) of the Stationary Phases Toward PAHs Differing in the Number of Carbon Atoms in their Molecule (Nos. 1–3) and to Isomeric PAHs (Nos. 4–6). Mobile Phase; Hexane (Upper Numbers) and Hexane Containing 10% Chloroform (Lower Numbers)

		$\alpha = k_2'/k_1'$			
No.	PAH	SiO_2	C_3CN	$C_{10}CN$	$i\text{-}C_{10}CN$
1	**Triphenylene**/Phenanthrene	1.80	1.84	2.74	2.15
		1.44	1.56	2.26	2.08
2	**1,2,7,8-Dibenzanthracene/**	1.71	1.83	2.68	2.22
	Tetraphene	1.33	1.49	2.15	1.96
3	**Benz[a]pyrene**/Triphenylene	1.04	1.11	1.62	1.49
		1.04	1.10	1.50	1.45
4	**Phenanthrene**/Anthracene	1.08	1.11	1.04	1.03
		1.12	1.07	1.04	1.04
5	**Triphenylene**/Tetracene/	1.11	1.13	1.12	1.02
		1.18	1.16	1.12	1.11
6	**Perylene**/Benz[a]pyrene	1.10	1.11	1.09	1.08
		1.11	1.13	1.11	1.08

TABLE 15 Coefficients B_1, B_2, B_3, B_4, B_0, the Correlation Coefficients (r),[a] and Fisher Coefficients (F) for Eq. (4). Stationary Phase: Silica Gel or Various Cyanoalkyl Phases. Mobile Phase: n-Hexane Containing 10 vol% Chloroform, Temperature $-29°C$

Parameter	SiO_2	C_3CN	$C_{10}CN$	$i\text{-}C_{10}CN$
B_1	0.0672	0.0714	0.0757	0.0757
r	0.8888	0.9211	0.8330	0.8492
B_2	0.0047	0.0126	0.0864	0.0713
r	0.8632	0.8796	0.9510	0.9431
B_3	0.0201	0.0226	0.0856	0.0746
r	−0.1600	−0.0423	−0.0212	−0.0203
B_4	0.1996	0.1454	0.1139	0.1067
r	0.7310	0.7131	0.8078	0.7937
B_0	−1.7742	−1.4852	−1.5516	−1.8271
Multiple correlation	0.9783	0.9946	0.9991	0.9989
F	50.19	208.04	1230.25	1039.48

[a] r is the correlation coefficient for B_i versus lg k'.

TABLE 16   Coefficients A_1, A_0, the Multiple Correlation Coefficient (R), and Fisher Coefficient (F) for Eq. (24). Stationary Phase: Silica Gel or Various Cyanoalkyl Phases, Temperature $-29°C$

Parameter	SiO_2	C_3CN	$C_{10}CN$	$i\text{-}C_{10}CN$
		Mobile phase; hexane		
A_1	0.0601	0.0685	0.1047	0.0867
A_0	−0.8253	−0.9824	−1.4866	−1.6439
R	0.9989	0.9996	0.9998	0.9995
F	2202.1	6717.3	13189.0	4721.8
		Mobile phase; hexane + 10 vol% chloroform		
A_1	0.0417	0.0460	0.0836	0.0759
A_0	−1.3976	−1.1772	−1.4643	−1.7033
R	0.9739	0.9963	0.9997	0.9994
F	92.0	679.2	7465.1	4080.4

coefficients and Fisher criteria (F) calculated with respect to the retention of naphthalene, anthracene, phenanthrene, tetracene, tetraphene, and 1,2,5,6-dibenzanthracene on the analyzed stationary phases upon elution by nondried hexane or hexane with 10% chloroform.

Table 16 shows that the interaction of the cata-condensed PAHs and the nitrile groups of cyanodecyl phases is stronger $(A_1 = 0.1047$ and $0.0867)$ than the interaction with the silanol groups on the silica gel surface $(A_1 = 0.0601)$. The values of coefficient A_1 for the cyanopropyl phase $(A_1 = 0.0685)$ are greater than those for silica gel but noticeably less than those for the cyanodecyl phases. It is of interest to note the values for the silica gel μ-Porasil and the cyanolkyl phase μBondapak CN $(A_1 = 0.063$ and 0.057, respectively) presented by Ref. 73, which are close to the values obtained by us. A comparison of coefficients B_1 in Eq. (4) and A_1 in Eq. (25) shows that they practically coincide for the cyanodecyl phases. In the case of silica gel and the cyanopropyl phase, the values of A_1 are less than the values of B_1, whereas they are greater than those for B_2, due to a significant difference in the interaction between these phases and the peripheral (type I) and inner (type II) carbon atoms of the PAH molecules.

Thus it follows that the application of Eq. (25) for series of cata-condensed PAHs also enables us to define the difference in the properties of the studied stationary phases toward PAHs. However, Eq. (25) is less informative than Eq. (4) and does not describe the behavior of pericondensed PAHs as well as the retention sequence of isometric PAHs.

VIII. RETENTION OF POLYPHENYLS AND SUBSTITUTED POLYCYCLIC AROMATIC HYDROCARBONS IN THE SYSTEM HYDROXYLATED SILICA-*n*-HEXANE

The study of the regularity of chromatographic retention of polyphenyls (PPH) and substituted (PPHs) and substituted PAHs is very important because this knowledge makes possible the use of NP HPLC for determination of individual PAHs (many of them carcinogenic) in different samples. In exhaust gases from internal combustion engines PAHs exist presumably as unsubstituted aromatic hydrocarbons or they can contain one or two methyl groups. The structure of PAH molecules in coal tar is also simple.

In this connection was studied the retention regularity of PPHs and unsubstituted and methyl-substituted PAHs on hydroxylated silica Silasorb 600 with *n*-hexane used as mobile phase.

Table 17 presents the structural parameters of the molecules and the values of the retention parameters (the capacity factor k') for the aromatic hydrocarbons (AHs) investigated.

Figure 4 shows the chromatogram obtained from a model mixture (CCl_4 and the AHs benzene, naphthalene, biphenyl, phenanthrene, chrysene, and *para*-terphenyl). All components of the mixture were completely separated (resolution $R_s > 1$ for all pairs of consecutive peaks) and peaks were reasonably symmetrical and narrow, indicating the high efficiency of the chromatographic system used.

It is apparent that to a first approximation the chromatographic retention of PAH is proportional to the number of aromatic cycles:

$$Z = f(n_{\hat{c}}) \tag{26}$$

where Z is the chromatographic retention parameter (e.g., lg t'_R, lg V' or lg k') and $n_{\hat{c}}$ is the number of aromatic cycles in a PAH molecule. This can also be expressed by a linear equation:

$$Z = A_1 n_{\hat{c}} + A_0 \tag{27}$$

where A_1 and A_0 are constants.

Equation (27), however, obviously only describes satisfactory the retention of substances belonging to one class of compound, e.g., unsubstituted linearly annellated polycyclic aromatic hydrocarbons, or to a series of polyphenyls (Fig. 5, Table 17).

As can be concluded from Fig. 5, the retention of PAH isomers (anthracene-phenanthrene, tetraphene-tetracene) mainly depends on the number of aromatic cycles, but the corresponding k' values for isomers are remarkably different. The presence of methyl groups in molecules of PAHs increases their retention in comparison with unsubstituted molecules. The retention of PAH molecules is

TABLE 17 Structural Parameters and Retention Characteristics of Polyaromatic Hydrocarbons

| No. | Substance | Chem. formula | Number of carbon atoms of type | | | | | n_π | $L(\text{Å})$ | k' |
			n_I	n_{II}	n_{III}	n_{IV}	n_{CH3}			
1	Benzene	C_6C_6	6	0	0	0	0	6	5.5	0.62
2	Naphthalene	$C_{10}H_8$	8	2	0	0	0	10	8.0	1.21
3	Acenaphthene	$C_{12}H_{10}$	6	2	0	2	2	10	8.0	1.55
4	Fluorene	$C_{13}H_{10}$	8	2	0	2	1	12	10.5	2.76
5	Diphenyl	$C_{12}H_{10}$	10	2	0	0	0	12	10.5	2.04
6	3-Methyldiphenyl	$C_{13}H_{12}$	9	2	0	1	1	12	10.5	2.42
7	3,3'-Dimethyldiphenyl	$C_{14}H_{14}$	8	2	0	2	2	12	10.5	2.76
8	Phenanthrene	$C_{14}H_{10}$	10	4	0	0	0	14	9.5	2.42
9	Anthracene	$C_{14}H_{10}$	10	4	0	0	0	14	10.5	2.24
10	2-Methylanthracene	$C_{15}H_{12}$	9	4	0	1	1	14	11.5	2.46
11	o-Diphenylbenzene	$C_{18}H_{14}$	14	4	0	0	0	18	10.5	5.23
12	m-Diphenylbenzene	$C_{18}H_{14}$	14	4	0	0	0	18	13.5	5.35
13	p-Diphenylbenzene	$C_{18}H_{14}$	14	4	0	0	0	18	15.5	5.79
14	1,3,5-Triphenylbenzene	$C_{24}H_{18}$	18	6	0	0	0	24	13.5	11.61
15	Perylene	$C_{20}H_{12}$	12	8	0	0	0	20	10.5	5.52
16	Tetracene	$C_{20}H_{12}$	12	6	0	0	0	18	11.8	4.54
17	Tetraphene	$C_{20}H_{12}$	12	6	0	0	0	18	11.8	4.76

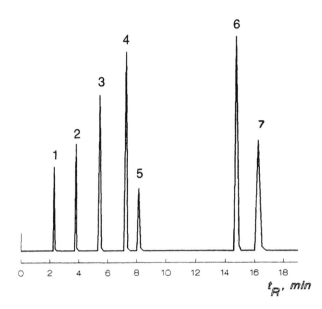

FIG. 4 Chromatogram of a model mixture: 1, CCl_4 (unadsorbed compound); 2, benzene; 3, naphthalene; 4, biphenyl; 5, phenanthrene; 6, chrysene; 7, *para*-terphenyl. Column: Silasorb 600 (62 × 2 mm, 5 μm). Mobile phase: *n*-hexane, 100 μL min^{-1}.

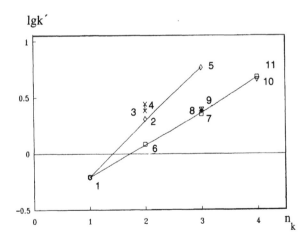

FIG. 5 The logarithm of the capacity factor (log k') as function of the number of aromatic rings (n_e) in an aromatic hydrocarbon molecule. 1, benzene; 2, diphenyl; 3, 3-methyldiphenyl; 4, 3,3'-dimethyldiphenyl; 5, *para*-diphenylbenzene; 6, naphthalene; 7, anthracene; 8, phenanthrene; 9, 2-methylanthracene; 10, tetraphene; 11, tetracene.

strongly dependent on the type of bond between aromatic rings in these molecules. So the biphenyl and *para*-terphenyl molecules are retained considerably more strongly than corresponding PAHs with two and three condensed rings (Fig. 5).

Obviously, the dependence of HPLC retention on PAH structure is multiparametric; consideration of one parameter only in a correlation equation does not provide adequate description of chromatographic performance unless the correct parameter is chosen.

It is possible to use Eq. (25) to correlate the retention of PAHs with the number of carbon atoms in the PAH molecule:

$$\lg k' = A_1 n_c + A_0 \tag{25}$$

where n_c is number of carbon atoms in a series of linearly or angularly annellated cata-condensed PAHs and A_1 and A_0 are constants. Equation (25) has been used to evaluate chromatographic properties of various stationary phases used for chromatography of PAH [45,70,72,73]. The value of A_1 can be used as an integrated characteristic of specific interactions between the stationary phase and a PAH molecule. The correlation Eq. (25) provides better results (Fig. 6) than Eq. (27) (see Fig. 5).

Figure 6 demonstrates that the slope of the linear dependencies $\lg k'$ on n_c is different for polyphenyls and polycyclic aromatic hydrocarbons; the *para*-terphe-

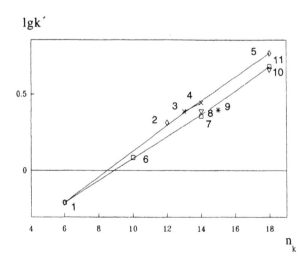

FIG. 6   The logarithm of the capacity factor ($\log k'$) as function of the number of carbon atoms (n_c) in an aromatic hydrocarbon molecule. Substance identification as for Fig. 5.

nyl molecule is retained more strongly than tetracene or tetraphene molecules with the same number of carbon atoms.

Assuming an identical NP retention mechanism for aromatic hydrocarbons in the polyphenyl series and for PAH, it is possible to describe their retention by a single or by a set of closely related correlation equations. In this case it is expedient to start to describe the additive retention of the aromatic hydrocarbon of each series AH (PAH or PHH) by the simplest equations. If necessary, it is possible to use more complex equations using more detailed description of molecular structure until an adequate model of chromatographic retention is obtained (an adequate retention model being characterized by small differences between calculated and experimental data and by high values of the correlation coefficient). For this purpose comparison of increments related to atoms of the same type in molecules of different aromatic hydrocarbons (PAH and PPH) should be performed as the equation becomes increasingly complex. The final aim is to achieve a good proximity of values of increments.

It is clear that from the perspective of construction of retention modes for individual PAHs in NP HPLC the use of the number of aromatic rings n or the number of carbon atoms n_c in an AH molecule as a simple parameter does not make possible an adequate description of retention.

It was shown that the retention of unsubstituted PAHs in NP HPLC (on hydroxylated silica gel and various cyanoalkyl stationary phases eluted with n-hexane) is described by the equation:

$$\lg k' = B_1 n_I + B_2 n_{II} + B_3 n_{III} + B_4(n_\pi/L) + B_0 \tag{4}$$

The values of B_i of Eq. (4) are given in Table 12.

With methyl-substituted PAH and PPH it is necessary to allocate four types of carbon atoms (Fig. 7). Similarly in terms of Eq. (4) it is also possible to describe the retention of methyl-substituted PAHs and PPHs by means of

$$\lg k' = A_1 n_I + A_2 n_{II} + A_3 n_{III} + A_4 n_{IV} + A_5(n_\pi/L) + A_0 \tag{28}$$

where A_1–A_4 are the equation coefficients reflecting the contribution to retention of different types of carbon atom (see Fig. 7).

Values of the coefficients A_i for PAH and PPH, calculated from the data in Table 17, are presented in Tables 18 and 19. It is apparent (see Tables 12 and 18), that the values of the coefficients A_i and B_i for different types of carbon atom from PAH molecules have the same sign and close values. The differences for the B_4 and A_5 values can be explained by the use of retention values of different PAH molecules when calculating the coefficients for Tables 12 or 18, and the introduction in Eq. (28) of an additional type (IV) of carbon atom, because of the presence of methyl substitution in PAH molecules. Thus, Eq. (28), similar to Eq. (4) offered earlier, describes the chromatographic behavior of PAHs of

(a)

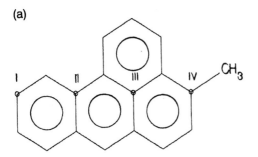

(b)

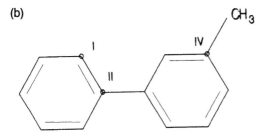

FIG. 7 Types of carbon atoms in molecules of methyl-substituted polycyclic aromatic hydrocarbons (a) and polyphenyls (b).

TABLE 18 Coefficients A_1, A_2, A_3, A_4, A_5, A_0 for Eq. (28) and Multiple Correlation (R) for Condensed Polyaromatic Hydrocarbons. Stationary Phase: Hydroxylated Silica Gel. Mobile Phase; n-Hexane. Temperature 22°C

	Equation coefficients	
Equation	A_i	Significance
$\lg k' = A_1 n_I + A_2 n_{II} + A_3 n_{III} + A_4 n_{IV} + A_5(n_\pi/L) + A_0$ (28)	A_1	0.122 ± 0.0036
	A_2	0.0211 ± 0.000
	A_3	0.021 ± 0.0043
	A_4	0.160 ± 0.0079
	A_5	-0.05 ± 0.029
	A_0	-0.8850 ± 0.0000
	R	0.9993

TABLE 19 Coefficients A_i, C_i, D_i, F_i, and Multiple Correlation (R) for Polyphenyl Hydrocarbons. Stationary Phase: Hydroxylated Silica Gel. Mobile Phase; n-Hexane. Temperature 22°C

		Equation coefficients	
Equation		X_i	Significance
$\lg k' = A_1 n_e + A_0$	(27)	A_1	0.40 ± 0.033
		A_0	-0.05 ± 0.088
		R	0.9806
$\lg k' = C_1 n_I + C_2 n_{II} + C_3 n_{IV} + C_0$	(30)	C_1	0.0189 ± 0.0000
		C_2	0.1500 ± 0.0000
		C_3	0.0349 ± 0.0009
		C_0	-0.1500 ± 0.0000
		R	0.9953
$\lg k' = D_1 n_I + D_2 n_{II} + D_4(n_\pi/L) + D_0$	(31)	D_1	-0.06 ± 0.02
		D_2	0.32 ± 0.053
		D_4	-0.09 ± 0.073
		D_0	0.4 ± 0.14
		R	0.9968
$\lg k' = F_1 n_I + F_2 n_{II} + F_3 n_{IV} + F_4(n_\pi/L) + F_0$	(32)	F_1	0.0184 ± 0.0036
		F_2	1.16 ± 0.0000
		F_3	0.0398 ± 0.0009
		F_4	-8.96 ± 0.09
		F_0	-0.0809 ± 0.0000
		R	0.99685

different classes (cata-condensed and peri-condensed) in NP HPLC and correctly predicts a retention order for isomeric PAH.

Comparison of coefficient values of Eq. (28) for PAH and Eqs. (30)–(32) for PPH (Tables 18 and 19) shows that they often have different values for nominally identical types of carbon atoms (Fig. 7). This confirms the difficulty of mathematical description of PAH and PPH retention by use of a single equation, although in the description of retention for each of these classes (PAH and PPH) the high values of multiple correlation coefficients are obtained for AH as a separate class. The noted discrepancy in coefficient values from equations, describing the chromatographic behavior of PAHs and PPHs, indicates the different mechanisms of adsorption of PAH and PPH molecules. This is a result of the different orientation of the molecules when adsorbed on a silica surfaces because flat orientation is hindered or impossible for polyphenyls (e.g., *ortho*-terphenyl) because of high potential energy barriers to internal rotation [74]. Although this difference complicates the optimization of chromatographic separation of PAH and PPH, it in-

creases the possibility of identifying individual AHs belonging to these classes
of organic molecule.

IX. OPTIMIZATION OF A MOBILE PHASE COMPOSITION AND IDENTIFICATION OF COMPONENTS OF PAH MIXES

The following two problems are most frequently found in liquid chromatography
of PAHs: (1) optimization (choice) of separation conditions and (2) identification
of the components of the analyzed complex mixture. Thus, in order to predict
the PAH retention times and choose the analysis conditions (in our case, the
mobile phase composition) for separating the mixture within a certain (given)
period of time, we can apply Eq. (5) (see Table 4). The precision of prediction
would meet the practical requirements. Equation (5) is found to be feasible for
the purpose of substance identification.

Besides the properties of the PAH molecules the concentration of more polar
addition to the mobile phase should be considered in order to optimize the mobile
phase composition. Therefore, in this case Eq. (8) appears to be more useful than
Eq. (5).

$$\lg k' = E_1 n_I + E_2(n_{II} + n_{III}) + E_3(n_\pi/L) + E_4 C + E_0 \qquad (8)$$

where E_i denotes the coefficients, C is the content of ethanol in the mobile phase,
vol% (in this case, for more convenient treatment of the experimental data, we
applied the concentration of ethanol in water, but not otherwise), as it enables
the prediction of the properties of the mobile phase of any (quantitative) composi-

TABLE 20   Values of Coefficients E_1, E_2, E_3, E_4, E_0, Correlation Coefficients,[a] and
Fisher Coefficients (F) for the Dependence Eq. (18) (the Retention Values of
Anthracene, Pyrene, Triphenylene, Tetraphene, Benz[a]pyrene, 1,2,7,8-
Dibenzanthracene Were Used for the Calculation); Column; MCH-10 "MicroPak,"
Water Content in Ethanol from 0 to 40 vol%, Temperature 39°C

Dependence coefficients					Multiple correlation	F
E_1	E_2	E_3	E_4	E_0		
0.05254	0.09754	−0.13558	−0.03195	1.87454	0.99649	884.66
(0.34055)[a]	(0.36186)	(0.14352)	(−0.92228)			

[a] Parentheses are used to give correlation coefficients E_i with $\lg k'$.

TABLE 21 Results of the Structural-Chromatographic Analysis of Polyaromatic Hydrocarbons [Calculations According to Eq. (5)]

Substance	n_I	$n_I + n_{II}$	n_π/L	$L = (n_I + n_{II} + n_{III})/(n_\pi/L)$
1. Phenanthrene	8.97	3.93	1.169	11.03
	(10)	(4)		(9.5)
2. Anthracene	9.89	3.79	1.289	10.56
	(10)	(4)		(10.5)
3. Pyrene	9.48	5.99	1.479	10.46
	(10)	(6)		(9.5)
4. Triphenylene	12.13	5.96	1.946	13.44
	(12)	(6)		(9.5)
5. Chrysene	10.51	6.44	1.260	13.48
	(12)	(6)		(11.8)
6. Tetraphene	12.72	6.48	1.806	10.63
	(12)	(6)		(11.8)
7. Perylene	12.86	8.02	2.250	9.28
	(12)	(8)		(10.5)
8. Benz[a]pyrene	12.58	8.02	1.919	10.73
	(12)	(8)		(11.8)
9. 1,2,7,8,-Dibenzanthracene	13.35	7.79	1.376	15.36
	(14)	(8)		(13.5)

[a] The actual values of the molecule's parameter are given in parentheses.

tion, even in the absence of relative experimental data in the composition. This equation is a specific case of the well-known dependency [41,75]:

$$\lg k' = A + BC \tag{29}$$

where A and B are the coefficients, and C is the concentration of one of the substances in the two-component mobile phase. The values of these coefficients are given in Table 20. The calculations have shown that Eq. (8) describes the retention on the six reference substances and other PAHs.

Three types of problems occur in analytical chemistry [76]: (1) when the qualitative composition of the test mixture is completely known and the reference samples are available, i.e., individual substances or the data on their retention; (2) when the object of the test is of known origin and when the reference samples of the supposed components or the data on their retention are available; (3) when information concerning the object composition is not available. When analyzing PAHs for solving the problems of the first and particularly of the second type, Eq. (5) is the most feasible. The proposed approach to the identification of the components (chromatographic zone) consists of the following: the analyzed mixture is separated on one and the same sorbent (column) with the mobile phases of various composition for which the coefficients of Eq. (5) were calculated. Then, using the experimental data, the logarithms of the capacity factors are calculated, and the system of linear equations of type (5) is constructed and solved with respect to n_I, n_{II} + n_{III}, and L (Table 21) with the use of a computer. Thus we obtain the point in three-dimensional space and can correctly identify the substance using the values of the point coordinates (comparing the values with the data bank for the given class of substances). Certainly another advantage of Eq. (5) is the discreteness and evenness of two (n_I and n_{II} + n_{III}) of three characteristic parameters, which simplifies rounding off to the nearest integer and increases the probability of correct identification.

REFERENCES

1. S. N. Lanin and Yu. S. Nikitin, Zh. Analit. Khimii (Rus). 42:1661 (1987).
2. S. N. Lanin and Yu. S. Nikitin, Abstr. IV All Union Symp. on Molecular Liquid Chromatography, Riga, 1987, p.11.
3. S. N. Lanin and Yu. S. Nikitin, Chromatographia. 25:272 (1988).
4. S. N. Lanin and Yu. S. Nikitin, Pure Appl. Chem. 61:2027 (1989).
5. S. N. Lanin, Yu. S. Nikitin. A. A. Pyatygin, and S. M. Staroverov, Chromatographia. 25:147 (1989).
6. S. N. Lanin and Yu. S. Nikitin, Zh. Analit. Khimii (Rus). 45:1939 (1990).
7. A. V. Kiselev, in Phizicheskaja Khimiya, Sovremennye problemy (Ya. M. Kolotyrkin, ed.), Khimija, Moscow, 1982, p. 180.
8. A. V. Kiselev, Chromatographia. 11:691 (1978).
9. A. V. Kiselev, Mezhmolekulyarnye vzaimodeistviia v adsorbtsii i khromatographii, Vysshaja Shkola, Moscow, 1986, p. 360.
10. A. V. Kiselev, D. P. Poshkus, and Ya. I. Yashin, Molekulyarnye osnovy adsorbtsionnoi khromatographii, Khimija, Moscow, 1986, p. 270.
11. A. Ya. Gorbachevskii, A. V. Kiselev, Yu. S. Nikitin, and A. A. Pyatygin, Vestnik Mosk. Univ. Ser. 2. Khimija. 26:364 (1985).
12. A. Ya. Gorbachevskii A. V. Kiselev, Yu. S. Nikitin, and A. A. Pyatygin, Chromatographia. 20:533 (1985).
13. B. A. Rudenko, Z. Yu. Bulycheva, and L. V. Dylevskaya, Zh. Analit. Khimii (Rus). 39:344 1984.
14. Z. Yu. Bulycheva, B. A. Rudenko, L. V. Dylevskaya, and V. F. Kutenev, Zh. Analit. Khimii (Rus). 40:330 (1985).
15. D. N. Grigorieva and R. V. Golovnya, Zh. Analit. Khimii (Rus). 40:1733 (1985).
16. R. Kaliszan, CRC Crit. Anal. Chem. 16:323 (1986).
17. R. Kaliszan Quantitative Structure–Chromatographic Retention Relationships, John Wiley and Sons, New York, 1987.
18. Ho Chu-Noi, D. L. Karlesky, J. R. Kennedy, and I. M. Warner, J. Liq. Chromatography 9:1 (1986).

19. K. Jinno, *Eighth International Symposium on Capillary Chromatography*. Riva del Garda, Italy, 1987, Pt.2, Pp. 1143–1152.
20. V. D. Shatz and O. V. Sachartova, *Vysokoeffektivnaya gidkostnaya chromatographya*, Zinante, Riga, 1988, p. 390.
21. P. L. Grizzle and J. S. Thomson, Anal. Chem. *54*:1071 (1982).
22. R. B. Sleight, J. Chromatography *83*:31 (1973).
23. A. Bylina, L. Gluzinski, K. Lesniak, and B. Radwanski, Chromatographia *17*:132 (1983).
24. R. E. Koopmans and R. Rekker, J. Chromatography *285*:267 (1984).
25. R. B. Hermann, J. Phys. Chem. *76*:2754 (1972).
26. F. S. Calixto and A. G. Raso, Chromatographia. *14*:596 (1981).
27. K. Jinno and A. Ishigaki, J. High Resolut. Chromatogr. Commun. *5*:668 (1982).
28. M. Randic, J. Chromatography *161*:1 (1978).
29. A. Radecki, H. Lamparczyk, and R. Kaliszan, Chromatographia *12*:595 (1979).
30. M. J. M. Wells, C. R. Clark, and R. Patterson, J. Chromatogr. Sci. *19*:573 (1981).
31. S. A. Wise, W. J. Bennett, F. R. Guenther, and W. E. May, J. Chromatogr. Sci. *19*: 457 (1981).
32. H. Hanai, Chromatographia *12*:77 (1979).
33. J. Sliwiok, A. Macioszczyk, and T. Kowalska, Chromatographia *14*:138 (1981).
34. J. Sliwiok, T. Kowalska, B. Kocjan, and B. Korczak, Chromatographia *14*:363 (1981).
35. K. Jinno, Anal. Lett. *15*:1533 (1982).
36. R. Kaliszan, J. Chromatogr. *220*:71 (1981).
37. W. Markowski, T. Dzido, and T. Wawrszynowicz, Pol. J. Chem. *52*:2063 (1978).
38. R. Kaliszan and H. Lamparczyk, J. Chromatogr. Sci. *16*:246 (1978).
39. R. J. Hurtubise, T. W. Allen, and H. F. Silver, J. Chromatography *235*:517 (1982).
40. N. Tanaka, Y. Tokuda, K. Iwaguchi, and M. Araki, J. Chromatography *239*:761 (1982).
41. T. Hanai and J. Hubert, J. Chromatography *291*:81 (1984).
42. J. F. Schabron, R. J. Hurtubise, and H. F. Silver, Anal. Chem. *49*:2253 (1977).
43. K. Jinno and K. Kawasaki, Chromatographia *17*:445 (1983).
44. K. Jinno and M. Okamoto, Chromatographia *18*:495 (1984).
45. A. N. Ageev, A. V. Kiselev, and Ya. I. Yashin, Chromatographia *13*:669 (1980).
46. J. E. Wilkinson, E. Struppe, and P. W. A. Jones, *Third International Symposium on Chemistry and Biology: Carcinogenesis and Mutagenesis*, Ann. Arbor Science, 1979, p. 217.
47. A. Matsunaga, Anal. Chem. *55*:1375 (1983).
48. M. Popl, V. Dolansky, and J. Mostecky, J. Chromatography *117*:117 (1976).
49. M. Popl, V. Dolansky, and J. Mostecky, J. Chromatography *59*:329 (1971).
50. O. A. Beiko, M. Yu. Isakov, V. Yu. Telly, and V. F. Novikova, Neftechimija, *22*: 702 (1982).
51. M. Popl, V. Dolansky, and J. Mostecky, J. Chromatography *91*:649 (1974).
52. H. Hemetsberger, A. Haac, and J. Kohler, Chromatographia *14*:341 (1981).
53. H. Hemetsberger, H. Klar, and H. Ricken, Chromatographia *13*:277 (1980).
54. W. Holstein, Chromatographia. *14*:468 (1981).
55. C. H. Lochmuller and C. W. Amoss, J. Chromatography *108*:85 (1975).

56. W. Holstein and D. Severin, Chromatographia *15*:231 (1982).
57. E. P. Lankmayer and K. Muller, J. Chromatography *170*:139 (1979).
58. J. Chmielowiec and A. E. George, Anal. Chem. *52*:1154 (1980).
59. K. G. Liphard, Chromatographia *13*:603 (1980).
60. V. Martiniand and J. Janak, J. Chromatography *65*:477 (1972).
61. V. M. Tatevskii, Zh. Fiz. Khimii. *68*:1157 (1994).
62. V. M. Tatevskii, Vestnik Mosk. Univ. Ser. 2. Khimija. *19*:635 (1978).
63. V. M. Tatevskii, *Teorija fiziko-khimicheskich svoistv molekul i veschestv*, MGU, Moscow, 1987, p. 238.
64. V. M. Tatevskii, Vestnik Mosk. Univ. Ser. 2. Khimija. *23*:315 (1982).
65. A. J. P. Martin, Biochem. Soc. Symp. *3*:4 (1949).
66. V. A. Palm, *Osnovy kolichestvennoi teorii organicheskikh reaktsii*, Khimija, Leningrad, 1977, p. 359.
67. L. R. Snyder, *Principles of Adsorption Chromatography*, Marcel Dekker, New York, 1968, p. 413.
68. H. B. Klevens, J. Phys. Chem. *54*:283 (1950).
69. O. W. Adams and R. L. Miller, J. Am. Chem. Soc. *88*:404 (1966).
70. M. C. Hennion, C. Picard, C. Combellas, M. Cande, and R. Rosset, J. Chromatography *210*:211 (1981).
71. L. R. Snyder, M. A. Quarry and J. L. Glaich Chromatograpia *24*:33 (1987).
72. W. R. Melander and Horvath, Chromatograpia *15*:86 (1982).
73. C. E. Ostman A. L. Colmsjo, Chromatograpia *25*:25 (1988).
74. W. J. Orville-Thomas (ed.), *Internal Rotation of Molecules*, Mir Press, Moscow, 1977, p. 510.
75. L. R. Snyder, J. W. Dolan and J. R. Gant, J. Chromatogr. *165*:3 (1979).
76. M. S. Vigdergauz, Zh. Analit. Khimii (Rus). *39*:151 (1984).

7

Polymer-Modified Silica Resins for Aqueous Size Exclusion Chromatography

YORAM COHEN, RON S. FAIBISH, and MONTSERRAT ROVIRA-BRU Department of Chemical Engineering, University of California, Los Angeles, Los Angeles, California

I. Introduction 263

II. Overview of Silica Resins Modification Techniques 267
   A. Coating of silica resins 267
   B. Silylation of silica resins 274
   C. Polymer-grafted silica resins 286

III. Graft Polymerization 292
   A. Process overview 292
   B. Graft polymerization and size exclusion chromatography 300

IV. Summary 308

   References 308

I. INTRODUCTION

Size exclusion chromatography (SEC) is an analytical method in which polymers, proteins, and polypeptides, dissolved in an appropriate solvent, are separated by their molecular size as they are eluted through a column packed with a porous support. The choice of the optimal mobile and stationary phases is essential for a successful separation process [1]. Common packing materials that are used for organic mobile phases are crosslinked polystyrene resins or silane-derivatized silica (e.g., with a variety of organosilanes), whereas for aqueous mobile phases the most common stationary phases are crosslinked hydroxylated polymethacrylate, polypropylene oxide gels, or glyceryl-derived silica [2]. Packing materials that are based on crosslinked polysaccharides (e.g., dextrans) have also been used extensively in aqueous SEC [3]. However, these soft resins exhibit poor mechanical stability when exposed to the high pressures in high performance SEC [3–5]. On the other hand, rigid SEC packings such as silica-based resins present several advantages such as mechanical, chemical, and thermal stability, all of which are very important in high-performance SEC [3,5,6]. Unfortunately, the most serious drawback of native silica packings is the presence of reactive silanol groups on the silica surface [5,7,8].

Various solute–silica surface nonsterical interactions, which are facilitated by the negatively charged silanol groups on the silica surface, lead to poor SEC performance of native silica packings. In essence, the silanol groups may be considered as weak ion exchangers that can interact with charged solute molecules (e.g., charged proteins and polymers) causing delayed retention (of anionic species) or early elution (of cationic species) [9–11]. Indeed, when native silica resins are used in SEC analyses, it is not uncommon to observe band shifts and broadening of elution chromatograms due to secondary retention phenomena. This secondary retention is especially noticeable with aqueous SEC, where mobile phase pH can dramatically affect the nature of the silica surface. Depending on the pH and ionic strength of the aqueous mobile phase, phenomena such as *ion exclusion* and *ion exchange* can be the leading causes of secondary retention with silica packings. Ion exclusion can occur when the pH of the mobile phase is greater than approximately 2, in which case the silica surface will be negatively charged (isoelectric point, pI, of silica is about 2). Thus, negatively charged solute molecules will have a shorter-than-expected retention time due to solute–silica repulsive forces. Aqueous SEC of anionic solutes, for example, will result in a higher apparent molecular weight (MW) than the true MW. On the other hand, if the solutes are positively charged, under the same pH conditions as above (pH > 2), ion exchange will take place when solute molecules are attracted to the silica surface. As a consequence, solute retention times will be longer than expected and a lower apparent MW than the true one will be obtained.

In addition to the above two phenomena, significant nonspecific surface adsorption of solutes to the silica surface may occur due to binding of solute molecules (in particular, biopolymers such as proteins) to surface silanol groups [3,8,12]. The effect of the above nonspecific adsorption can be illustrated by size exclusion chromatograms for the adsorption of Ficoll (a densely branched polysaccharide of a spherical shape) onto a silica gel packing (Fig. 1) [13]. The net difference between the areas of chromatogram A (Ficoll solution only) and chromatogram B (Ficoll + silica) is a clear indication that some of the eluting Ficoll adsorbed onto the silica. Variations in the elution volume, V_e, due to the

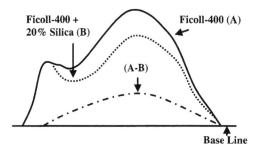

FIG. 1 Adsorption of Ficoll on silica. (A) Ficoll without silica. (B) Ficoll with silica. (A-B) The difference between A and B (indicative of Ficoll adsorption onto silica). (From Ref. 13.)

above secondary retention phenomena can be quantified by considering the relationship between the elution volume and changes in K, the equilibrium partition coefficient:

$$V_e = V_0 + KV_i \tag{1}$$

where V_0 is the void volume of the SEC column (i.e., volume of the macropores) and V_i is the pore volume of the column packing material. Total exclusion of solute molecules from the resin's internal pores will occur when $K = 0$, whereas $K = 1$ is the condition of total permeation of solute molecules through the resin's pores. A value of K greater than 1 results when the silica support acts as a weak cationic exchanger, while a value of K less than 1 indicates ion exclusion phenomena [14].

Ion exchange, ion exclusion, and nonspecific solute–silica interactions, discussed above, represent the most common nonsterical interactions that lead to poor SEC column performance. Typical SEC calibration curves that illustrate good and poor performance of a stationary phase are shown in Fig. 2. Improved resolution of solutes of various molecular weights is better when most of the silica silanol groups are unavailable for interactions with the solutes.

In order to minimize solute–silanol interactions in aqueous SEC with silica-based resins, researchers have explored several techniques to "deactivate" the reactive silanol groups on the native silica surface. The "deactivation" techniques rely on masking of the silanol groups with species that will not adsorb the solutes to be separated. As summarized in Table 1, the presently available surface modification methods include coating or adsorbing of polymers onto the silica resins with or without polymer crosslinking, covalent bonding of organosilanes to the silanol groups, and covalent bonding of polymer chains to the silanol groups via polymer grafting and graft polymerization. Schematic representation

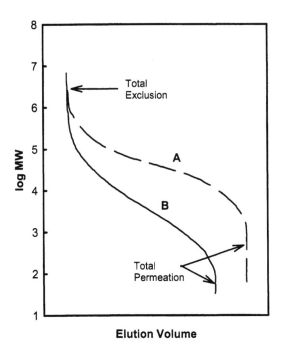

FIG. 2   SEC calibration curves for poor (A) and good (B) column performance.

of the three main methods of silica modification is given in Fig. 3. In the case of surface modification with polymers, the degree of crosslinking of the coating polymer layer or the density of the grafted polymer layer allows for control of the silica resin pore size (see Secs. II and III). An overview of the various surface modification techniques is given in the subsequent sections followed by a detailed

TABLE 1   Various Modification Techniques for SEC Packings

Method	Type of surface modifier bonding	Typical size of polymeric modifier
Silylation	Covalent bond	Short chains
Polymer coating/adsorption with or without crosslinking	Physisorption and/or chemisorption	Low to high MW ($\leq 10^5$) chains
Polymer grafting	Covalent bond	High MW chains
Graft polymerization	Covalent bond	Monomer to high MW chains

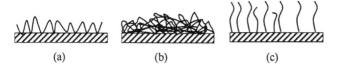

(a)   (b)   (c)

FIG. 3   Different silica surface modification methods. (a) Polymer coating w/o crosslinking. (b) Polymer coating with crosslinking. (c) Polymer grafting or graft polymerization.

description and analysis of the graft polymerization techniques, which is the focus of this chapter.

II. OVERVIEW OF SILICA RESINS MODIFICATION TECHNIQUES

A. Coating of Silica Resins

1. General Aspects

Coating of the native silica is usually carried out by deposition of a polymer from a solution that contains a crosslinking agent onto the silica surface. The solvent is then evaporated and polymer crosslinking, if desired, is often thermally induced [8]. Polymers that have been commonly used in this technique include polyethylene oxide, PEO (also known as polyethylene glycol, or PEG), dextran, and polystyrene. The degree of polymer coating (i.e., mg polymer/m^2) and the extent of crosslinking have been shown to be of paramount importance in controlling undesired silica–solute interactions [3,4,8,12]. One method to quantify the number of exposed silanol groups on different coated silica resins is to titrate the charged silanol groups on the surface of the resins with an appropriate reagent (e.g., tetramethylammonium hydroxide) [8]. This method allows for the selection of optimal coating conditions to be used in order to completely mask the silica surface silanol groups by the polymer coating layer. The study of Frere and Gramain [8], for example, utilized the above titration method to determine that the number of silanol groups on the surface of bare silica resins can be dramatically reduced by coating the resins with methacrylates of PEO. The study demonstrated that complete masking of the silanol groups was achieved with the coating polymer when the titrated solution pH was between 5 and 8. At a higher titrated solution pH [9–10], silanol group masking was incomplete and a maximum of about 86% reduction in the number of exposed silanol groups was obtained. Variations in the masking efficiency of the silanol groups were determined by coating bare silica resins with different amounts of polymer containing different amounts of a crosslinker agent. The study showed that there was an optimum crosslinker content of 8–10%, which yielded the best coated resins for aqueous SEC applica-

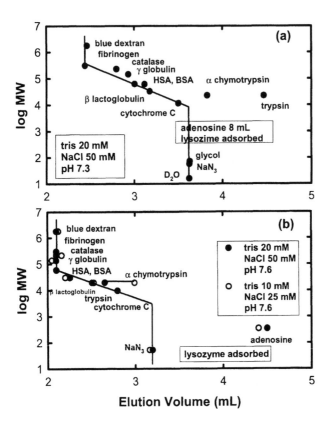

FIG. 4   Calibration curves for proteins with the poor resin A (a) and the good resin B (b) methacrylate PEO-coated silica resins. (From Ref. 8.)

tions. To illustrate the effective masking of surface silanol groups, the performance of the PEO-coated silica resins in SEC of proteins, displayed in terms of standard SEC calibration plot, is given in Fig. 4. Resin A (Fig. 4a) exhibited a much better SEC performance than resin B (Fig. 4b), which possessed the highest number of exposed silanols than any of the tested resins.

In a more recent study, Matthijs and Schacht [3] reported that high molecular weight dextran coatings (MW = 7×10^4) led to improved masking of the silanol groups and thus reduction in silica–solute interactions. These investigators used diethylaminoethyl (DEAE) in order to introduce a positive charge on the silica surface for better masking of the negatively charged silanol groups. However, careful choice of the amount of DEAE used has to be exercised, as was demonstrated in another study by Santarelli et al. [12]. They showed that if the percentage of dextran units bearing DEAE groups is below a certain value (4% in their study), the neutralization of the negatively charged silanol groups is incomplete. On the other hand, above a certain percentage of the DEAE-bearing dextran units (10% in this case) excess positive charge on the resin's surface can lead to undesired cationic exchange as observed with various proteins such as thyroglobulin, β-amylase, pepsin, trypsin, and cytochrome c.

Smaller pore size polymer-coated supports can result in a drastic loss of pore volume and, consequently, a decrease in column resolution [3]. Indeed, the choice of pore size of the native, unmodified, porous silica to be coated is a very important consideration. Zhou et al. [4], for instance, showed that an agarose-coated silica resin of a larger pore size exhibits better protein SEC performance than a smaller pore size coated resin, as illustrated in Fig. 5. On the other hand, Letot et al. [15] compared SEC performance of dextran- and PEO-coated silica resins ranging in pore size from 500 to 4000 Å. They have showed that the smaller

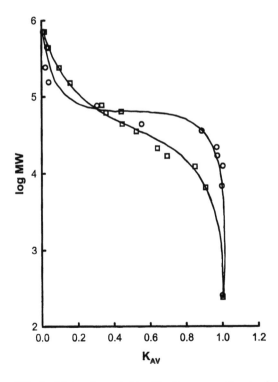

FIG. 5 Molecular weight calibration curves for standard proteins on agarose-coated silica columns of two different pore sizes (mobile phase: aqueous solution containing 0.02 M Tris-HCl and 0.15 M NaCl; pH = 7.4). (○) 300 Å and (□) 1250 Å. (From Ref. 4.)

pore size resins (500–1000 Å) produced SEC calibration curves with a higher resolution than the modified resins with larger pore size with PEO and dextran standards (Fig. 6). It should also be noted that nonsterical interactions between dextran and the coated silica resin were reduced as the resin pore size decreased, as illustrated in Fig. 6. It can be concluded, therefore, that the choice of a suitable resin pore size cannot be decoupled from the effects of the resin coating material and the solute to be separated on the SEC performance of the modified resin.

In order to overcome insufficient masking of the silica's silanol groups, some studies have suggested aqueous SEC operations under high ionic strength conditions (high salt concentrations) [3,8,11,13,16]. This approach is illustrated in Fig. 7 for the elution of the protein α-chymotrypsin using a Tris buffer solution with sodium chloride, which was used to vary solution ionic strength, with PEO-coated silica as the SEC packing material [8]. The elution chromatograms, with the various coated silica resins, are observed to be narrower when the ionic strength of the eluent is higher. The narrower chromatograms at higher ionic strengths can

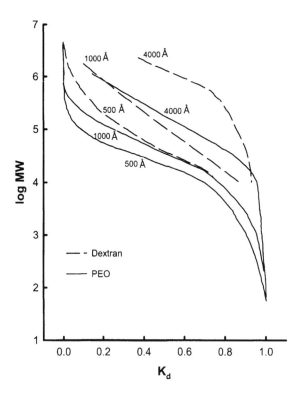

FIG. 6 Molecular weight versus the equilibrium partition coefficient calibration curves for PEO and dextran standards using 500-, 1000-, and 4000-Å PVP-coated silica resins. (From Ref. 15.)

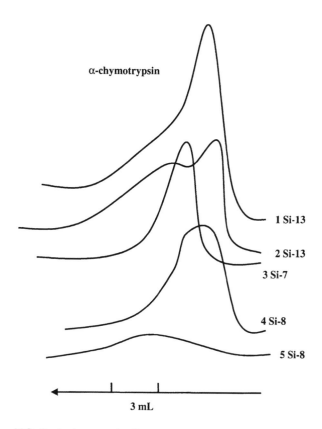

α-chymotrypsin

1 Si-13

2 Si-13

3 Si-7

4 Si-8

5 Si-8

3 mL

FIG. 7 Ionic strength effect on SEC elution curves of α-chymotrypsin for a variety of PEO-coated silica resins. (Aqueous mobile composition: (1,3) 20 mM Tris, 50 mM NaCl; (2) 10 mM Tris, 25 mM NaCl; (4) 20 mM Tris, 500 mM NaCl, and (5) 20 mM Tris, 150 mM NaCl; pH = 7.6.) (From Ref. 8.)

be explained by the charge masking effect of the salt cations in solution. The cations are attracted to the negatively charged silanol groups and effectively mask these groups and prevent any specific interactions with the solute molecules. The above method of charge masking, which relies on increasing the ionic strength of the solution, requires the controlled addition of salts; this approach may be unnecessary if the surface charge of the modified resin used is sufficiently masked by the coating (or covalently bonded) material.

2. Polyvinylpyrrolidone-Coated Silica Resins

In the mid-1980s, studies have been conducted to illustrate the potential use of polyvinylpyrrolidone (PVP) as a coating material for aqueous SEC silica resins [6,15]. PVP was first prepared by Reppe [17] from acetylene, formaldehyde, and ammonia. The protein-like structure of PVP renders the polymer biocompatible with possible uses in a variety of medical applications (such as blood plasma extender and contact lenses material) [6,10,18]. Some of the earliest uses of PVP in chromatographic separations were made in the separation of aromatic acids, aldehydes, and phenols with crosslinked PVP supports [6]. However, the permeability of these crosslinked support materials was very low resulting in poor column separation capabilities. An alternative approach was proposed in which chromatographic separation resins were prepared by adsorbing PVP onto porous silica resins [6]. Given the strong adsorption of PVP onto the silica surface, it was argued that porous silica resins with adsorbed PVP would be a stable packing material for high-performance liquid chromatography (HPLC) applications [15]. The application of the PVP-coated silica resins to aqueous SEC was especially of interest because PVP is biocompatible and thus could be used to separate proteins and water-soluble polymers.

The performance of the PVP-coated silica resins in aqueous SEC was investigated in several studies [6,10,15]. It was found that columns packed with the PVP-coated silica resins were able to separate a variety of proteins and highly polar compounds with minimal adsorption by the modified silica packings. Of particular interest was the observation that highly polar solute molecules, which are normally adsorbed and retained by native silica columns, were not retained

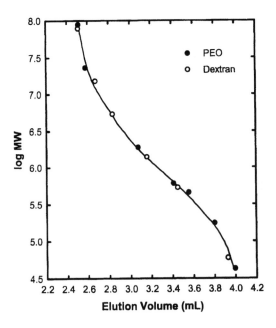

FIG. 8 Calibration curve for PEO and dextran using a PVP-coated resin with a pore size of 500 Å. (From Ref. 15.)

by the PVP-coated resins. In fact, it was found that as the polarity of the solute molecules increased, their adsorption onto the native silica resin was greater than onto the PVP-coated resin. As shown in the illustration of Fig. 8, the overlap of the universal calibration curves for the PEO and the dextran samples, with the adsorbed PVP-silica resins, demonstrates the absence of nonsteric effects in the elution of these solutes, which are present with native silica resins.

In summary, it is also important to note that an optimal coated silica resin must not only perform the desired separation but should also exhibit satisfactory long-term chemical stability. Polymer coating of silica has been shown effective in reducing solute adsorption onto the silica; however, most of the coated resins exhibited short-term stability of the coating polymer layer [5,11]. It has been demonstrated, for example, that some coated resins lose a portion of the coating layer in the initial elution stages of SEC [15]. In addition, the incomplete coating of the native silica resins can lead to serious nonsteric effects such as solute retention or repulsion due to solute-exposed silanol group interactions, as was discussed above. Indeed, the instability of PVP-coated silica resins was evaluated by Cohen and Eisenberg [10] over many elution cycles, as illustrated in Fig. 9.

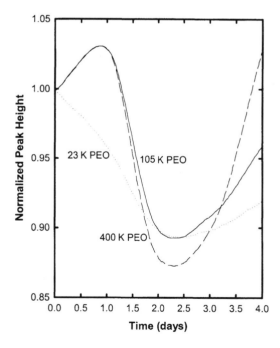

FIG. 9 Column instability for adsorbed PVP-silica resin for different injected molecular weight samples of PEO using pure water as the mobile phase. (Peak height is normalized with respect to the initial peak height.) (From Ref. 10.)

The chromatograms resulting from successive injections of three polymer samples over a period of 4 days demonstrated significant variability in peak height. Such behavior suggests simultaneous adsorption/desorption of the PVP layer and, thus, the instability of the coating layer with time.

B. Silylation of Silica Resins

1. The Silylation Procedure

Despite the intensive effort to develop modified silica resins by adsorbing polymer onto the silica surface, the instability of the coating layer at different temperatures, flow rates, and compositions of the mobile phase has hindered the commercial acceptance of such resins. In aqueous SEC, silanol groups on the silica surface contribute to the adsorption of hydrophilic polymers. Also, surface silanols can contribute in part to the adsorption and denaturalization of proteins and enzymes, a behavior that makes the presence of silanol groups undesirable in aqueous SEC with silica resins.

An alternative and popular way to reduce solute–silica surface interactions is by chemically reacting the surface silanol groups with organosilanes. The silylation method results in an organosilane that is covalently bonded to the silica surface, resulting in stable and robust SEC silica resins. A particularly effective class of organosilanes is of the type R_nSiX_{4-n}, where R is a nonhydrolyzable organic group and X is a hydrolyzable functional group (e.g., halogen, amine, alkoxy, or akyloxy). The chemistry of the nonhydrolyzable organic group provides the desired surface chemical functionality, whereas the hydrolyzable groups react with the surface silanol groups. Thus, silylation reactions allow the introduction of a variety of functional groups and hence offer a range of possibilities for tailoring chromatographic supports with the desired chemical characteristics. Although a wide range of commercial silylated silica SEC resins are available, the actual surface chemistry (and thus specific silanes used) of these resins is often proprietary. However, the most common silanes employed in the synthesis of hydrophilic bonded silica are listed in Table 2 and a list of selected commercial hydrophilic silica columns is presented in Table 3.

There are a number of basic approaches of chemically bonding organosilane groups onto the active surface of silica via organosilane reactions (Figs. 10 and 11). In general, there are two distinct classes of silylation reactions: aqueous phase silylation and anhydrous silylation. The aqueous phase silylation reaction may be divided into three main separate steps as illustrated in Fig. 10i–iii for an alkoxysilane: hydrolysis, silane–silane condensation, and silane–substrate condensation. In the presence of water, alkoxysilane molecules hydrolyze to form hydroxysilanes [19,20]. Furthermore, hydroxysilanes with two or three hydroxyl groups can undergo a condensation-polymerization reaction that can lead to the formation of polysiloxanes (Fig. 10ii), which then compete with the monosilanes

TABLE 2 Silanes Employed in the Synthesis of
Hydrophilic Bonded Silicas

No.	Formula
1	$CH_3COO(CH_2)_3$—$Si(OC_2H_5)_3$
2	$CH_3COO(CH_2)_3$—$Si(CH_3)_2(OC_2H_5)$
3	$CH_3COO(CH_2)_5$—$Si(OC_2H_5)_3$
4	$CH_3COO(CH_2)_5$—$Si(CH_3)_2(OC_2H_5)$
5	$\underset{\diagdown O \diagup}{CH_2\text{-}CH}$—$(CH_2)_2$—$Si(OC_2H_5)_3$
6	$\underset{\diagdown O \diagup}{CH_2\text{-}CH}$—$CH_2$—$O$—$(CH_2)_3$—$Si(OCH_3)_3$
7	$\underset{\diagdown O \diagup}{CH_2\text{-}CH}$—$CH$—$CH_2$—$O$—$(CH_2)_3$—$Si(OCH_3)_3$
8	$NH_2(CH_2)_3$—$Si(OC_2H_5)_3$
9	$NH_2(CH_2)_2NH$—$(CH_2)_3Si(OC_2H_5)_3$
10	$NH_2(NHCH_2CH_2)_2(CH_2)_3Si(OC_2H_5)_3$
11	$\underset{L_{NH}}{HN}$—$(CH_2)_3Si(OC_2H_5)_3$
12	$NH_2CONH(CH_2)_3Si(OC_2H_5)_3$
13	$CH_3CONH(CH_2)_3Si(OC_2H_5)_3$
14	$CF_3CONH(CH_2)_3Si(OC_2H_5)_3$
15	$CH_3SO_2NH(CH_2)Si(OC_2H_5)_3$
16	$CH_3CONHCH_2CONH(CH_2)_3Si(CO_2H_5)_3$
17	C_2H_5NHCO—$O(CH_2)_3(CH_2)_3Si(OC_2H_5)_3$
18	$NH_2(CH_2)_3Si(OC_2H_5)$
19	$R(CH_2$—CH_2—$O)_n$—$O(CH_2)_3Si(OC_2H_5)_3$
	($R = CH_3, C_2H_5, nC_4H_9; 1 < n < 3$)

Source: Ref. 7.

for surface reactive groups. Thus, in the presence of a substrate with silanol groups, the condensation reaction results in the formation of covalent bonds between the silane and the silica surface (Fig. 10iii). Silane groups already bonded on the surface can still react with oligomeric or polymeric hydroxyorganosiloxanes in the bulk (Figs. 10iv and 10v). The presence of various silane species, including chemically bonded polysilanes, on a glass surface after aqueous silylation was demonstrated via Raman spectroscopy in the early study of Koenig and Shih [21]. Furthermore, the aqueous phase silylation was shown to lead to the formation of multilayer coverage due to the presence of polysilanes. In practice, however, sterical hindrance severely limits the surface silane yield. The reaction scheme presented above is clearly an oversimplification. In reality, partial hydrolysis of the alkoxy groups of the unreacted silane can result in mono-, di-, and

TABLE 3 Examples of Commercial Hydrophilic Bonded Silicas

Trade name	Composition	Fractionation range for native globular proteins[a] (kDa)	Mean particle size (μm)	Suppliers
TSK-gel SW type (analytical)	Silica with bonded hydrophilic polar groups (exact composition not known)	1–1000	10,13 ± 2	Toyo Soda Manufacturing (Tokyo)
TSK-gel SW type (preparative)	Silica with bonded hydrophilic polar groups	1–1000	13,17 ± 2	Toyo Soda Manufactureing (Tokyo)
LiChrosorb Diol	Silica with bonded glycerolpropyl groups	10–100	5,7,10	E. Merck (Darmstadt)
Protein-Pak	Polymerized glycerolpropyl–bonded silica	1–500	10	Waters Associates (Milford, MA)
μBondagel	Ether-bonded silica	2–2000	10,20	Waters Associates (Milford, MA)
SynChropak GPC	Silica with bonded glycerolpropyl groups	NG	10	Synchrom Inc. (Linden, IN)
Synchropak CATSEC	Silica with polymerized polyamine coating	SEC for cationic polymers	10	Synchrom Inc. (Linden, IN)
Si-Polyol	Silica with hydophilic bonded groups	NG	3,5,10,30	Serva Feinbiochemica GmbH & Co. (Heidelberg)
SP (Daltosil Polyol)	Silica with hydrophilic bonded groups	NG	40–200	Serva Feinbiochemica GmbH & Co. (Heidelberg)

[a] Fractionation over the indicated range is possible with one or combination of different columns available from the manufacturer.
Source: Ref. 7.

(i) Hydrolysis of 3-glycidoxypropyl triethoxysilane to the corresponding hydroxy derivatives

$$R-Si(OR')_3 + 3 H_2O \rightarrow R-Si(OH)_3 + 3 R'OH$$

(ii) Intermolecular condensation of hydroxy derivatives, forming siloxanes

$$n(R-Si(OH)_3) \xrightarrow{-H_2O} (R-\underset{\underset{OH}{|}}{\overset{\overset{OH}{|}}{Si}}-O)_nH \qquad n > 1$$

(iii) Intermolecular condensation between hydroxy derivatives and the surface hydroxyl group

$$\equiv Si-OH + HO-Si(OH)_2R \rightarrow \equiv Si-O-Si(OH)_2R + H_2O$$

(iv) Oligomeric or polymeric phase formation due to the reaction of hydroxyorganosiloxanes with the surface hydroxyl groups

$$\equiv Si-OH + HO(\underset{\underset{OH}{|}}{\overset{\overset{OH}{|}}{Si}}-O)_nH \rightarrow \equiv SiO(\underset{\underset{OH}{|}}{\overset{\overset{OH}{|}}{Si}}-O)_nH + H_2O$$

(v) Reaction of monomeric bonded 3-glycidoxypropyl triethoxysilane with oligomeric or polymeric hydroxyorganosiloxanes

$$\equiv Si-O-\underset{\underset{R}{|}}{\overset{\overset{OH}{|}}{Si}}-OH + HO(\underset{\underset{R}{|}}{\overset{\overset{OH}{|}}{Si}}-O)_mH \rightarrow \equiv Si-O\underset{\underset{R}{|}}{\overset{\overset{OH}{|}}{Si}}-O(\underset{\underset{R}{|}}{\overset{\overset{OH}{|}}{Si}}-O)_mH$$

FIG. 10 Silylation reaction in aqueous media. R is an organic residual group, that contains the desired functionality to be introduced on the silica surface (e.g., $-(CH_2)_3-O-CH_2-CH\text{-}CH_2$). OR' can be any hydrolyzable group (e.g., $-O-CH_3$). (From Ref. 7.)

trihydroxy derivatives; di- and trihydroxy derivatives can then lead not only to linear condensation but to formation of branched and cyclic siloxanes. Crosslinking can also occur between adjacent surface-bonded groups. Therefore, silylation can be considered to be a complex reaction, which in general is difficult to control.

Surface silylation and thus the resulting surface properties are highly dependent on the silane bulk concentration, pH, and the isoelectric point of the sub-

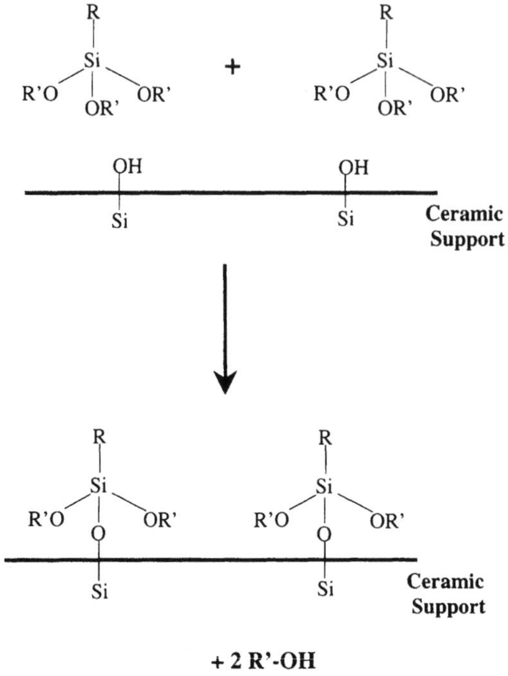

FIG. 11   Anhydrous silylation reactions. R is an organic residual group that contains the desired functionality to be introduced on the silica surface (e.g., —CH=CH$_2$). OR′ can be any hydrolyzable group (e.g., —O—CH$_3$).

strate. Studies have shown, for example, that in polymer-silylated glass filler composites the adhesive forces between silica/glass filler and the polymer matrix varied markedly with the pH of the reaction medium [22,23]. It is important to note that higher sensitivity to the pH of the reaction medium was found for ionic functional silanes relative to nonionic functional silanes.

The surface concentration of hydroxyl groups on a fully hydroxylated silica is about 4.6 OH groups/nm^2 [24]. Thus, the monolayer surface concentration of bonded silanes is about 7.5 μmol/m^2. In practice, however, steric hindrance severely limits the surface silane yield. The effect of the size of the specie approaching the surface can be studied by changing the silane hydrolyzable functional group. In the study of Unger et al. [25], silane surface coverage was shown to decrease by approximately 20% when one of the methyl groups of trimethylchlorosilane was replaced by an n-butyl or n-octyl group. Furthermore, replacing the methyl groups by bulky phenyl groups resulted in a decrease in the silane surface coverage by approximately 50–70%, depending on the number of methyl groups replaced by the phenyl groups. In a later study, Schomburg et al. [26] showed that the resulting surface silane concentration is affected by the presence of bulky side groups of the organosilanes due to sterical hindrance effects. The above study showed, for example, that the silane surface coverage for bulky octa-decyldimethylsilylenolate was approximately 40% that of trimethylsilylenolate.

The formation of polysilanes in aqueous phase silylation reactions leads to sterical effects that can reduce the uniformity of the silylated layer and the surface coverage. As a result, anhydrous silylation reactions, in which the formation of polysilanes is eliminated, are often employed. Anhydrous silylation reactions may be carried out with pure silane or in an anhydrous organic solvent. In this method the silica surface must be dried before the reaction. However, since surface silanol groups are essential for silylation, extreme caution must be taken to ensure that the drying step does not lead to condensation of the silanol groups to form siloxane bonds, which occurs at temperatures above approximately 190 ± 10°C [27].

In the absence of water, polysilanes are not formed, and the reaction scheme is simplified resulting in the attachment of monosilanes onto the surface (Fig. 11). As a consequence, silylation yields (e.g., μmol silane/m^2) reached with anhydrous reactions are lower than with hydrosilylation; however, the fraction of reacted surface hydroxyls is expected to be higher than in aqueous silylation. Bulk poly-condensations during aqueous reactions can lead to low silane yield due to steric hindrance effects associated with the polysilanes that diffuse to the surface to react with surface silanols. It is noted, however, that it is possible with the presence of a small and controlled quantity of water on the silica surface to enable condensation reactions between already bonded silanes and bulk siloxanes, thereby increasing the silane surface coverage obtained with anhydrous reactions (Table 4). The role of water is nicely illustrated in the early study of Majors and Hopper [29] who conducted a silylation reaction by introducing a predetermined amount of water vapor into the reaction vessel. Their results revealed that the surface concentration of octadecyltrichlorosilane was about 2 μmol/m^2 in the presence of water relative to about 0.5 μmol/m^2 without the use of water. Similarly, the silane yield of 2-cyanoethyltriethoxysilane increased from approximately 1.8 without water to 11 μmol/m^2 in the presence of water vapor. Majors and Hopper [29] proposed that silylation coverage can be increased by using a two-step reaction. In the first step, the primary silane (e.g., dimethyldichlorosi-lane) is bonded to the silica surface using the anhydrous silylation technique. Subsequently, the substrate is exposed to water vapor, resulting in replace-ment of the unreacted chlorines by hydroxyl groups. In the next step, a large excess of the desired secondary silane is reacted with the primary silane-bonded silica in the presence of water vapor. In the above approach, the primary silane serves to extend the silanol groups away from the silica surface and thereby decrease sterical hindrance effects, which may occur if the secondary silane

TABLE 4 Performance of Silylation on Silica Surfaces

Silane	Silica	Silylation method	Surface coverage ($\mu mol/m^2$)	Ref.
Octadecyltrichlorosilane	LiChrosorb (Si-100-5)	Bulk	3.7	Schomburg et al., 1983
Octadecyltriethoxysilane		Vapor	3.0	
Octadecyltrichlorosilane		Phase	4.4	
Octadecyldimethylsilylenolate	Nucleosil (100-5)	Silylation	3.9	
Allyldimethyldilylenolate			5.6	
			6.6	
Trimethylchlorosilane	Porous	Bulk	4.5	Unger et al., 1976
Butyldimethylchlorosilane		Vapor	3.6	
Octyldimethylchlorosilane		Phase	3.8	
Hexadecyldimethylchlorosilane		Silylation	3.4	
Phenyldimethylchlorosilane			2.6	
Butyldiphenylchlorosilane			1.8	
Triphenylchlorosilane			1.5	
Sulfonamidetriethoxysilane	Merck (Si-100)	Anhydrous silylation	5.25	Engelhardt and Mathes, 1977
Aminetriethoxysilane			4.97	
Amidetriethoxysilane			4.02	
Trifluoroamidetriethoxysilane			3.86	
Glycinamidetriethoxysilane			3.01	
Octadecyltriethoxysilane			2.78	
Glycoltriethoxysilane			2.17	
Octadecyldimethyltrifluoroacetoxysilane	Porous	Anhydrous silylation	3.32	Kinkel and Unger, 1984
Octadecyldimethylchlorosilane			3.34	
Octyldimethyltrifluoroacetoxysilane			3.45	
Octyldimethylchlorosilane			3.48	
Ethyldimethyltrifluoroacetoxysilane			3.93	
Ethyldimethylchlorosilane			4.01	
Trimethyltrifluoroacetoxysilane			4.48	
Trimethylchlorosilane			4.40	

Source: Ref. 28.

TABLE 5 Effect of Surface Interactions on Exclusion Chromatography of Water-Soluble Polymers with Water as Effluent

Stationary phase	Characteristic retention of solute in SEC columns		
	Dextrans	Polyethylene glycols	Proteins
Silica	Excluded	Strongly retarded	—
RP-C_{18}	Retarded	Strongly retarded	Strongly retarded
"Trifluoroamide"	Excluded	Retarded	Strongly retarded
"Sulfonamide"	Excluded	Retarded	Retarded
"Glycol"	Excluded	Retarded	Retarded and excluded
"Glycinamide"	Excluded	Excluded	Weakly retarded
"Amide"	Excluded	Excluded	Excluded

Source: Ref. 31.

groups are bulky. Using the above two-step procedure, the silane yields obtained for vinylmethyldichlorosilane and allylphenyldichlorosilane were 66 and 28 $\mu mol/m^2$, respectively, which are significantly above the monolayer coverage, which is no higher than 6–10 $\mu mol/m^2$.

An important criterion for the applicability of silanes for SEC resin modification is the stability of the silane–substrate bond. In all of the aforementioned silylation studies, the silylated substrates were thoroughly washed with an organic solvent, sometimes at elevated temperatures, to remove excess nonbonded silanes. The bonded silane layer was shown to be stable under these conditions. However, the stability of the silane-modified surfaces in the presence of water is often not ascertained in most silylation studies. Yet it should be apparent, that since water molecules may hydrolyze the bonded silane, it is essential to determine the stability of the silylated substrate in aqueous media [30]. Indeed, in composite adhesion studies, where silanes are used to improve filler-polymer matrix adhesion, it is a common practice to test the strength of the material after prolonged immersion in boiling water [22].

2. Silylated Silica Resins for Size Exclusion Chromatography

Over the last two decades, with developments in the synthesis of organosilanes, silane-modified silica resins have emerged as a class of versatile SEC resins. One of the classic studies that documents the diverse properties of silane-modified silica SEC resins was published by Engdelhard and Mathes [31]. The above study teaches that hydrophilic organic groups grafted onto silica SEC resins by silylation reactions can be effective in SEC separation of proteins and water-soluble polymers such as dextran and polyethylene glycols (Table 5), while virtually

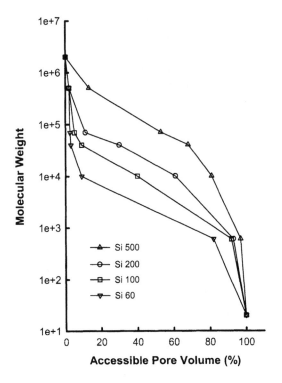

FIG. 12   Effect of pore size on calibration curves for dextrans on amide-bonded silicas.
Pore size: 60 Å (Si-60), 100 Å (Si-100), 200 Å (Si-200), and 500 Å (Si-500). Samples:
dextran standards MW 500,000 to 10,000 and raffinose (MW = 595). (From Ref. 36.)

eliminating surface adsorption in some cases. For example, SEC silica resins with
surface-bonded amide groups enable the elution of protein standards and polymers such as PEO without solute adsorption. The amide group is not neutrally
charged, hence it is unclear as to whether electrostatic interactions affect the
separation. However, given that polyacrylamide gels were shown to be successful
for protein separations [32–35], it should not be surprising that the amide-bonded
silica SEC resin is similarly a good resin for protein SEC analysis.

Interactions of solutes with silylated surfaces are expected to vary with the
silylation surface coverage. This effect was noted by Engelhardt and Mathes [36],
who argued that significant differences in the interaction of PEG with similar
functional end-groups attached to the surface silane (amine, diamine, and tria-
mine) are attributable to differences in surface coverage of the specific functional
groups. In the same study, dextran SEC using an amide-bonded silica resin demonstrated the importance of pore size on MW resolution. As Fig. 12 shows, silica

with pore size smaller than 100 Å is not suitable for polymer SEC since the effective size of the pores is reduced after silylation. Thus, if the initial pore size is too small, total exclusion of low MW polymer may occur after silylation. Improved polymer MW resolution is obtained for pore sizes in the range of 300–1000 Å; therefore, when preparing silylated resins the initial pore size must be carefully considered relative to the expected pore size reduction upon silylation.

The effect of controlling the hydrophobicity of the silica surface was demonstrated by Krasilnikov and Borisova [37] for the purification of viruses via SEC using silylated silica columns. The above study analyzed the adsorption of several kinds of viruses representative of their parent families (tick-borne encephalitis and west Nile -flaviviridae-, rabies -rhabdoviridae-, polio 1 -entheroviridae-, influenza -ortomyxoviridae-, and hepatitis B -hepadnoviridae-) in 12 different modified silica supports prepared either by surface silylation, coating, or grafting. The modified silica surface included, amine, tris-carboxyl, benzoyl, hydroxybenzoyl, salicyloyl, glycerol, hydroxypropyl functional groups and polyvinylpyrrolidone and albumin. Strong adsorption was reported for modified silica with hydrophobic surface properties. Thus, the results obtained in the study of Kraslinikov and Borisova [37] indicated that the existence of hydrophobic interactions is the primary reason for virus adsorption. Adsorption was decreased by inducing a negative charge on the silica surface and was virtually eliminated by modifying the surface with hydrophilic groups such as grafted polyvinylpyrrolidone chains (see Sec. III).

For most commercial silylated silica supports, the precise chemical and physical composition of the surface bonded phases is proprietary. Nevertheless, there are several published studies with commercial columns that document the effect of different parameters such as ionic strength, pH, composition of the mobile phase, and type of silane bonded to the silica. The available literature teaches that even with the "best columns" that are marketed as nonadsorbing columns, ionic and hydrophobic interactions can affect solute retention. Therefore, control of pH, ionic strength, and type of salt used is essential to optimize SEC operation. For example, ionic interactions contribute to retention, mainly through the variation of eluent pH in the separation of proteins on diol-modified silica columns [38]. At the same time it has been reported that peak shifting toward higher retention times, for the analysis of protein hydrolyzates with a poly(2-hydroxyethylaspartamide)–silica column, occurs with increasing ionic strength [39]. Hydrophobic interactions can also be relevant in SEC of proteins with silylated silica resins, as shown by Corradini et al. [40]. Depending on the nature and concentration of the salt used, SEC with chemically bonded silica columns can exhibit a sieving effect, electrostatic or hydrophobic interactions. In addition, SEC of polymers is affected by ionic and hydrophobic interactions, even with hydrophilic SEC resins. For example, Mori [41] showed that the addition of a simple electrolyte suppressed the ion exclusion effect in SEC of anionic and nonionic water-soluble polymers on silica gel with bonded hydrophilic groups, but the addition of an excess electrolyte promoted hydrophobic interactions.

Changes in solvent conditions often lead to changes of the predominant solute–surface interactions. This can be used not only to improve difficult separations but to enable the use of a given column for different types of chromatographic separations. For example, silylation in an aqueous medium was used by Miller et al. [42] to modify silica resins for applications in SEC and hydrophobic interaction chromatography (HIC). In the above study, five different ether-bonded phases of the structure $= Si\!\!=\!\!(CH_2)_3\!\!-\!\!O\!\!-\!\!(CH_2\!\!-\!\!CH_2\!\!-\!\!O)_n\!\!-\!\!R$, where n = 1, 2 or 3 and R = methyl, ethyl or n-butyl, were evaluated for protein analysis. Under high ionic strength conditions (3 M ammonium sulfate), hydrophobic interactions with the protein were observed with the polyether-bonded phases, and the observed behavior was characteristic of HIC. It is worth noting that the use of weakly hydrophobic phases in HIC provides a milder adsorptive surface, leading to elution of proteins in their active state. At moderate (0.5 M ammonium acetate) to low (0.05 M ammonium acetate) ionic strength, the bonded ether linkages were found to be noninteracting and thus usable for SEC. In the study of Miller et al. [42], it was found that as the length of the alkyl group at the end of the bonded ether chains increased, the degree of hydrophobic interactions also increased. Calibration curves for both protein and polystyrene were in close correspondence (Fig. 13), which indicates that ether-bonded phases with nonionic weakly hydrophobic properties have the desired properties for SEC where separation is due to molecular size. Therefore, the elution behavior was shown to depend only on the size of the solute and not on solute–support interactions. SEC behavior for those phases was not affected by the ionic strength of the mobile phase in a range between 0.05 and 0.5M ammonium acetate, which was interpreted as proof of the minimization of silanol accessibility on the ether phase.

In practice, even with the most efficient silylation procedure, residual hydroxyl groups will remain on the silica surface. This may be due to the use of silanes with bulky organic groups that limit the ability of the silane to react with the less accessible hydroxyl groups on the surface. Residual hydroxyls also result from the hydrolysis of a chloro- or alkoxysilane (e.g., $RSiCl_3$ or $RSiOR'_3$). As discussed previously, interactions between the solute and OH groups on either the native silica or the hydrolyzable silane organic group should be avoided in SEC. One can, however, take advantage of such interactions in nonideal SEC. For example, several authors have noted that, under certain conditions, glycerol propylsilane–modified silica resins exhibit latent hydrophobic [43,44] and anionic [38,45–47] properties that can be manipulated to increase the resolving power of the column. The anionic charge of a resin can be suppressed with 0.1 M or higher ionic solutions of a nonbuffering salt [48]. In contrast, Ovalle [49] demonstrated that zwitterion buffer can be used to amplify electrostatic interac-

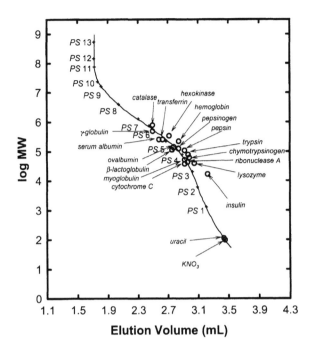

FIG. 13 Universal calibration curve for SEC of proteins and polystyrene standards using an ether-bonded resin [≡Si—(CH₂)₃—O—(CH₂—CH₂—O)₃—CH₃]. (From Ref. 42.)

tions between proteins and glycerol propylsilane–modified silica support. This latter approach was shown to be effective in separating potato homogenate–containing proteins of similar molecular weights but varying pI values. It is noted that these proteins were not resolved by SEC methods or by standard anion or cation exchange chromatography. Between proteins of different sizes, pI was the dominant factor in determining the elution order of the proteins tested. When pI values were similar, the larger protein eluted first. The degree of protein adsorption in the study of Ovalle [49] was high (60% of the injected sample) and the level of recovery was found to vary depending on the protein. Despite the high degree of protein retention, the column did not appear to leach adsorbed proteins, and peak shapes remained constant over a 4-year period.

Clearly, silane-modified silica resins have shown a wide range of applicability in SEC. However, as with polymer-coated resins, the residual silanol groups on the silica surface can lead to deterioration of SEC performance. Therefore, other alternatives, which include polymer grafting and graft polymerization, have been proposed as described in the following sections.

C. Polymer-Grafted Silica Resins

1. Overview

Attachment of polymer chains to the silica surfaces by polymer grafting involves the chemical binding of a preformed, and usually anionically polymerized, monodispersed polymer onto active support sites. Alternatively, polymer chains can be grown monomer by monomer from surface active sites by free radical polymerization or polycondensation reactions. This latter method of surface modification is known as graft polymerization. The basic idea of the above techniques is to bind to the silica surface polymeric chains with the desired chemical functionality. The first step in both procedures often involves surface activation by silylation, where the organic group of the organosilane provides the reactive surface species.

It is important to note that silylated SEC resins are also modified by bonding of short chains (usually not longer than C_{18}) to the substrate surface (as discussed above in Sec. B). In contrast, polymer grafting and graft polymerization allows the attachment of much longer chains. Polymer grafting has the advantage of enabling the formation of a polymer phase of narrow molecular weight distribution. This approach, however, results in a lower surface density, relative to graft polymerization, due to sterical hindrance associated with the large polymeric chains that must reach the inner porous matrix of the silica. On the other hand, graft polymerization results in a denser polymer brush layer on the silica surface because the small monomers that diffuse to the surface can react with active surface sites. The ability to manipulate surface chain density is essential for SEC resin performance, and thus graft polymerization is a preferred method of surface modification for SEC applications.

2. Silica Resins Prepared by Polymer Grafting

The surface characteristics of modified silica SEC resins are affected by the size and surface density of the grafted polymer chains. Due to the difficulty in measuring surface density and molecular weight of the grafted chains, most studies report the degree of grafting in terms of polymer graft yield (mg polymer/m^2). The polymer graft yield provides a useful measure of the success of the grafting procedure, especially relative to polymer adsorption. However, for SEC applications, information of the surface graft density ($\mu mol/m^2$) is a more informative parameter. The compilation of polymer grafting data (Table 6) reveals a number of trends. In general, for higher molecular weight polymers, as the graft yield increases, the surface polymer chain density decreases. One possible explanation for the above trend is that once the larger polymer molecules react with the surface groups, the remaining surface groups can become shielded from the reactive bulk polymer chains due to increased sterical hindrance.

Polymer graft yield is strongly dependent on both molecular weight and con-

TABLE 6 Polymer Grafting on Silica Surfaces

Polymer MW	Silica	Graft yield (mg/m²)	Surface coverage (μmol/m²)	Ref.
PS-Cl[a]	Aerosil 200			Laible and Hamann, 1980
1500		0.60	0.40	
2000		0.75	0.40	
3100		1.20	0.40	
4200		1.40	0.35	
6200		1.80	0.30	
9400		2.10	0.20	
14600		2.20	0.15	
PS-OCH₃[b]	Nonporous			Edwards et al., 1984
13800		1.20	0.09	
25600		3.40	0.13	
PS[c]				Papirer and Nguyen, 1972
4000	Aerosil-130-O⁺	0.63	0.16	
	Aerosil-130-O⁺	0.63	0.16	
	Aerosil-130-O⁺	0.58	0.15	
	Aerosil-130-O⁺	0.61	0.15	
27800	Aerosil-130-O⁺	0.43	0.02	
PS[c]				Bridger et al., 1979
2000	Nonporous-O⁺	1.89	0.95	
5000		3.80	0.76	
7000		4.70	0.67	
22000		2.89	0.13	
51000		1.33	0.03	
PS[c]				Edwards et al., 1984
34400	Nonporous-O⁺	4.6	0.13	
PS[c]	Aerosil 100-Cl[#]			Hamann et al., 1975
104		0.10	5.60	
520		0.20	2.00	
1040		0.30	1.30	
2080		0.50	1.20	
4170		0.80	1.00	
6250		1.20	0.95	
8330		1.50	0.90	
16660		2.80	0.85	
PS[c]				Bridger et al., 1979
13000	Nonporous-Cl[#]	4.33	0.33	
PVPy-Cl[d]	Aerosil 200			Hamann et al., 1975
105		0.08	3.95	

TABLE 6 Continued

Polymer MW	Silica	Graft yield (mg/m^2)	Surface coverage (μmol/m^2)	Ref.
210		0.15	3.70	
315		0.20	3.50	
525		0.30	2.80	
630		0.30	2.20	
735		0.30	2.00	
1050		0.30	1.45	
2100		0.55	1.35	
4200		1.05	1.25	
8400		1.90	1.15	
16800		3.50	1.05	
PVPy-Cl[e]	Aerosil 200			Hamann et al., 1975
105		0.10	5.20	
210		0.20	5.10	
315		0.30	4.80	
525		0.40	3.70	
735		0.40	2.70	
1050		0.40	1.85	
4200		1.20	1.45	
8400		2.30	1.35	
PEG[f]				Papier et al., 1987
62		0.27	4.35	
106		0.33	3.11	
150		0.43	2.87	
194		0.53	2.50	
2000		1.87	0.94	
4000		2.40	0.60	
100000		2.50	0.25	

[a] Chloride-terminated polystyrene.
[b] Methoxy-terminated polystyrene.
[c] Polystyrene.
[d] Polyvinylpyridine-Cl ($T = 20°C$).
[e] Polyvinylpyridine-Cl (T = 45°C).
[f] Polyethylene glycol.
[+] Siloxane-modified silica.
[#] Chloride-modified silica.
Source: Ref. 28.

centration of the polymer chains in the treatment solution. Early studies have shown that for dilute polymer solutions at identical polymer mass concentrations the graft yield is independent of molecular weight [50]. For high initial mass concentrations, however, the graft yield is inversely proportional to the polymer molecular weight. In contrast, at low polymer concentration but for equal initial molar concentrations, the graft yield increases with molecular weight. At high polymer molar concentrations, for high molecular weight polymers, the polymer graft yield reaches a plateau faster than in dilute solutions. Therefore, at high initial polymer concentrations, higher polymer graft yield is achieved with lower molecular weight polymer.

The study of Lecourtier et al. [51] reported on the performance of SEC resins with low (MW = 1000 and 8000) and high (MW = 50,000) molecular weight polystyrene grafted on porous chlorinated silica (Table 7). The above study illustrated that when separating nonadsorbing solutes, the upper and lower exclusion limit increased with the molecular weight of the grafted polymer. At the same time, a significant decrease in the pore volume and the specific surface area of the resin was reported for the grafted polystyrene SEC resins. Based on a series of SEC studies with polystyrene and alkanes it was concluded that the grafting procedure resulted in a polystyrene phase bonded to the exterior of the silica SEC resins. The study of Lecourtier et al. [51] teaches that SEC of large molecules is best achieved when grafting occurs outside the silica pores. Clearly, there is an interplay between the molecular weight of the polymer approaching the surface and the space that the grafted chain occupies. The use of long chains results in a higher surface mass density of grafted polymer while increasing the fraction of inaccessible surface area (or micropore volume). The above two effects suggest the existence of an optimum polymer molecular weight for a maximum column SEC performance.

The observed dependence of polymer graft yield on molecular weight must be interpreted with consideration of the internal pore size of the silica support relative to the chain size. For example, polymer grafting onto silica with large

TABLE 7 Effect of the Molecular Weight on the Graft Yield

Support	Specific surface (m^2/g)	Porous volume (cm^3/g)	Graft yield (mg/m^2)
Silica	345	0.58	—
Silica-PS 1000	60	0.15	3.40
Silica-PS 8000	50	0.09	3.54
Silica-PS 50,000	145	0.37	3.20

Source: Ref. 56.

pores (i.e., $L \ll D_p$, where L is the chain size and D_p is the pore diameter) and where the chains are in the dilute regime (i.e., negligible interactions among grafted chains) will result in graft yield that increases with polymer molecular weight. On the other hand, when the molecular size of the polymer chain is large relative to the pore size, pore diffusion limitations will reduce the graft yield as the molecular weight increases. Similarly, for large chains sterical hindrance effects will also lead to decreasing polymer graft yield with increasing polymer molecular weight. Clearly, as the chain size approaches the distance between surface anchoring sites, an already grafted polymer chain can shield adjacent reactive sites from the reactive polymer species in solution. The larger grafted polymer chains, with larger radii of gyration, tend to shield reactive sites over a greater area than smaller grafted chains. In addition, large polymer chains in solution require a larger unhindered area for grafting to occur. Clearly, for porous SEC silica-based resins, the effects of sterical hindrance should be more pronounced, where the size of the polymer modifier limits its diffusion into the porous resin's matrix. For example, when the mean pore diameter (D_p) of the starting silica resin falls in the mesopore range and below ($D_p < 50$ nm), polymer chains with a radius of 1.4 nm (e.g., polystyrene-vinylmethyldiethoxysilane copolymer of MW $= 10,000$ [52]) will have limited access to the pores. Therefore, the obtained polymer graft yield will be lower than calculated based on a freely accessible surface. In order to reduce the effects of sterical hindrance and obtain a higher and more uniform polymer surface density, it is more appropriate to use the method of graft polymerization for SEC resin preparation, which has distinct advantages, as discussed in Sec. III.

3. Polymer Grafting and Size Exclusion Chromatography

The synthesis of SEC resins by polymer grafting was reported in the literature for protein [53] and polymer [51–54] separations. These studies have shown that the performance of polymer-grafted resins, as a function of the molecular weight of the polymer chains, depends on the quality of the solvent. For example, Lecourtier et al. [51] evaluated silica-polystyrene-grafted resins and found that, under good solvent conditions, the exclusion limit and solute retention increased as the molecular weights of the grafted molecules increased (Fig. 14a). Conversely, for solvents having low affinity for both solute and grafted polymer, the influence of the bonded molecule size was reversed, namely, the exclusion limit and solute retention decreased as the molecular weights of grafted molecules increased (Fig. 14b). The size, degree of swelling, and grafting location of the grafted polymer on the silica resin are responsible for the above observations. Studies of swelling of the polymer layer under good solvent conditions and measurements of the apparent pore volume by liquid nitrogen adsorption showed that polystyrene was mainly grafted outside the pores, preventing direct access of the solute into the pores. Thus, under good solvent conditions, there are two processes

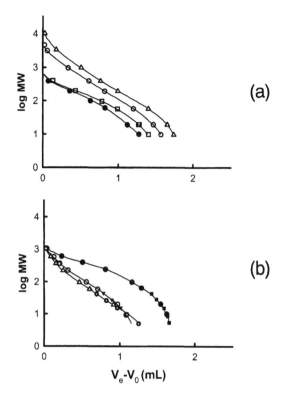

FIG. 14 Effect of the polymer molecular weight and solvent quality on SEC calibration curves. (a) Elution volume (V_e) of polystyrenes (molecular weight M) by chloroform on pure silica Si-60 (○) and silicas bonded with polystyrene MW = 1000 (●), MW = 8000 (□), and MW = 50,000 (△). (b) Elution of polystyrenes (PS) and alkanes (Alk) by dimethylformamide on silica bonded with polystyrene MW = 1000 [PS (●), Alk (■)], MW = 8000 [PS (○), Alk (▼)], and MW = 50,000 [PS (△), Alk (○)]. (From Ref. 51.)

controlling the chromatographic separation: steric exclusion effect in the silica pores and partition of the solute between the grafted phase and the bulk phase. At sufficiently high molecular weight of the grafted polymer chains, pore entrances may be blocked by the polymer, thus preventing the penetration of larger solutes into the interior of the SEC resin (Fig. 14a). Small solutes, however, can partition into the grafted layer and this process largely depends on the difference between solute affinity toward the grafted molecules and the solvent. It is worth noting that reverse chromatography occurs when solutes exhibit high affinity toward the stationary phase. Under these latter conditions, the retention time is affected by the interaction parameters between the solute and the grafted layer, which in turn also depend on polymer size. For example, dimethylformamide is a better solvent for polystyrene MW = 1000 than for polystyrene of MW = 50,000. Thus, elution volume for a grafted polystyrene of MW = 50,000 should be lower than for a grafted polystyrene of MW = 1000 reverse phase (Fig. 14b).

Polymer chains grafted onto silica can be used to screen specific solute–silica interactions. However, solvent parameters such as hydrophobicity or ionic strength must be optimized to minimize solute–SEC resin interactions. The effect of solvent quality on electrostatic and hydrophobic solute–surface interactions is illustrated in the study of Petro et al. [53]. These authors demonstrated the effect of solvent power on electrostatic and hydrophobic interactions between proteins and both diol-grafted and dextran-grafted silica resins. Protein retention was found to be higher on the diol-silica resins than on the dextran-silica resins, which indicated that protein–surface silanol interactions were significantly suppressed by the dextran surface phase (MW = 40,000). As expected, an optimal salt concentration was found at which protein–surface interactions were at minimum. An even more effective shielding of residual silanols was achieved by grafting aminoethyl-derivatized dextran (AE-dextran) onto the silica SEC resin. However, the performance of the latter modified resin was affected by the content of the aminoethyl groups, which essentially reduced the cation exchange effect of silanols in the presence of positively charged amino groups. Nevertheless, with increasing content of amino groups on the dextran graft, anion exchange behavior of the resin emerges, resulting in increased retention of acidic proteins at low salt concentration. Although residual electrostatic interactions can be minimized, by increasing the ionic strength of the solvent, hydrophobic resin–protein interactions also increase with increasing salt concentration. Thus, despite the improvement of modified SEC resins, ionic strength is also a parameter that needs to be optimized for aqueous SEC applications.

In closure, grafted polymers can be effective in controlling specific surface–solute interactions (e.g., minimizing surface adsorption of proteins). However, it is necessary to optimize solvent conditions under which SEC is performed. Additionally, the resin's pore structure can be significantly altered by the grafting process [54]. Thus, the initial pore size distribution of the SEC resin must be carefully selected, in relation to the size of the grafted chains, in order to produce a SEC resin of the desired molecular weight resolution.

III. GRAFT POLYMERIZATION

A. Process Overview

Free radical graft polymerization procedure typically consists of two separate steps: a silylation reaction producing a vinylsilane-modified surface, and a subsequent surface polymerization of a vinyl monomer that results in a chemically bonded polymer phase (see Fig. 15). In order to achieve the desired characteristics for the grafted polymer phase it is necessary to understand the various controlling reaction steps involved in the production of that material. An overview of the typical experimental procedure is presented focusing on the techniques of pretreatment, silylation, hydrolysis, and graft polymerization.

1. Surface Pretreatment

Surface activation is a crucial step for subsequent graft polymerization. Most surface activation procedures rely on surface silylation. However, prior to silylation, the silica resin must be cleaned with a dilute acid to remove metal ions (e.g., iron) and any traces of organic solvents from the surface. The resin is then washed in deionized water for a prescribed period of time in order to hydrolyze surface siloxane groups into silanol groups. A final surface preparation step consists of drying the resin in order to remove excess surface water.

In the silylation step silane compounds with the desired functionality are reacted with surface hydroxyls. For example, amorphous silicas possess, depending on the temperature history of silica, surface silanol ($\equiv$Si—OH) and/or siloxane $\equiv$Si—O—Si$\equiv$) groups. Hydroxyl concentration for native hydroxylated silica is reported to be 4.6 hydroxyls per nm^2 [24]. Since the silanes react only with the surface hydroxyl groups, the maximum possible surface coverage with surface silanes is therefore 4.6 molecules/nm^2.

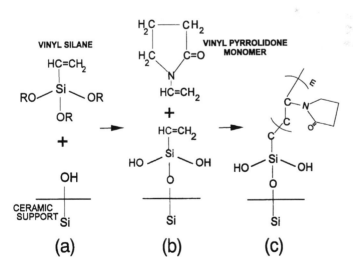

FIG. 15 Steps in vinylpyrrolidone (VP) grafting onto silica: (a) surface silylation; (b) VP surface grafting; (c) resultant attached PVP surface chains.

2. Surface Silylation

Prior to graft polymerization, the silica surface is activated as illustrated in Fig. 11 for the attachment of a vinyltrialkoxysilane onto a hydroxylated surface. Since the vinyl groups of these attached silane molecules provide the surface anchoring sites for the grafted polymeric chains, the ability to regulate the silylation coverage provides a method for controlling the surface density (chains/area) of the resulting polymer chains. It is noted that, in the above example, the use of trialkoxysilane allows for multiple surface attachments and/or silane crosslinking, which produces multilayer coverage [55] as shown schematically in Fig. 16. Additional details regarding silylation of the silica resins are provided in Sec II.B. In the sections that follow the discussion is restricted to alkoxyvinylsilane, which can be bonded to silica in a controlled manner and under extremely mild reaction conditions.

Chemical bonding of alkoxysilanes onto the hydroxylated silica surface may be accomplished by either aqueous phase or anhydrous silylation methods as described in Sec. II.B. Briefly, aqueous phase silylation in the presence of water molecules leads to the formation of polysilanes, which may limit the development of a uniform surface coverage because of sterical effects. In addition, the silane yield is strongly affected by the pH of the reaction medium. Anhydrous silylation reactions may be performed using pure silane or in anhydrous organic solvents with properly dried substrates. The selection of the appropriate organic solvent is essential for the formation of a uniform brush-like surface coverage. On the other hand, high-yield multilayer coverage may be achieved with the addition of a small amount of surface water to the dried substrates [29,56].

Subsequent hydrolysis of the alkoxysilane-modified silica surface is necessary when the graft polymerization process is conducted in an aqueous solution due to the hydrophobic nature of the remaining alkoxy groups. Hydrolysis of the unreacted alkoxy groups increases the hydrophilicity of the silylated surface, thus facilitating a more efficient wetting of the support by the aqueous monomer solu-

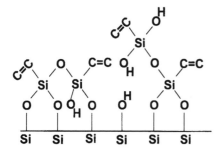

FIG. 16 Multilayer silane coverage of the silica surface.

tion. Hydrolysis can be carried out using a basic solution (typically at pH ≈ 9.5) as illustrated in Fig. 17.

3. Graft Polymerization Process

(a) General Aspects. The final part of the surface modification procedure is the graft polymerization reaction itself, as illustrated in the example of Fig. 15, in which an initiator is used to initialize the polymerization by attacking the —C=C— bond of both the monomer and the surface silane groups forming free radicals. During graft polymerization, the attachment of polymer to the surface results from one of three possible processes: polymer grafting, graft polymerization, and termination (Fig. 18). Polymer grafting refers to chemical bonding between living polymer and the silica surface sites. Graft polymerization is the process of attaching monomeric units, one by one, to active groups on a chain growing from the substrate surface. Termination results from the reaction of a live homopolymer chain and a growing surface chain, and is often grouped with polymer grafting defined previously. Due to sterical hindrance and diffusion limitations of the large polymer chains near the surface and in the pores of the resins, polymer grafting typically results in low graft density. On the other hand, as was mentioned in Sec. II.C, since graft polymerization involves the diffusion of small

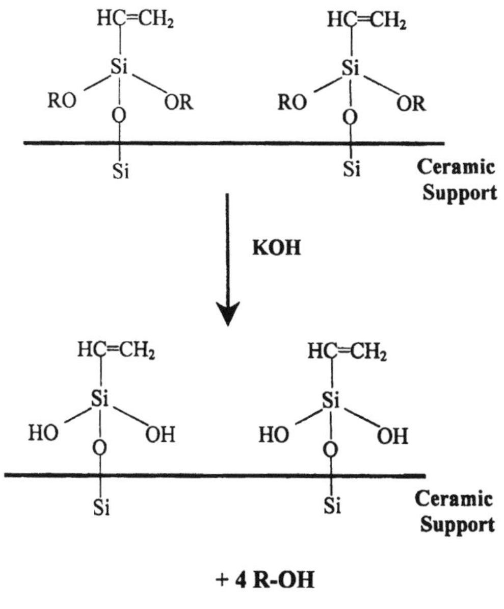

FIG. 17 Hydrolysis of the alkoxysilanes-modified silica surface with potassium hydroxide.

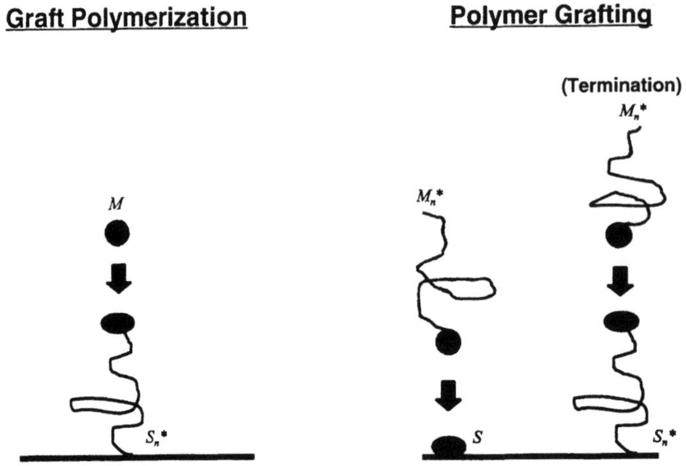

Graft Polymerization

Polymer Grafting

FIG. 18 Comparison between graft polymerization and polymer grafting. M: Monomer; M_n^*: active polymer chain of size n; S: surface site; S_n^*: active surface chain of size n.

monomer molecules to the silica surface, graft polymerization results in relatively high polymer yield and more uniform surface coverage than that obtained via polymer grafting. In order to ensure that graft polymerization is the primary process by which polymeric chains are grown on the silica surface, it is necessary to understand the kinetic mechanism of the graft polymerization process.

(b) The Kinetics of Graft Polymerization. The kinetics of free radical graft polymerization involves the following species: solvent, reaction initiator, free radicals, monomers in solution, surface-bound monomers, live growing homopolymer chains (i.e., chains in solution), live surface chains, terminated homopolymer, and terminally anchored surface chains. Consumption of monomer occurs by two processes, homopolymerization in solution and graft polymerization (which includes polymer grafting) on the silica surface, with the sum of these two consumption rates representing the overall rate of monomer consumption. The reaction sequence for free radical graft polymerization, as in traditional free radical polymerization mechanisms, can be divided into four different categories: initiation, propagation, transfer, and termination. A summary of the various reaction steps is given in Fig. 19. The various species that appear in the reaction

(I)
$$I_2 \overset{k_1}{\underset{k_1'}{\rightleftharpoons}} (I_2) \overset{k_2}{\rightarrow} (I_2^*) \overset{k_d^*}{\rightarrow} 2I\bullet \qquad\qquad (I_2^*) \overset{k_r}{\rightarrow} I_2$$

(II)
$$(I_2^*) + M \overset{k_{am}^*}{\rightarrow} (IMI) \ (Monomer-Initiator\ Associate)$$

$$(I_2^*) + S \overset{k_{as}^*}{\rightarrow} (ISI) \ (Surface\ Site-Initiator\ Associate)$$

(III)
$$(IMI) \overset{k_{bm}}{\rightarrow} I\bullet + M_1\bullet + H_2O$$

$$(ISI) \overset{k_{bs}}{\rightarrow} I\bullet + S_1\bullet + H_2O$$

$$(IMI) \overset{k_{cm}}{\rightarrow} Q_m \ (Nonreactive\ Products)$$

$$(ISI) \overset{k_{cs}}{\rightarrow} Q_s \ (Nonreactive\ Surface\ Species)$$

(IV)
$$I\bullet + S \overset{k_{im}}{\rightarrow} S_1\bullet \qquad\qquad I\bullet + M \overset{k_{im}}{\rightarrow} M_1\bullet$$

(a)

FIG. 19 Initiation and graft polymerization reaction schemes. (a) Initiation reaction scheme. (I) formation of a cage hierarchy. (II) Formation of monomer–and surface–initiator associates. (III) Dissociation of monomer–and surface–initiator associates (monomer-enhanced initiation). (IV) Primary chain initiation reactions. (b) Graft polymerization reaction scheme. (I) Homopolymer propagation (in solution). (II) Polymer grafting onto surface sites. (III) Graft polymer propagation on surface sites. (IV) Chain transfer to monomer. (V) Chain transfer to active surface sites. (VI) Termination reactions. I_2 = "compact cage"; I_2^* = "diffuse cage"; $I\cdot$ = primary initiator radical; M = monomer; S = surface site; $S\cdot$ = activated surface site; $M\cdot$ = single monomer radical; M_i and S_i ($i = 2, n, n + 1$, or m) = growing homopolymers (in solution) and surface chains, respectively; H_i and G_i ($i = n, m$, or $n + m$) = dead homopolymer and grafted polymer chains, respectively. (From Ref. 57.)

$$M_1\bullet + M \xrightarrow{k_{pmm}} M_2\bullet \qquad\qquad M_1\bullet + S \xrightarrow{k_{pms}} S_2\bullet \qquad\qquad S_1\bullet + M \xrightarrow{k_{psm}} S_2\bullet$$

(I) $\vdots$ \qquad\qquad\qquad (II) $\vdots$ \qquad\qquad\qquad (III) $\vdots$

$$M_n\bullet + M \xrightarrow{k_{pmm}} M_{n+1}\bullet \qquad M_n\bullet + S \xrightarrow{k_{pms}} S_{n+1}\bullet \qquad S_n\bullet + M \xrightarrow{k_{psm}} S_{n+1}\bullet$$

$$M_n\bullet + M \xrightarrow{k_{trmm}} H_n + M_1\bullet \qquad\qquad M_n\bullet + S \xrightarrow{k_{trms}} H_n + S_1\bullet$$

(IV) \qquad\qquad\qquad\qquad\qquad\qquad (V)

$$S_n\bullet + M \xrightarrow{k_{trsm}} G_n + M_1\bullet \qquad\qquad S_n\bullet + S \xrightarrow{k_{trss}} G_n + S_1\bullet$$

$$M_m\bullet + M_n\bullet \xrightarrow{k_{tdmm}} H_m + H_n$$

$$M_m\bullet + M_n\bullet \xrightarrow{k_{tcmm}} H_{m+n}$$

$$M_m\bullet + S_n\bullet \xrightarrow{k_{tcms}} G_{m+n}$$

(VI)

$$M_m\bullet + S_n\bullet \xrightarrow{k_{tdms}} H_m + G_n$$

$$S_m\bullet + S_n\bullet \xrightarrow{k_{tcss}} G_{m+n}$$

$$S_m\bullet + S_n\bullet \xrightarrow{k_{tdss}} G_m + G_n$$

(b)

FIG. 19 Continued

steps are defined in the figure legends and a detailed discussion of the reaction mechanisms can be found elsewhere [57,58].

When the substrate resin is exposed to the monomer reaction mixture, both graft polymerization and polymer grafting occur. Therefore, it is prudent to determine optimal conditions that would suppress the contribution of polymer chains that are produced by polymer grafting. Such an investigation was recently performed by Cohen et al. [57], and an illustrative set of results is presented in Fig. 20. The figure illustrates that both initial monomer concentrations and reaction temperature are key factors in controlling polymer grafting contribution to the overall grafting process. The percent contribution of polymer grafting to the total graft yield is seen to increase with increasing reaction temperature and decreases with increasing initial monomer concentration. At high reaction temperature, the mobility of the larger polymer chains in solution increases, thus promoting the transport of the live chains to the silica surface. On the other hand, when the initial monomer concentration is high, the rate of graft polymerization dominates chain formation (see Fig. 19b). In conclusion, for the vinylpyrrolidone-silica sys-

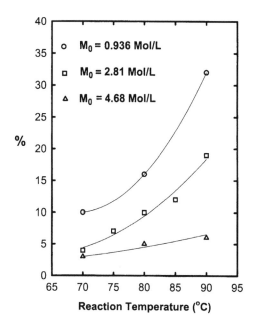

FIG. 20   Percent contribution of polymer grafting of polyvinylpyrrolidone (PVP) to overall polymer graft yield as a function of temperature and initial monomer concentration (M_0). Graft polymerization of VP onto impermeable silica particles in an aqueous suspension.

tem, the grafting of PVP is best carried out at low temperature and high monomer concentration in order to minimize the contribution of polymer grafting. It is worth noting that in the example of Fig. 20 the practical initial monomer concentration is 2.81 mol/L because at the higher concentration of 4.68 mol/L the viscosity of the reaction mixture increases rapidly, making it difficult to maintain a well-mixed suspension.

B. Graft Polymerization and Size Exclusion Chromatography

Modification of SEC silica resins via graft polymerization is an efficient method of masking surface silanol groups on silica surfaces. This approach, as mentioned previously, results in covalently bonded polymeric chains that are grown monomer by monomer on the silica surface, thus yielding modified resins with potential for long-term chemical and mechanical stability. The first reported modification of a silica resin with a grafted pyrrolidone-type surface group (i.e., 1-N-pyrrolidyl-2-dimethylchlorosilylethane) for aqueous SEC of various polymeric chains was reported in the mid 1980s [6]. However, these resins demonstrated marginal chemical stability relative to PVP-coated resins [6].

In order to improve the stability of PVP-modified silica resins, direct grafting methods of PVP onto silica resins via graft polymerization or polymer grafting have been proposed as alternatives to coated silica resins to yield stable silica-supported polymer (SSP) resins for SEC [6,10,37,59–61]. Polymer grafting, which involves the covalent bonding of live polymer chains to the silica surface, can be used to produce chemically stable resin with monodispersed polymeric surface chains [10]. However, grafting of large polymeric chains onto the silica surface and inside the pores is problematic due to diffusional limitations and steric hindrance effects (see Sec. II.C.2 and II.C.3 and [10]). In contrast, the free radical graft polymerization method (described previously in Sec. III.A.3) promotes a more uniform surface coverage with a higher surface chain density, albeit with a wide molecular weight distribution that is typical for free radical homopolymerization.

There are several major operating conditions that affect the performance of SSP/SEC columns. These include grafted polymer surface density, chain length, solvent power of the eluent, shear stress (or shear rate) at the interstitial pore walls, and column temperature. When the length of the anchored chains is long, relative to the distance between the anchoring points on the surface, the chains extend away from the surface forming a so-called brush layer (Fig. 21). A simple criterion for the formation of a brush layer was proposed by de Gennes [62]:

$$\left(\frac{\alpha}{D}\right)^2 > N^{-6/5} \tag{2}$$

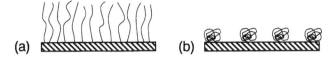

FIG. 21 Polymer chain conformation in a good solvent at (a) high and (b) low surface chain density. High surface chain density corresponds to the so-called polymer brush layer.

where α is the monomer size, D is the average distance between grafted sites, and N is the number of monomers per grafted chain. If the inequality in Eq. (2) is satisfied, the grafted polymer layer is in the dense brush regime. In this type of grafted layer conformation, the chains extend away from the surface with partial overlap between neighboring chains. When the inequality of Eq. (2) is reversed, the grafted chains collapse against the surface in a coiled mushroom-type conformation. The surface density of the grafted polymer phase, and in turn the conformation of the polymer chains, affects the permeability of the packed SEC column. A dense polymer surface layer, consisting of high molecular weight polymer chains, will reduce the permeability of the column, resulting in increased operational pressure. At the same time, it is expected that reduction in resin size and column porosity will reduce the lower MW exclusion limit. It is important to note that as the quality of the solvent increases, the polymer surface phase swells to a greater degree, lowering the column permeability [59,63]. On the other hand, as the solvent quality (for the surface polymer) decreases, the chains retract to the surface resulting in higher column permeability [59].

The variation of permeability of SEC columns packed with SSP resins can be utilized as a measure of the effective length of the chains. Hydrodynamic permeability measurements are expected to be sensitive to long chains, a fact that is well known from measurements of the hydrodynamic thickness of adsorbed polymers [64,65]. The permeability, which is obtained from simple flow rate–pressure drop measurements, can be related to the effective hydrodynamic thickness of the polymer surface layer via the following equation [64,65]:

$$\frac{L_H}{R_E} = 1 - \left(\frac{k_G}{k}\right)^{1/4} \tag{3}$$

where L_H is the effective hydrodynamic thickness of the grafted polymer chains, k and k_G are the permeabilities of the packed column packed with the native silica resin and the silica-grafted polymer resin, respectively, and R_E is the effective pore radius given by [66]:

$$R_E = \frac{\epsilon d_p}{12(1 - \epsilon)} \tag{4}$$

where d_p is the particle diameter and ϵ is the porosity of the packed column. Equation (3) assumes that the pores in the column are represented by equivalent cylindrical pores. Thus, a decrease in column permeability due to a grafted polymer phase (relative to the native silica resin) is equivalent to a decrease in the pore diameter or, equivalently, to an increase in the effective hydrodynamic thickness of the polymer surface layer. As an illustration, the flow rate–pressure drop behavior for a PVP-grafted silica resin is shown in Fig. 22 for the flow of different solvents. The native silica was Nucleosil 1000-10 (N1000-10) with an average particle diameter of 10 mm, an average pore diameter of 1000 Å, and a surface area of 50 m²/g. It is evident that there are significant differences in the permeabilities (observed from the slopes of the flow curves) between the two solvents

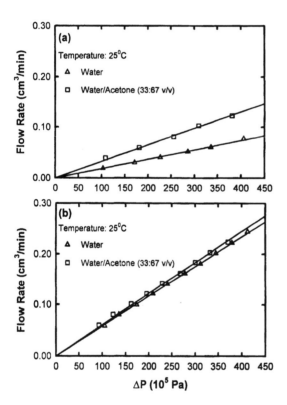

FIG. 22   (a) Flow rate versus pressure drop for column packed with PVP-grafted silica resin G1 (with 121 monomers per grafted chain). (b) Flow rate versus pressure drop for column packed with PVP-grafted silica resin G2 (with 63 monomers per grafted chain). Same elution conditions are used for both columns as indicated in a and b. (From Ref. 59.)

(water and water-acetone) for the resin with the longer grafted chains containing 121 monomers per chain (Fig. 22a). This behavior is indicative of the swelling of the chains as the solvent power increases from a theta solvent (water-acetone (33:67 v/v) to a good solvent (pure water). On the other hand, the column packed with the resin of shorter chains (containing 63 monomer units per chain; Fig. 22b) exhibits a much smaller variation in column permeability with solvent power. This behavior is consistent with the discussion above and suggests that for the silica-grafted PVP resin there is an additional operational consideration, namely, the variation in pore size with solvent power. Additional flow rate–pressure drop curves for the N1000-10 resin with grafted PVP chains of about 130 monomers/chain are given in Fig. 23. The curves illustrate nicely the increasing

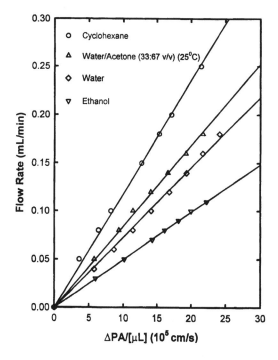

FIG. 23 Flow curves for a PVP-grafted N1000-10 column for different solvents. The slope of the curves is the column permeability $\{k = Q/[\Delta PA/(\mu L)]$, where $Q =$ flow rate, $\Delta P =$ pressure drop across the packed column, $A =$ column cross-sectional area, $\mu =$ viscosity of mobile phase, and $L =$ column length$\}$. Swelling of the grafted PVP phase is evident by the decrease in the slope of the flow curves with increasing solvent power. The resin was prepared by graft polymerization with initial VP concentration of 5% by volume. (From Ref. 63.)

column permeability with decreasing solvent power. When a very hydrophilic eluent such as ethanol is used, significant polymer swelling decreases the column permeability below that obtained with pure water. On the other hand, when a very hydrophobic eluent such as cyclohexane is used, polymer swelling is minimal and column permeability is observed to be highest (Fig. 23). The curve for the theta solvent water-acetone (33:67 v/v) gives intermediate column permeability.

As noted earlier, if the surface density of the grafted chains is sufficiently high, one should expect the permeability of the column to be invariant with the shear stress in the interstitial pores of the column. Indeed, as illustrated in Fig. 24 for a silica resin with grafted PVP polymer chains in the brush regime and with chain length/resin diameter ratio of about 9.3×10^{-4}, the column permeability for water is independent of the column shear stress (or shear rate). In addition, the column packed with silica resins grafted with longer chains (121 monomers per

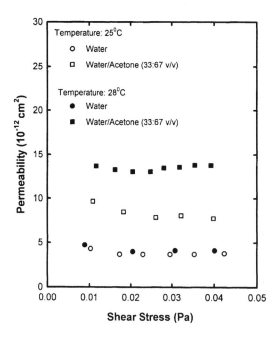

FIG. 24 Permeability of PVP grafted-silica column G1 (with 121 monomers per grafted chain). Effect of solvent power on pore–wall shear stress, $\tau = [\Delta P/(L_e/L)(L/R_E)]$, where ΔP = pressure drop across column, L_e/L = the ratio of the effective fluid path length to column length (=1.57 for spherical packings), and R_E = effective pore radius of pores in the packed column, which is given by Eq. (4). (From Ref. 59.)

chain; curves in Fig. 22a) exhibits a lower permeability (compare Fig. 22a and b). Clearly, the length of the polymer chains and solvent power affect the resin's pore size; therefore, these variables are important to consider when designing SEC applications with silica-grafted polymer resins.

The effect of grafted surface chains on SEC separation of polyvinylpyrrolidone and polyethylene oxide standards is shown in Fig. 25. The grafted resins yield a linear calibration of log(MW) versus elution volume over a considerable molecular weight range. Moreover, it is apparent that the PVP-30, which has longer chains than the PVP-10 resins, has a lower resolution at the lower MW region (i.e., near the lower MW exclusion limit) and also a lower upper MW exclusion limit. Although the above two resins have not been optimized for aqueous SEC, the results do suggest that the length of the surface chains (in the brush regime) can be manipulated to synthesize resins for SEC analysis of different MW ranges. It is also worth noting that the calibration curve for an adsorbed PVP-10K resin (prepared by adsorbing a 10K MW PVP onto the surface of the silica resin) has a much narrower range of MW applicability with the added disadvantage of poor resin stability (see Fig. 25).

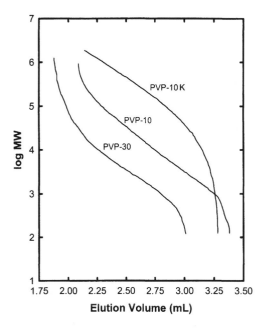

FIG. 25 Molecular weight versus elution volume calibration curves of PEO and PVP standards for adsorbed PVP-silica (PVP-10K) and PVP grafted-silica (PVP-10 and PVP-30). (From Ref. 10.)

As implied by the above discussion, the resulting performance of silica-grafted polymer resins can be tailored by adjusting the length of the polymer chains relative to the initial silica resin pore size. An illustration of this type of control is provided in Fig. 26, where resins with initial pore size of 4000 and 1000 Å were used in SEC of PEO standards. For this type of silica resin an upper MW exclusion limit of 7×10^5 was obtained with a lower MW exclusion limit of about 1000. Of the three resins shown in Fig. 26, the N4000-grafted resin, which was prepared by graft polymerization at an initial monomer concentration of 30% (VP 30%) (by volume), had the lowest upper exclusion limit. This is consistent with the fact that the contribution of graft polymerization (i.e., sequential addition of monomer to growing surface chains) to the formed grafted polymer phase is higher at the higher initial monomer concentration, resulting in longer grafted chains with measurable reduction in pore size for the N4000-10 (VP 30%) resin. A native silica resin with smaller initial diameter can also be used to achieve a separation capability in the range displayed for the N4000 resins. Such a case is illustrated in Fig. 26 for the 1000-Å pore silica resin (N1000-10) onto which vinylpyrrolidone was polymerized with initial monomer concentration of 5%. At

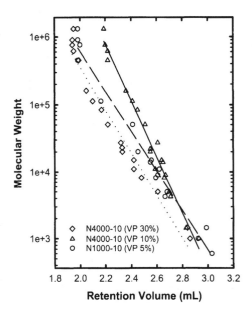

FIG. 26 Molecular weight versus elution volume calibration curves of PEO standards for the larger pore size (4000 Å) PVP-grafted silica resins N4000-10 (VP 10% and VP 30%), and the smaller pore size PVP-grafted silica resin (1000 Å) N1000-10 (VP 5%). (From Ref. 63.)

the lower initial monomer concentration the contribution of polymer grafting (i.e., attachment of living polymer chains to the surface) to the grafted polymer phase increases and thus a layer of shorter chains results. The grafting of the shorter PVP chains on the smaller pore size resin (N1000-10) produced a modified resin with similar SEC performance as the larger pore size resin (N4000-10). The ability to control the size of the chains is crucial to designing effective grafted polymer resins. Clearly, this can only be accomplished with a detailed understanding of the graft polymerization kinetics as discussed in Sec. III.A.

Changes in SEC column operating temperature can also significantly affect the permeability of the PVP-grafted resin. For example, the permeability of PVP-grafted resin-packed columns increases with eluent temperature, as illustrated for the water-PVP systems in Fig. 27. This result is consistent with the typical trend that the solvent power of a good solvent is expected to decrease with increasing temperature because the swelling of PVP in a good solvent is an exothermic

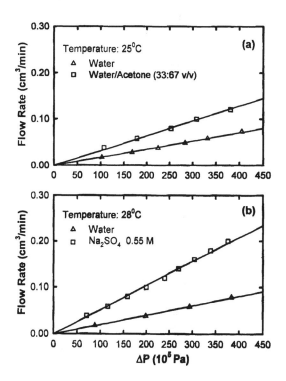

FIG. 27 Temperature effect on flow curves in columns packed with PVP grafted-silica resin G1: (a) at 25°C and (b) at 28°C. (Note: column permeability is given by $k = [Q/\Delta PA/(\mu L)]$, as defined in Fig. 23.) (From Ref. 59.)

process [67]. The decrease in solvent power results in lower swelling of the polymer chains and, as a consequence, higher column permeability [59]. It is also interesting to note that the permeability of the PVP-grafted silica resin for water-acetone (33:67 v/v) at 25°C was lower than that for the 0.55 M aqueous sodium sulfate solution at 28°C (Fig. 27), even though both eluents are theta solvents. This interesting difference between the two solvents is consistent with estimates of the end-to-end distance of the PVP chains in these solvents as described in [67]. Similar trends in column permeability with changes in column temperature can also be seen in Fig. 24.

IV. SUMMARY

The grafting of polymers onto silica resins using free radical graft polymerization for aqueous SEC applications is a relatively recent development. The motivation behind the development of this technique was to produce modified SEC-silica resins that would not interact with the solutes to be separated. Thus, the degree of masking of the silica surface silanol groups is by far the most important property that modified silica resins for SEC must possess. In addition, the modified silica resins should exhibit long-term chemical stability. Clearly, the graft polymerization approach presented in this chapter for PVP-grafted silica resins results in resins with the above desired properties. Moreover, graft polymerized silica SEC resins have been shown to present better long-term chemical stability than polymer-coated resins.

Grafted polymers can also be used to alter the porosity of native silica with large pores and thus change the exclusion limit of the resins. This can be achieved by controlling the density and length of the polymer chains. Additionally, solvent power and column temperature were shown to markedly affect the conformation of the grafted polymer chains and thus alter column permeability. This may also be used to obtain desired column permeabilities and better SEC performance for aqueous as well as nonaqueous SEC systems.

Although the focus of this chapter was on the graft polymerization of PVP onto silica resins for aqueous SEC, there may be many other polymers suitable for this particular resin modification technique. Indeed, the very promising results with the PVP-grafted silica resins should encourage future studies of other polymeric materials that may be suitable for grafting onto silica resins for use in a variety of SEC systems.

REFERENCES

1. R. Eksteen, and K. J. Pardue, in *Handbook of Size Exclusion Chromatography* (Wu Chi-san, ed.), Marcel Dekker, New York, 1995, pp. 47–101.

2. E. G. Malawer, in *Handbook of Size Exclusion Chromatography* (Wu Chi-san, ed.), Marcel Dekker, New York, 1995, pp. 1–24.
3. G. Matthijs, and E. Schacht, J. Chromatography A *755*:1 (1996).
4. F. L., Zhou, D. Muller, X. Santarelli, J. Jozefonvicz, J. Chromatogr. *476*:195 (1989).
5. D. Horák, J. Hardil, and M. J. Beneš, in *Strategies in Size Exclusion Chromatography*, (M. Potschka, and P. L. Dubin, eds.), American Chemical Society, Washington, D. C., 1996, pp. 190–210.
6. J. Köhler, *Chromatographia 21*:573 (1986).
7. K. K. Unger, and J. N. Kinkel, in *Aqueous Size-Exclusion Chromatography*, (P. L. Dubin, ed.), Elsevier, Amsterdam, 1988, pp. 193–234.
8. Y. Frere and Ph. Gramain, J. Reactive Polym. *16*:137 (1992).
9. V. A. Romano, T. Ebeyer, and P. L. Dubin, in *Strategies in Size Exclusion Chromatography* (M. Potschka, and P. L. Dubin, eds.), American Chemical Society, Washington, D. C., 1996, pp. 88–102.
10. Y. Cohen and P. Eisenberg, in *Polyelectrolyte Gels: Properties, Preparation, and Applications* (R. S. Harland, and R. K. Prud'homme, eds.), American Chemical Society, Washington, D.C., 1992, pp. 254–268.
11. P. L. Dubin, in *Aqueous Size-Exclusion Chromatography* (P. L. Dubin, ed.), Elsevier, Amsterdam, 1988, pp. 55–75.
12. X. Santarelli, D. Muller, and J. J. Jozefonvicz, J. Chromatography *443*:55 (1988).
13. G. Shah and P. L. Dubin, J. Chromatography *693*:197 (1995).
14. P. Pfannkoch, K. C. Lu, and F. E. Regnier, J. Chromatogr. Sci. *18*:430 (1980).
15. L. Letot, J. Lesec, and C. Quivoron, J. Liq. Chromatogr. *4*:1311 (1981).
16. H. D. Crone, R. M. Dawson, and E. M. Smith, J. Chromatography *103*:71 (1975).
17. W. Reppe, *Poly(vinylpyrrolidone)* Verlag Chemie: Weinheim, FRG, 1954.
18. L. Blecher, D. H. Lorenz, H. L. Lowd, A. S. Wood, and D. P. Wyman, in *The Handbook of Water-Soluble Gums and Resins* (R. L. Davidson, ed.), McGraw-Hill, New York, 1980, pp. 21.1–21.21.
19. B. Arkles, Chemtech *7*:766 (1977).
20. Petrarch Systems, Inc. *Silicon Compounds Register and Review*, Bristol, Pennsylvania 1987.
21. J. L. Koenig and P. T. K. Shih, J. Colloid Interface Sci. *36*:247 (1971).
22. E. P. Plueddemann, in *Interfaces in Polymer Matrix Composites, Composite Materials* (E. P. Plueddemann, ed.), Academic Press, New York, p. 6, 1974.
23. E. P. Plueddemann and G. L. Stark, Modern Plastics *55*:74 (1974).
24. R. K. Iler, *The Chemistry of the Silica*, John Wiley and Sons, New York, 1979.
25. K. K. Unger, N. Becker, P. Roumeliotis, J. Chromatography *125*:115 (1976).
26. G. Schomburg, A. Deege, J. Kohler, and U. Bien-Vogelsang, J. Chromatography *282*:27 (1983).
27. L. T. Zhuravlev, Reac. Kinet. Catal. Letters *50*:15 (1993).
28. M. Chaimberg, *Free Radical Graft Polymerization of Vinyl Pyrrolidone*, Ph.D. dissertation, University of California, Los Angeles, 1989.
29. R. E. Majors, and J. M. Hopper, J. of Chromatogr. Sci. *12*:767 (1974).
30. M. Chaimberg and Y. Cohen, J. Colloid Interface Sci. *134*:576 (1990).
31. H. Engelhard and D. Mathes, J. Chromatography *142*:311 (1977).

32. L. R. Snyder and J. J. Kirkland, *Introduction to Modern Liquid Chromatography*, John Wiley and Sons, New York, 1979.

33. T. Kremer and L. J. Boros, Wiley and Sons 1979.

34. C. J. Van Oss, Sep. Puri. Meth. *11*:131 (1982/83).

35. D. R. Absolom, Sep. and Purif. Meth. *10*:239 (1981).

36. H. Engelhard and D. Mathes, J. Chromatography *185*:305 (1979).

37. I. Krasilnikov and V. Borisova, J. Chromatography *446*:211 (1988).

38. P. Roumeliotis, and K. K. Unger, J. Chromatography *218*:535 (1981).

39. M. P. C. Silvestre, M. Hamon, and M. Yvon, J. Agric. Food Chem. *42*:2778 (1994).

40. D. Corradini, R. Filippetti, and C. Corradini, J. Liq. Chromatography *16*:3393 (1993).

41. S. Mori, J. Chromatography *471*:367 (1989).

42. N. T. Miller, B. Feibush, and B. L. Karger, J. Chromatography *316*:519 (1985).

43. D. E. Schmidt, R. W. Glese, D. Conron, and B. L. Karger, Anal. Chem. *52*:177 (1980).

44. L. R. Meyerson, and K. I. Abraham, Peptides *7*:481 (1986).

45. F. E. Regnier and K. M. Gooding, Anal. Biochem. *103*:1 (1980).

46. H. D. Crone, and R. M. Dawson, J. Chromatography *129*:91 (1976).

47. P. Roumenliotis, and K. K. Unger, J. Chromatography *185*:445 (1979).

48. F. E. Regnier, Meth. Enzymol. *91*:137 (1983).

49. R. Ovalle, Anal. Biochem. *229*:1 (1995).

50. E. Papirer, and V. T. Nguyen, Angew. Makromol. Chem. *28*:31 (1973).

51. J. Lecourtier, R. Audebert, and C. Quivoron, J. Liq. Chromatography *1*:479 (1978).

52. A. Kurganov, O. Kuzmenko, V. A., Davankov, B. Eray, K. K. Unger, and U. Trudinger, J. Chromatography *506*:391 (1990).

53. A. Kurganov, V. Davankov, T. Isajeva, K. Unger, and F. Eisenbeiss, J. Chromatography A *660*:97 (1994).

54. M. Petro, P. Gemeiner, and D. Berek, J. Chromatography A *665*:37 (1994).

55. K. M. R. Kallury, P. M. Macdonald, and M. Thompson, Langmuir *10*:492 (1994).

56. R. P. Castro, *Development of ceramic-supported polymeric membrane for filtration of oil emulsions*, Ph.D. dissertation, University of California, Los Angeles, 1997.

57. Y. Cohen, J. D. Jou, W. Yoshida, and L. J. Bei, in *Oxides Surfaces* (J. A. Wingrave, ed.), Marcel Dekker, New York, 1997.

58. M. Chaimberg and Y. Cohen, AIChE J. *40*:294 (1994).

59. Y. Cohen, P. Eisenberg, and M. Chaimberg, J. Colloid Interface Sci. *148*:579 (1992).

60. A. E. Ivanov, L. Zigis, M. F. Turchinskii, V. P. Kopev, P. D. Reshetov, V. P. Zubov, L. N. Kastrikina, and N. I. Lonskaya, Mol. Genet., Mikrobiol. Virusol. *11*:39 (1987).

61. K. Komiya, Y. Kato, Can. Pat. 1987, 1,293,083; Chem. Abstr. *117*:163,125 (1991).

62. P. G. de Gennes, Macromolecules *13*:1069 (1980).

63. C. Chen, Behavior of polyvinylpyrrolidone-grafted silica resin in size exclusion chromatography, M.S. thesis, University of California, Los Angeles, 1992.

64. R. Varoqui, and P. Dejardin, J. Chem. Phys. *66*:4395 (1977).

65. P. Gramain, and P. Myard, Macromolecules *14*:180 (1981).

66. Y. Cohen, and A. B. Metzner, AIChE J. *27*:705 (1981).

67. P. J. Flory, *Principles of Polymer Chemistry*, Cornell University Press, Ithaca, NY, 1953.

8

Polycation-Modified Siliceous Surfaces for Protein Separations

YINGFAN WANG and PAUL L. DUBIN Department of Chemistry, Indiana University—Purdue University, Indianapolis, Indiana

I.   Introduction   311

II. Experimental   314
     A. Materials   314
     B. Methods   315

III. Results and Discussion   317
     A. Selectivity   317
     B. Protein binding   317
     C. Polyelectrolyte adsorption layer   318
     D. Chromatography   319
     E. Quasi-elastic light scattering   320
     F. Capillary electrophoresis   322

IV. Conclusions   326

     References   326

I. INTRODUCTION

Chromatographic separation of proteins is based on their specific or nonspecific interactions with the stationary phase of chromatographic columns. Specific interactions have been utilized in affinity chromatography, in which protein separation results from the interaction of protein binding sites with immobilized ligands [1]. Various ligands including protein G, biotin, and glucose have thus been immobi-

311

lized to provide highly selective separation of their complementary proteins [1]. Nonspecific interactions based on electrostatic, hydrophobic, and dipole–dipole effects also provide the basis for several types of protein chromatography. For example, ion exchange chromatography depends on the competitive interaction of proteins and simple ions with the packing material, and proteins are separated primarily according to their net charge. In reverse phase liquid chromatography (RPLC) [2], the elution sequence of proteins depends on the overall hydrophobicity. Lastly, elution in size exclusion chromatography (SEC) [3] depends on protein molecular dimensions. These examples indicate the diversity of protein chromatography.

The stationary phases of chromatography columns are normally supported by a core of inorganic or polymeric material [2]. Generally, chromatographic supports must be mechanically stable, uniform with respect to size and pore distribution, and chemically inert to both solvent and solute molecules [2]. Although crosslinked polysaccharides, polyacrylamide, and other polymeric materials are still widely used for protein separations, the chromatographic applications of these materials tend to be constrained by their limited mechanical strength, low column efficiency, and swelling [4]. On the other hand, inorganic materials such as silica, alumina, and glass beads exhibit good mechanical stability. Silica and glass beads are particularly useful in chromatographic applications due to the ease of surface modification and commercial availability [2,4].

Silica and glass beads typically require surface modification for protein separation. This modification may involve a chemically stable bonded stationary phase that prevents undesired interactions between proteins and active silanol groups, or it may involve introduction of desired active sites that can interact with proteins. Siliceous surfaces have been modified with both small molecules and macromolecules. Petro and Berek [4], who summarized recent developments in polymer-modified silica gels, pointed out that polymer-modified silica stationary phases are more chemically stable than monomer-modified stationary phases. It can also be noted that a substantial proportion of silanol groups remain unaltered after surface modification with low molecular weight ligands. These uncovered silanol groups often show undesirable interaction with solutes. This problem can be reduced by macromolecule modification, which provides more complete surface coverage.

Macromolecular modification of the silica surface has also been proven to be an effective method to prevent protein adsorption in capillary electrophoresis, and many polymers have been employed for this purpose. Regnier et al. [5] modified fused silica capillaries with polyethyleneimine (PEI). The subsequent crosslinking of adsorbed PEI provided a stable polymer layer on silica surface. Their results show that PEI-modified capillary minimized adsorption of positively charged proteins onto the capillary surface. Neutral polymers have also been used in capillary electrophoresis to prevent protein adsorption. Kubo [6] used a polyacrylamide-coated capillary to reduce the adsorption of bovine serum albumin (BSA). Generally, the techniques used to modify silica surfaces with polymeric materials are the same.

Interest in polymer modification of silica surface has recently been stimulated by potential applications to capillary electrophoresis (CE) [5,7–9]. Previous study demonstrated that polymers can be immobilized either through chemical modification or through physisorption [4,10]. In the first class of immobilization, polymers are covalently bound to the silica surface by reaction with silanol groups or with other functional groups introduced onto the silica surface [4,7,8]. Although chemical modification appears to yield a relatively stable polymer stationary phase, the cost and inflexibility of this method impede further application, and recent attention has been focused on physisorption [8,9]. The second of these methods involves the adsorption of polymer onto a silica surface without chemical reaction [4,9]. In this method, both ionic and nonionic polymers can be directly deposited onto silica and may remain on the silica surface permanently. Physisorption may also be used to bind monomers or oligomers, followed by polymerization, with or without crosslinking, on the silica surface [5]. Since polyelectrolytes are readily physisorbed onto oppositely charged surfaces through electrostatic interactions, they have been used to modify the chromatographic properties of silica materials [9].

Two approaches toward the examination of polymer-modified silica can be visualized, either by examination of chromatographic properties or by elucidation of the structure of the adsorbed polymers. Chromatographic properties include column efficiency, resolution, and relative retention; by "structure of adsorbed polymers" we refer to phenomena at the molecular level, such as dimensions, configuration, and local environment of the immobilized molecules. These two approaches should complement each other in the interpretation of chromatographic results.

While extensive chromatography studies have been carried out on polymer-modified HPLC columns, few have focused on the effect of the structure of adsorbed polymers on chromatographic separations. A similar problem is evident in the field of CE. The absence of such studies may point to a need for a more complete understanding of how polymers, especially polyelectrolytes, adsorb on HPLC substrates and the concomitant relationship between the configuration of the adsorbed polymers and resultant chromatographic properties.

The configuration of adsorbed polyelectrolyes has been described in terms of trains, loops, and tails. Fleer et al. [11] provide a detailed review on polymer adsorption at interfaces. They pointed out that the configurations of adsorbed polymers are determined by polymer adsorption conditions, such as solvent properties, polymer molecular weight, polymer concentration, polydispersity, adsorption time, surface geometry, surface chemical properties, and substrate chemical heterogeneity. Because chromatographic properties of proteins are determined by the interactions between proteins and binding sites in the stationary phase, it would be expected that the nature and number of loops and tails of the adsorbed polymers in the stationary phase should strongly influence the subsequent retention of proteins.

Although numerous polymers can be immobilized on columns for protein separation, we are especially interested in immobilized polyelectrolytes. Siliceous surfaces can be readily coated by physisorption of polyelectrolyte. To some extent, these coated phases should retain the binding properties of the free polyelectrolyte. Previous studies indicate that poly(diallyldimethylammonium chloride) can selectively bind β-lactoglobulin in the presence of BSA, even though these two proteins have similar isoelectric points. It is expected that the separation of proteins on polyelectrolyte-immobilized stationary phases should resemble the separations of proteins by polyelectrolyte coacervation, because both phenomena involve the same electrostatic interactions. We have attempted to establish some empirical relations for the efficiency and selectivity of protein separation via polyelectrolyte coacervation [12]. If the structure of the polyelectrolyte is partially conserved after immobilization onto the silica surface, similar efficiency and selectivity is expected to be observed in chromatographic separations.

The primary question in the current study is whether the selectivity of a polyelectrolyte for protein separation is retained after it is immobilized onto a siliceous surface. A more general goal is to establish the relation between the polyelectrolyte adsorption process and subsequent chromatographic properties of the modified silica surface. To accomplish this, two types of polyelectrolyte-modified siliceous surfaces were examined. First, polyelectrolyte-modified chromatography grade glass, prepared under different adsorption conditions, was used to bind proteins. Second, polyelectrolytes were immobilized onto capillary inner surfaces and the modified capillaries were used for protein CE. Both size exclusion chromatography (SEC) and dynamic light scattering were employed to study the structure of the adsorbed polymer layer, and the results were used to interpret the protein binding experiments.

II. EXPERIMENTAL

A. Materials

Poly(diallyldimethylammonium chloride) (PDADMAC) (Merquat 100, Calgon Corp., Pittsburgh, PA) with a nominal MW of 2×10^5 and $M_w/M_n > 5$ was dialyzed (molecular weight cutoff = 12,000–14,000) and freeze-dried before use. PDADMAC L-120 ($M_n = 35,000$) was a gift from Dr. W. Jaeger (Fraunhofer-Institut, Teltow, Germany). Monodisperse silica particles, with Stokes' radius $(R_s) = 40 \pm 0.5$ nm (Shokubai Co., Osaka), were kindly supplied by Dr. Y. Morishima (Macromolecular Chemistry Dept., Osaka University). Bovine serum albumin was purchased from Boehringer Mannheim (Indianapolis, IN) (Lot 100062), and β-lactoglobulin was obtained from Sigma (Lot L2506). Proteins were used without further purification. Controlled pore glass (BioRan-CPG) (30–60 μm grain size, pore diameter 29.4 nm, and 136 m^2/g specific surface area) was obtained from Schott Gerate (Mainz, Germany). Pullulan standards (Shodex standard, P-82, Lot 50501) were obtained from Showa Denko K.K. (Tokyo, Japan).

B. Methods

1. Preparation of PDADMAC-CPG

Prior to polyelectrolyte adsorption, CPG was first cleaned using the procedure recommended by the manufacturer, namely, washing with pH 9.0, 1% sodium dodecyl sulfate (SDS) at room temperature for 2 h, then with deionized (DI) water until no foaming was observed, and drying at 89°C for more than 12 h.

PDADMAC was adsorbed onto cleaned CPG at different pH, ionic strength, and adsorption times. CPG was added slowly into the PDADMAC solution with stirring for 24 h to reach equilibrium. The solution was centrifuged and the solid was washed 5 times with DI water before drying at 89°C for more than 24 h. The pH, ionic strength, and adsorption time for polyelectrolyte adsorption are designated as pH_0, I_0, and t_0.

2. Protein Binding

All protein binding experiments were performed at $pH_1 = 9$ and $I_1 = 0.1$. About 100 mg of PDADMAC-CPG was added into 0.1 g/L protein solution with stirring for 1 h. The solution was centrifuged and the protein concentration in the supernatant was measured by UV at 278 nm. Figure 1 schematically illustrates the procedure for polyelectrolyte adsorption and subsequent protein binding.

Polyelectrolyte desorption was not monitored during protein binding experiments due to lack of applicable detection methods. However, we found that pullulan elution volumes from PDADMAC-CPG packed columns, measured with the chromatographic procedures described in the next section, did not change during 48 h of chromatographic elution. Since pullulan elution volumes were significantly different for PDADMAC-treated versus nontreated columns, this result strongly suggests the absence of polyelectrolyte leaching for PDADMAC-CPG. Similar results were also found in capillary electrophoresis experiments where the electrophoretic mobility of protein on a polyelectrolyte-coated capillary was stable during multiple runs [8,9].

3. Chromatography

The chromatographic system included a Minipump (Milton Roy, St. Petersburg, FL), a 100 μL sample loop, an R401 differential refractometer (Waters, Milford, MA), and a Kipp and Zonen recorder (Model BD 112, Delft, Holland). A stainless steel column (25 cm × 0.5 cm ID) was dry-packed with PDADMAC-CPG (pH_0

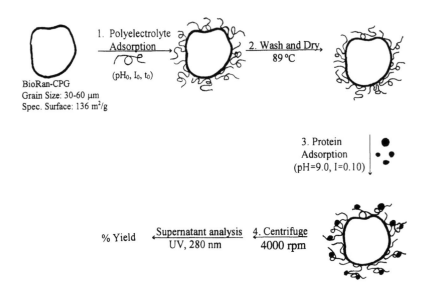

FIG. 1 Schematic process of batch adsorption experiment.

= 9, t_0 = 12 h, C_p = 20 g/L, and I_0 = 0, 0.10, 0.50, and 1.00, respectively).
The column efficiency measured with D_2O was 2.4 × 10³ plates/m. Boric acid–
NaOH buffer at selected pH and ionic strength was used as the mobile phase.
The flow rate was maintained between 0.5 and 0.6 mL/min.

4. Quasi-Elastic Light Scattering

PDADMAC was adsorbed on silica particles by slowly adding 0.002 wt% silica
into a solution of 0.2 wt% PDADMAC at preadjusted pH and ionic strength. The
solution was filtered through 0.45-μm Whatman filters before light scattering
measurement. The apparent stokes radius R_s^{app} of PDADMAC-silica was determined by quasi-elastic light scattering with a Brookhaven (Holtsville, NY) 72-
channel BI-2030 AT digital correlator, using a 100-mW argon ion laser.

5. Capillary Electrophoresis

PDADMAC adsorption was conducted at pH 8, ionic strength 0.5 by washing
polymer solution through preconditioned fused silica capillary (Restek,
Bellefonte, PA) under alternating high (5–10 min) and low (200–540 min) pressure for 16 h. After adsorption, the column was washed with pH 8, I = 0.05 Tris
buffer for 30 min before mobility measurement. The electroosmotic flow (EOF)
was measured with mesityl oxide (Aldrich Chemical Company, Inc.). The value
of electroosmotic mobility (μ_{EOF}) was determined experimentally using Eq. (1):

$$\mu = \frac{L_d\,L_t}{V t_{MO}} \qquad (1)$$

where L_d (cm) is the length of the capillary length from the inlet to the detector, t_{MO} (s) is the migration time of electrically neutral marker (mesityl oxide), V is the voltage applied across the capillary, and L_t is the total capillary length. After PDADMAC coating, EOF was reversed (in the opposite direction of uncoated capillary) and the polarity of electrodes was switched for detection.

III. RESULTS AND DISCUSSION

A. Selectivity

Figure 2 shows selective protein binding by PDADMAC-CPG. β-lactoglobulin is more favorably bound to PDADMAC-CPG than BSA, which is consistent with the coacervation selectivity of PDADMAC for these proteins [12]. Thus, the selective protein binding properties of PDADMAC in solution are retained after it is immobilized on the CPG surface.

B. Protein Binding

The degree of protein binding (yield) on polyelectrolyte-treated glass was studied as a function of pH_0. Figure 3 shows the yield upon batch mixing of BSA with polyelectrolyte-treated glass at $pH_1 = 9.0$ ($pH_0 = 0.9$ and 10.0) and $I = 0.1$. It is clear that protein binding on polyelectrolyte-treated glass is affected by the

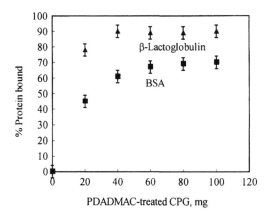

FIG. 2 % BSA and β-lactoglobulin bound on PDADMAC-CPG. PDADMAC adsorption conditions: $pH_0 = 9.0$, no added salt; $C_p = 1.0$ g/L. Protein binding conditions: $pH_1 = 9.0$, $I_1 = 0.10$, $C_{pr} = 0.10$ g/L.

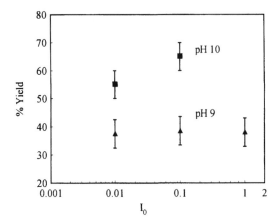

FIG. 3   Effect of preparation pH (pH_0) and ionic strength (I_0) on subsequent BSA bind-ing. PDADMAC adsorption conditions: $pH_0 = 9.0$; $C_p = 0.10$ g/L. BSA binding condi-tions: $pH_1 = 9.0$, $I_1 = 0.10$, $C_{pr} = 0.10$ g/L, 10 g/L PDADMAC-CPG.

initial conditions of polyelectrolyte adsorption. According to Fleer [13], the amount of adsorbed polyelectrolyte increases with the substrate surface charge density. The pH titration of CPG [14], shows that its charge density is doubled from pH 9.0 to pH 10.0. Consequently, more polyelectrolyte is adsorbed at pH 10, and the resultant PDADMAC-CPG binds more protein.

   Our preliminary results also show that polyelectrolyte adsorption conditions (I_0, t_0) also affected subsequent protein binding. The effect of I_0 and t_0 on protein binding has been explained by Wang and Dubin [15].

C. Polyelectrolyte Adsorption Layer

The amount of adsorbed polyelectrolyte is a key parameter in the interpretation of protein binding results. Commonly used techniques for the quantitation of adsorbed polymer include adsorption isotherms, ellipsometry, and reflectometry [16]. However, these techniques are not applicable to the PDADMAC-CPG sys-tem for several reasons. First, a relatively concentrated polymer solution was used in the current experiments to ensure adsorption leading to a loop-and-tail type of configuration. The amount of polymer adsorbed was therefore insignifi-cant compared to the bulk polymer concentration. Attempts to generate adsorp-tion isotherms through elemental analysis failed. Other techniques, such as ellip-sometry and reflectometry, are normally applied only to planar surfaces. Recently, the absorption of polymers tagged via fluorescence [17] or radio active labels [18] has been effectively used to monitor polymer adsorption. However, the lack of such labeled polymers impeded the use of these two powerful techniques. Therefore, we employed two semiquantitative methods to determine polymer adsorption layer thickness on PDADMAC-CPG and the results were used to interpret protein binding experiments.

D. Chromatography

SEC was employed to measure the pore radius of polyelectrolyte-treated and native CPG using the cylindrical pore model [19]. According to this model,

$$K_{SEC} = \left(1 - \frac{R}{r_p}\right)^2 \tag{2}$$

where K_{SEC} is the measured partition coefficient, R is the solute radius, and r_p is the pore radius. Although Eq. (2) rests on an unrealistically well-defined pore geometry, it has been empirically verified by several groups [20,21], who have observed that plots of $K_{SEC}^{1/2}$ versus R yield straight lines with slopes of $1/r_p$. With this method, the pore radius of CPG was obtained both before and after polyelectrolyte adsorption, and the difference between these two values (Δr_p) was used to estimate polyelectrolyte adsorption layer thickness (δ_H).

Figure 4 shows a plot of $K_{SEC}^{1/2}$ versus R for pullulan standards eluted at $I_1 = 0.001$ on both native and polyelectrolyte-treated CPG (pH$_0$ = 9.0, C_p = 20 g/L, no salt added), giving $r_p = 11.4 \pm 0.4$ and 8.9 ± 0.3 nm, for native and

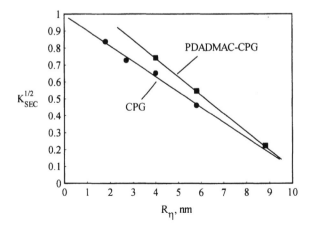

FIG. 4 Dependence of SEC chromatographic partition coefficient on pullulan Stokes radius, used to determine CPG effective pore size. SEC mobile phase: pH 9.0, 0.001 M Tris buffer. PDADMAC-CPG prepared at pH = 9.0, $I = 0$, C_p = 20 g/L.

polyelectrolyte-treated CPG, respectively. Therefore, $\delta_H = \Delta r_p = (11.4 \pm 0.4)$ $- (8.9 \pm 0.3) = 2.5 \pm 0.5$ nm. The error limits shown here are obtained from the standard deviation of the slope. Additional measurements at $I_1 = 0.01$ and 0.10 both gave $\delta_H = 1.7 \pm 1.5$ nm. The relatively large error in δ_H led us to seek other methods to quantitate δ_H.

E. Quasi-Elastic Light Scattering

Quasi-elastic light scattering (QELS) was chosen as an alternative method to examine the structure of the adsorbed polyelectrolyte. Since CPG particles were too large to be characterized by QELS, we used small silica particles (KE-E10) and low molecular weight polyelectrolyte (L-120) to model PDADMAC adsorption on CPG. The system was chosen to minimize sedimentation and bridging flocculation. It is expected that the influence of pH_0, I_0, and t_0 on polyelectrolyte adsorption thickness should be similar to that of CPG/PDADMAC.

The hydrodynamic radius of silica was measured both before and after polyelectrolyte adsorption, and the difference between the two ΔR_s^{app} was taken as the hydrodynamic polyelectrolyte adsorption layer thickness (HLT), with the results shown in Fig. 5. Numbers in parentheses are the conditions used for the QELS measurement, and the filled circles represent results obtained at pH values other than adsorption conditions (pH $\neq$ pH_0). HLT is seen to increase with pH_0 and to be virtually independent of the measurement pH. HLT for weakly adsorbed chains would be expected to change with measurement pH. Therefore, this initial

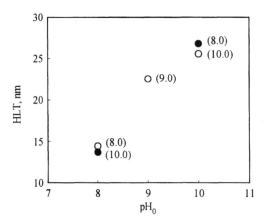

FIG. 5 Effect of adsorption pH on hydrodynamic layer thickness (HLT) on silica (by QELS). Silica concentration = 0.02 g/L, C_p = 2 g/L, I = 0.01 Numbers in parentheses indicate the measurement pH.

pH dependence suggests a relatively strong adsorption of PDADMAC on the silica surface.

The dependence of protein binding to polyelectrolyte-treated CPG on pH_0, I_0 and t_0 suggests that the structure of the adsorbed polyelectrolyte, namely, the arrangement of loops and tails, affects subsequent protein binding. Therefore, it should be possible to relate protein binding to PDADMAC-CPG to δ_H and HLT obtained from SEC and QELS, respectively.

An explanation for the effect of pH on polyelectrolyte adsorption layer thickness and subsequent protein binding is illustrated schematically in Fig. 6. Let us assume that n binding sites per polymer chain are required for polyelectrolyte retention on the CPG surface. At low pH, where the surface charge density of

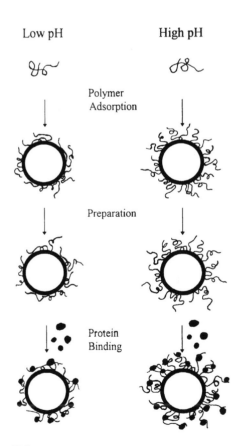

FIG. 6 Schematic depiction of effect of polyelectrolyte adsorption pH on subsequent BSA binding.

CPG is low, a relatively flat configuration is needed to produce the requisite number of ionic contacts, and fewer polyelectrolyte molecules are adsorbed. On the other hand, an increase in surface charge density at high pH means that a shorter length of adsorbed chain can provide energetic stabilization of the bound state, resulting in a subsequent increase in the adsorption amount. Furthermore, repulsion between adsorbed chains yields a thicker δ_H. QELS results show that HLT increases with pH_0 (Fig. 5); therefore, more sites are available for subsequent protein binding, as verified by the protein binding results in Fig. 3.

Although Fig. 6 rationalizes the effect of pH_0 on protein binding with QELS results, it should be noted that a significant discrepancy exists between the silica model system and polyelectrolyte-treated CPG. First, the discrepancy between δ_H (2.5 nm) and HLT (≥ 12 nm) suggests that the model system does not represent the true polyelectrolyte adsorption process on CPG. Silica is more highly curved than CPG, so that the configurational entropy of train-adsorbed polyelectrolyte on silica is more unfavorable than on CPG. If loops and especially tails are more predominant for PDADMAC on silica, HLT could be much higher than δ_H. Second, the surface charge density of silica is expected to be higher than that of CPG, which is likely to affect the polyelectrolyte adsorption layer thickness. Third, CPG has a porous surface whereas silica is relatively smooth. An oversimplified picture of polyelectrolyte adsorption within the pore used to obtain δ_H from SEC might contribute to the discrepancy as well. Finally, QELS is primarily sensitive to the tails of adsorbed polyelectrolytes, whereas both loops and tails contribute to the protein binding. All of these problems remain as challenges for future research.

F. Capillary Electrophoresis

Mixtures of BSA and β-lactoglobulin were separated on both coated and uncoated capillaries. As shown in Fig. 7, the PDADMAC-coated capillary better resolves the two proteins. Table 1 summarizes the data of protein titration charge (Z) and apparent protein mobility (μ). The calculated $|\mu|/|Z|$ are all close to 0.06 except the one for BSA on coated capillary (0.024). The comparable μ_{EOF} at two pHs suggests a similar surface charge density for both coated and uncoated capillaries. Therefore, the relatively low $|\mu|/|Z|$ for BSA on a coated capillary is more likely due to the interactions between BSA and the loops and tails of adsorbed polyelectrolytes, as suggested in the protein binding experiment with PDADMAC-coated CPG.

The interactions between proteins and polyelectrolyte-coated capillaries were studied by comparing apparent protein mobility (μ_{app}) with true mobility (μ). If $\mu_{app}/\mu = 1$, there is no protein–polyelectrolyte interaction. The smaller the mobility ratio μ_{app}/μ, the more strongly the proteins interact with the polyelectrolyte-modified surface. Electrophoretic mobility values for BSA are available in the

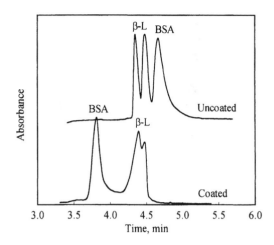

FIG. 7 Electrophoregram of BSA and β-lactoglobulin on both PDADMAC-coated (pH = 4.0, I = 0.05) and uncoated (pH = 8.0, I = 0.05) capillary.

literature, but most of the data were obtained by the moving boundary method, and substantial inconsistencies are noted. As discussed by Douglas et al. [23], these inconsistencies are likely due to the limitation of moving boundary method and the quality of the protein used. Protein mobility can be estimated using Henry's theory [24] of electrophoresis in the limits of low ζ potential ($\zeta e/kT < 1$):

$$\mu = \frac{Z_a e}{6\pi\eta r} \frac{H(\kappa r)}{(1 + \kappa r)} \tag{3}$$

where Z_a is the actual charge of protein, η is the viscosity of the medium, $H(\kappa r)$ is the Henry's function, κ is the Debye parameter, and r is the Stoke's radius of

TABLE 1 Protein Titration Charge [22] (Z) and Mobility (μ, $\times 10^4$ cm^2 V^{-1} s^{-1})[a]

β-Lactoglobulin				BSA			
pH 4.0[b]		pH 8.0[c]		pH 4.0[b]		pH 8.0[c]	
Z	μ_{app}	Z	μ	Z	μ_{app}	Z	μ
19	−1.02	−16	1.19	27	−0.64	−16	1.12

[a] EOF for all the capillaries are between 3.5 and 3.63.
[b] PDADMAC-coated capillary.
[c] Uncoated capillary.

the protein. Xia et al. [25] demonstrated as expected a low ζ of BSA at pH near PI. Therefore, we attempted to calculate the true mobility of BSA from reported literature values, with the ionic strength effect corrected using Eq. (3). Douglas and others [23] measured the electrophoretic mobility of BSA as a function of pH. Using their data and Eq. (3), the electrophoretic mobility of BSA at pH 4 and ionic strength 0.05 was calculated as 1.3×10^{-4} cm^2 V^{-1} s^{-1}. Hence, μ_{app}/μ of BSA was estimated as 0.49, which indicates an interaction between BSA and coated PDADMAC. However, we were unable to evaluate μ_{app}/μ for β-lacto-globulin due to insufficient literature data at pH 4.

An alternate method to estimate the electrophoretic mobility of protein from Eq. (3) has been proposed by Compton et al. and others [26,27]. First, they calculated the structural net protein charge, Z_c, via the Henderson-Hasselbalch equation:

$$Z_c = \sum_{n=1,4} \frac{P_n}{1 + 10^{pH-pK(P_n)}} + \sum_{n=1,5} \frac{N_n}{1 + 10^{pK(N_n)-pH}} \tag{4}$$

where P_n and N_n refer to the respective positively and negatively charged amino acids. Then, they used a semiempirical expression for protein mobility:

$$\mu = \frac{C_1 Z_a}{C_2 M^{1/3} + C_3 M^{2/3} I^{1/2}} \tag{5}$$

where C_1, C_2, C_3 are constants, Z_a is the electrophoretic protein charge, and M is the protein molecular weight. Compton et al. assumed that Z_a/Z_c is independent of pH at any given ionic strength. Therefore, Eq. (5) suggests that

$$\frac{\mu_1}{\mu_2} = \frac{Z_{C1}}{Z_{C2}} \tag{6}$$

Consequently, the electrophoretic mobility at one pH can be calculated from the mobility measured at another pH. As Compton et al. pointed out, there are many factors influencing the theoretical charge of proteins, such as specific ion adsorption, as well as structural and environmental influences on the pK$_a$ of individual amino acids. Therefore, the above equations can only be used for a first approximation of protein mobility. Douglas et al. [23] suggested that protein charge titration data, if available, should be used in Eq. 6 to calculate protein mobilities. In this study, electrophoretic mobilities of protein were measured on a coated column at low pH where no protein–polyelectrolyte interaction existed. The extrapolation of these measured mobilities as a function of protein titration charge gave μ at the desired conditions.

Electrophoretic mobilities of BSA at pH 3.2, 3.4, 3.6, 3.8, 4.0, and 4.2 are shown in Fig. 8 as a function of protein charge [22]. The extrapolated μ_{BSA} at pH 4 ($Z = 27$) is 1.35×10^{-4} cm^2 V^{-1} s^{-1}. Therefore, μ_{app}/μ_{BSA} at this pH and ionic strength is estimated as 0.47, which is consistent with the value of 0.49

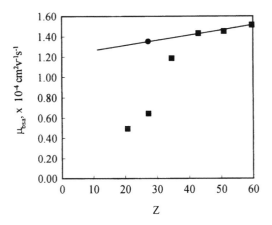

FIG. 8 Electrophoretic mobility of BSA (■) as a function of protein charge on PDADMAC-coated capillary. Line shows extrapolation from low pH to pH = 4.0 (●).

calculated from Eq. (3). Similar experiments were also carried out to obtain $\mu_{\beta\ell}$ as 1.20×10^{-4} cm^2 V^{-1} s^{-1} at pH 4.0, as shown in Fig. 9, leading to an estimate for $\mu_{app}/\mu_{\beta\ell}$ as 0.85. The smaller value μ_{app}/μ for BSA suggests a stronger interaction between BSA and the PDADMAC surface, accounting for the better separation achieved using coated capillaries.

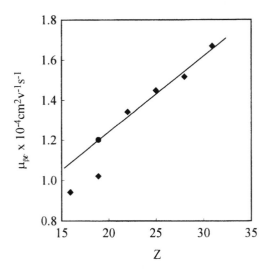

FIG. 9 Electrophoretic mobility of β-lactoglobulin (◆) as a function of protein charge on PDADMAC-coated capillary. Line shows extrapolation from low pH to pH = 4.0 (●).

Li et al. [28] measured the electrophoretic mobility of β-lactoglobulin as a function of pH on a Polybrene-coated capillary. They found that $\mu_{\beta\ell}$ was a linear function of pH between pH 3.3 and 4.3 (note that the protein titration charge is a linear function of pH in this pH range [29], and the extrapolation of $\mu_{\beta\ell}$ against pH yielded a correct isoelectric point. Contrary to our results, Li's data suggest that no protein–polyelectrolyte interactions exist on a Polybrene-coated capillary. This discrepancy may be attributed to the different polyelectrolyte adsorption conditions. In Li's method, the low polyelectrolyte concentration and low ionic strength of coating solution, approximately $C_p = 0.05\%$ and $I = 0.02$, would result in relatively flat polyelectrolyte adsorption. On the other hand, our high C_p and high ionic strength coating solution resulted in more loops and tails of polyelectrolyte in the adsorption layer, leading to better resolution in the coated capillary.

IV. CONCLUSIONS

We have demonstrated that PDADMAC can be immobilized on silica surfaces through physisorption. By properly controlling polyelectrolyte adsorption conditions it is possible to change the structure of immobilized polyelectrolyte and thus change the subsequent amount of protein binding. Most notably, the protein binding selectivity of PDADMAC is maintained after immobilization. This can be attributed to the interaction between proteins and the loops and tails of the immobilized polyelectrolytes. We have used modified silica surfaces to enhance protein separation in both chromatography and capillary electrophoresis. This study opens up the possibility for the design of novel stationary phases in both chromatography and capillary electrophoresis.

ACKNOWLEDGMENTS

The authors thank Dr. Y. Morishima for the silica samples, Dr. W. Jaeger for the PDADMAC samples, and the National Science Foundation for support via CHE 9505953.

REFERENCES

1. P. D. G. Dean, W. S. Johnson, and F. A. Middle, *Affinity Chromatography: A practical Approach*, IRL Press, Washington, D.C., 1985.
2. C. F. Poole, and S. K. Poole, *Chromatography Today*, Elsevier, New York, 1991.
3. P. L. Dubin, (ed.), *Aqueous Size-Exclusion Chromatography*, Elsevier, Amsterdam, 1988.
4. M. Petro, and D. Berek, Chromatographia *37*(9/10):549 (1993).
5. J. K. Towns, and F. E. Regnier, J. Chromatography *516*:69 (1990).

6. K. Kubo, Anal. Biochem. *241*:42 (1996).
7. A. M. Dougherty, N. Cooke, and P. Shieh, *Handbook of Capillary Electrophoresis* (J. P. Landers ed.), CRC Press Boca Raton, 1997, Chap. 24.
8. Q. Liu, F. Lin, and R. A. Hartwick, J. Chromatogr. Sci. *36*:126 (1997).
9. E. Cordova, J. Gao, and G. M. Whitesides, Anal. Chem. *69*:1370 (1997).
10. R. K. Iller, *The Chemistry of Silica*, John Wiley and Sons, New York, 1979, Chap. 5.
11. G. J. Fleer, M. A. Cohen Stuart, J. M. H. M. Scheutjens, T. Cosgrove, and B. Vincent, *Polymers at Interfaces*, Chapman and Hall, London, 1993, Chap. 7.
12. Y. Wang, J. Y. Gao, and P. L. Dubin, Biotechnol. Prog., *12*:356 (1996).
13. G. J. Fleer, Spec. Publ. Roy. Soc. Chem. *82*:34 (1991).
14. G. Shah, P. L. Dubin, J. I. Kaplan, G. R. Newkome, C. N. Moorefield, and G. R. Baker, J. Colloid Interface Sci. *183*:397 (1996).
15. Y. Wang and P. L. Dubin, J. Chromatogr. A, in press.
16. M. A. Cohen Stuart and G. J. Fleer, Annu. Rev. Mat. Sci. *26*:463 (1996).
17. M. A. Bos, and J. M. Kleijn, Biophys J. *68*:2566 (1995).
18. C. Huguenard, J. Widmaier, A. Elaissari, and E. Pefferkorn, Macromolecules *30*: 1434 (1997).
19. E. F. Casassa, and Y. Tagami, Macromolecule, *2*:14 (1969).
20. H. Waldmann-Meyer, J. Chromatography. *350*:1 (1985).
21. S. L. Edwards, and P. L. Dubin, J. Chromatography, *648*:3 (1993).
22. C. Tanford, S. A. Swanson, and W. S. Shore, J. Am. Chem. Soc. *77*:6414 (1955).
23. N. G. Douglas, A. A. Humffray, H. R. C. Pratt, and G. W. Stevens, Chem. Eng. Sci. *50*:743 (1995).
24. D. C. Henry, Proc. Roy. Soc. London A, *133*:106 (1931).
25. J. Xia, P. L. Dubin, Y. Kim, B. B. Muhoberac, and V. J. Klimkowski, J. Phys. Chem., *97*:4528 (1993).
26. S. Micinski, M. Gronvald, and B. J. Compton, Meth. Enzymol. *270*:342 (1996).
27. R. A. Mosher, P. Gebauer, and W. Thormann, J. Chromatography. A, *638*:155 (1993).
28. Y. J. Yao, K. S. Khoo, M. C. M. Chung, and S. F. Y. Li, J. Chromatography. A, *680*:431 (1994).
29. Y. Nozaki, L. G. Bunville, and C. Tanford, J. Am. Chem. Soc. *81*:5523 (1959).

9

Adsorption Processes in Surface Area Exclusion Chromatography

EMILE PEFFERKORN   Institut Charles Sadron, Strasbourg, France

ABDELHAMID ELAISSARI   Chemistry of Supports, CNRS–Biomérieux (UMR-103), Lyon, France

CLARISSE HUGUENARD   Laboratoire RMN et Chimie du Solide, Université Louis Pasteur, Strasbourg, France

I.	Introduction	330
II.	Kinetics of Polymer Adsorption	331
	A. Rate-controlling process in lattice adsorption of one species	332
	B. Rate-controlling process in off-lattice models for adsorption of one species	333
III.	Numerical Simulation of Chromatography	337
	A. Probability of adsorption in chromatography	338
	B. Size effects in adsorption chromatography	340
	C. Composition effects in adsorption chromatography	348
IV.	Experimental Systems	355
	A. The adsorbents	355
	B. The polymer–solvent systems	355
	C. Relevance of the adsorbing disc model to adsorbed polymer systems	361
V.	Surface Area Exclusion Chromatography	364
	A. Experimental device and methodology	364
	B. Chromatographic separation of unimers and micelles from micellar solutions of the diblock copolymer polystyrene-polyvinyl-2-pyridine	366
	C. Chromatographic separation of the protonated polyvinyl-4-pyridine	377

VI. Conclusion   384

       References   385

I. INTRODUCTION

Diffusion of polymer chains in random media controls a great number of phenomena such as exclusion chromatography, membrane separations, and ultrafiltration. The fundamental issue in these separation processes is the transport of a polymer chain from a region of larger volume to another region of smaller volume where the chain entropy is lower. The separation is thus induced by effects of entropic barriers on chain dynamics [1]. Adsorption phenomena of polymer chains at solid–liquid interfaces control the interfacial phenomena in this new approach of chromatographic separation mainly based on surface area availability effects. For one-component systems, different models have been developed to determine the relationships between the rate of surface coverage and the decrease in surface area availability [2]. When the size of the adsorbent becomes comparable with the remaining available surface areas, the effect of the excluded surface area should be taken into account [3,4]. This first quasi-ballistic model has been modified to take into account the random walk of the particle near the surface prior adsorption [5]. More recently, the initial random sequential adsorption (RSA) model has been modified to take into account the mobility of the adsorbent in the adsorbed state (in-plane mobility) [6]. Actually, this last feature has contributed in the past to determine two main systems. Statistical thermodynamic treatments of adsorption allow an exact calculation of the entropy of the sorbed phase to be made, provided that the model (equation of state, adsorption isotherm) is defined. The *localized* model (lattice theory, no nearest neighbors) considers that the sorbed molecule encounters potential wells deeper than kT and so becomes caged [7]. The *mobile* model considers the two-dimensional mobility in the adsorption plane to be governed by the mutual interactions existing between the adsorbent and the sorbed species. Obviously, the actual behavior of the adsorbent with regard to localized or mobile adsorption may depend on the correlation between adsorption energy and degree of surface coverage because in mobile systems a limited in-plane mobility may set in at high coverage. These different situations of chain adsorption are first presented from the theoretical and numerical points of view.

The adsorption models should be combined with chromatographic separation models. The plate model used to describe the solute partition between solid and liquid phases during flow of the suspension or solution through the permeable medium was introduced by Mayer and Tompkins [8] and later modified by Glueckauf [9,10]. Since in chromatographic columns of stacked beads the adsorbent is discontinuous whereas the solution flow is continuous, in the initial model of Mayer and Tompkins thermodynamic equilibrium was assumed to be attained at each plate between the solute and the sorbent phases. Glueckauf considered the lack of equilibrium arising from fast rates of mass transfer through the column. On the basis of a Langmuir-type process in which both adsorption and desorption contribute to establish thermodynamic equilibrium, the concentration ratio of the adsorbed $N_{s,i}$, and solute species N_i is found to be related to the adsorption energy. In irreversible adsorption and in many reversible adsorption processes, as the solute remains a long time in the adsorbed state, the progress of surface filling further modifies the thermodynamic partition. Numerical studies relative to these aspects are presented as a second part.

The results of numerical studies of polymer separation resulting from interfacial phenomena are confronted with experiments of polymer adsorption and chromatographic separation of polymers. The aim of the work was to clearly establish the correlation between some interfacial phenomena related to adsorption and the quality of the resulting separation.

II. KINETICS OF POLYMER ADSORPTION

The rate of polymer adsorption has to be investigated under conditions where the solution to surface transport does not constitute the limiting factor. Actually, when the polymer is injected into the chromatographic column, it is expected that the molecule will immediately interact with the surface (the stationary phase) due to the very small average thickness of the elutant (the mobile phase). Therefore, the diffusional higher mobility of polymers of small molecular weight is not expected to favor the faster adsorption of this class of polymer with respect to that of polymer of higher molecular weight. Moreover, since the chromatographic separation usually constitutes a fast process, it is important to only consider the initial step of fast surface filling and not the supplementary period of establishment of the adsorption equilibrium. With these assumptions, the elementary equation of the adsorption kinetics is expressed as follows:

$$\frac{dN_s}{dt} = N(t)K(N_s) \tag{1}$$

where $N(t)$ represents the average polymer concentration during the period dt, dN_s is the increase in surface coverage during dt, and $K(N_s)$ is the kinetic coefficient of the adsorption. The validity of the kinetic model applied to chromatographic conditions requires $K(N_s)$ to be a function of N_s and independent of $N(t)$.

A. Rate-Controlling Process in Lattice Adsorption of One Species

The expected decay of $K(N_s)$ as a function of N_s results from the difficulty encountered by the polymer (1) to systematically seek an unoccupied surface area and (2) from the necessity to dispose of the expected number of sets of contiguous sites to bind the macromolecule to the surface.

Using the relation of Arrhenius to express the kinetic coefficient $K(N_s)$ as a function of the adsorption energy $\Delta\mu(\text{ads})$, one obtains the following equation:

$$K(N_s) = K_0 \exp(-\Delta\mu(\text{ads})/kT) \tag{2}$$

with

$$\Delta\mu(\text{ads}) = \mu(\text{surf}) - \mu(\text{sol}) \tag{3}$$

The adsorption enthalpy $h(\text{surf}) - h(\text{sol})$ results from the exchange in the interface of adsorbed molecules of solvent with polymer segments and from the exchange in the solution of polymer–solvent interactions for solvent–solvent interactions. Since the same scheme is encountered throughout the surface covering, the enthalpic balance is assumed to be constant,

The entropy of the polymer in the solution $s(\text{sol})$ does not vary greatly in very dilute solution and may also be taken as being constant.

To a first approximation, the term $\Delta\mu(\text{ads})$ in Eq. (3) is determined by the variation of $Ts(\text{surf})$, which is evaluated as follows. The entropy S of the polymeric interface can be estimated from the usual law $S = k \ln W$, where W represents the number of ways of putting the N_s polymer molecules at the N_0 sites of the plane lattice [11,12]:

$$W = \frac{N_0!}{(N_0 - xN_s)!(N_s)!} \left(\frac{z-1}{N_0}\right)^{N_s(x-1)} \tag{4}$$

where z is the lattice coordination number and x the mean degree of polymerization. The entropy $s(\text{surf})$ of the isolated macromolecule in the situation $[N_0, N_s]$ is given by Eq. (5):

$$s = dS/dN_s \tag{5}$$

With the usual assumption of Flory, Eq. (5) is expressed as follows:

$$K(N_s) = K_0 \left(\frac{z-1}{z}\right)\frac{1}{N_s}\{v_{N_s+1}\} \tag{6}$$

where $\{v_{N_s+1}\}$ represents the expected number of sets of x contiguous sites available to the $(N_s + 1)$th molecule, if N_s molecules are previously adsorbed at random. Equation (6) can be developed into:

$$K(N_s) = K_0 x \frac{(N_0/x) - N_s}{N_s} \tag{7}$$

In this situation, the adsorption mechanism can be easily described if the adsorbent surface is assumed to contain N_0/x fictive adsorption sites and if there is no supplementary constraint to the planar distribution of the adsorbed segments. To approximate more exactly the term $\{v_{N_s+1}\}$ in Eq. (6), Flory proposed a better evaluation of the function determining the expectancy that a given cell adjacent to a previously vacant one is occupied:

$$\{v_{N_s+1}\} = (N_0 - xN_s)z(z - 1)^{x-2} (1 - f_i)^{x-1} \tag{8}$$

where

$$1 - f_i = \frac{N_0 - xN_s}{N_0 - 2(x - 1)N_s/z)} \tag{9}$$

Thus, Eq. (7) is modified into:

$$K(N_s) \propto K_0 \, x \frac{(N_0/x) - N_s}{N_s} \times \left[\frac{N_0 - xN_s}{N_0 - 2(x - 1)N_s/z} \right]^{x-1} \tag{10}$$

Equation (10) indicates that $K(N_s)$ strongly decreases with surface coverage when all of the x contiguous sites should be available for adsorption [13]. This situation has been revisited by Poland who takes into account the variety of polymer interfacial conformation [14]. Figure 1 shows the values of $K(N_s)$ as a function of (xN_s/N_0) derived from Eqs. (7) and (10) for adsorption of 1175 chains of 10 segments adsorbing on a square lattice of side equal to 114. Figure 2 shows the situation for adsorption of 588 chains of 20 segments. It is shown that the kinetic coefficient $K(N_s)$ strongly decreases with the lattice coordination number z and the polymerization index x for a given surface coverage.

B. Rate-Controlling Process in Off-Lattice Models for Adsorption of One Species

1. Random Sequential Adsorption Model (Localized Adsorption)

The random sequential adsorption (RSA) model without interaction is expected to generate a kinetic behavior for adsorption that is similar to that resulting from the lattice model. Actually, Eq. (4) assumes random distribution of the adsorbed species on targets, whereas in RSA the adsorption sites are not determined. The

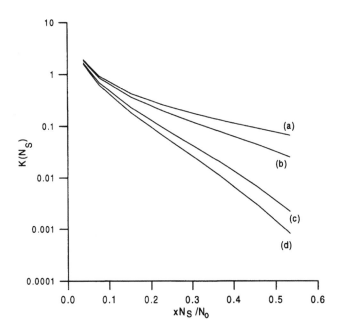

FIG. 1 Representation of the kinetic coefficient $K(N_s)$ as a function of the relative coverage xN_s/N_0 calculated using the model of Flory for 1176 (N_s) polymer chains of polymerization index x equal to 10, adsorbing onto a lattice of 13,100 adsorbing sites (N_0). (a) is obtained using Eq. (7); (b), (c), and (d) are obtained using Eq. (10) and the following values of the lattice coordination number 2, 3, and 4, respectively.

RSA model is usually employed to determine the deposition rate and position correlation of spheres or discs onto a plane surface and the main information is relative to the surface blocking effects of previously adsorbed species [15]. The model has been applied to colloidal particles [16] and was found to be valid for proteins [17]. This simple model may be applied to diluted systems of macromolecules strongly interacting with the adsorbent, so that irreversible localized adsorption is established after deposition. In the present RSA process, hard discs are placed randomly and sequentially on a surface without overlap. When the randomly selected position is free the disc is deposited on that position and when the selected position is already occupied the attempt is abandoned.

2. Random Sequential Adsorption Plus Surface Diffusion Model (Mobile Adsorption)

The RSA algorithm is modified to take into account that the adsorbed disc is continuously moving in the adsorption plane. For adsorption of one species, the

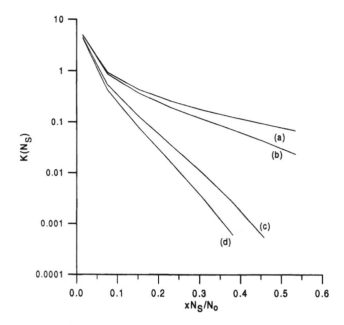

FIG. 2 Representation of the kinetic coefficient $K(N_s)$ as a function of the relative coverage xN_s/N_0 calculated using the model of Flory for 588 (N_s) polymer chains of polymerization index x equal to 20, adsorbing onto a lattice of 13,100 adsorbing sites (N_0). (a) is obtained using Eq. (7); (b), (c), and (d) are obtained using Eq. (10) and the following values of the lattice coordination number 2, 3, and 4, respectively.

model is similar to the model of random walk near the surface, for which interactions at large distances between a disc and the plane may modify the trajectory of the adsorbing disc [5]. In this situation, when the position corresponding to the randomly selected coordinates is already occupied, the adsorbing disc is able to move near the interface to search for a free area in the immediate vicinity of the first selected position. It is further assumed that a disc wandering around a selected position operates over the time required in the pure RSA process in agreement with the fact that attractive or repulsive forces modify the trajectory in the second situation. Obviously, this is equivalent to our major assumption that an adsorbed disc continuously moves in the adsorption plane while the solution to surface transfer is instantaneous.

The unidirectional diffusion path that allows a disc to escape from an already occupied area is given by:

$$(2r)^2 = 2Dt = 2(v/6)L^2 \tag{11}$$

where L expressed by r/λ is the length of the elementary jump. The probability for a random walker in the plane, which encounters an occupied area to reach a *closely* located and nonoccupied area, is expressed by Eq. (12) between the frequency and the length of the jumps:

$$96\lambda^2 > v > 24\lambda^2 \tag{12}$$

Clearly, the disc does not systematically explore the total allowed diffusion path, but the first free area will be occupied. Equation (12) implies that there is no possibility for the disc to investigate at large distances from the first impact point, in order to preserve a total diffusion path of similar length. If no free area is available for adsorption at the end of the random walk, the adsorption attempt does not succeed definitely.

The influence of the jump length on the fate of the adsorbing disc during this walk around the coordinates of an already adsorbed disc has been investigated by visual observation that indicated that the probability of finding a closely located free area increases when the jump length L decreases. This leads to the best positioning when the still-free area located between a set of adsorbed discs is close or equal to the disc area. Nevertheless, the running of this subroutine is rapidly time consuming, so that a value of λ equal to 10 was found to be a good compromise.

3. Kinetics of the Adsorption Process

The numerical simulation is employed to determine the validity of Eq. (1), which implies that $K(N_s)$ is independent of $N(t)$. Equation (1) may be modified in Eq. (13) to hold for simulation:

$$\frac{\Delta N_{cs}}{\Delta t_c} = K(N_{cs}) \times N_c \tag{13}$$

where ΔN_{cs} represents the number of discs adsorbed during the time Δt_c, when N_c discs are in front of the adsorbing plane.

The number of discs that are transferred from the solution to the adsorbent, that of discs that are in front of the plane and that being adsorbed per time unit, permit determination of the computer time and adsorption rate, both being scaling parameters. The simulation is carried out as follows.

At time $t = 0$, ΔN_c discs are placed in front of the plane. Each disc carried out one attempt to adsorb before the time is incremented by Δt_c. Simulation indicates that after the first period Δt_c, ΔN_{cs} discs succeed in adsorption whereas N_c discs fail. During a second period Δt_c, the constant number of attempts ΔN_c is increased by the number N_c, which failed in the first period. Successively, the nonadsorbed discs and the discs that are regularly supplied again attempt to adsorb during each period Δt_c. The resulting adsorption is calculated by summing the successive values of ΔN_{cs}. This procedure is implemented up to full-surface coverage of the plane with discs.

Three situations are considered:

1. $\Delta N_c \gg N_c$. During each period Δt_c, ΔN_c tentatives are made to place a disc and generally succeed. This corresponds to the early stage of the adsorption. Taking $\Delta t_c = 1$ implies that the time is increased by one unit when all of the discs tried to adsorb on the plane once every period.

2. $\Delta N_c \ll N_c$. Each period Δt_c allows N_c tentatives and adsorption almost never happens. This corresponds to the final stage of adsorption, where the rate of supplying the discs does not play a role. The time is increased by one unit when, as previously, all the discs have tried once to adsorb.

3. $\Delta N_c \cong N_c$. The same definition is adopted.

As one may expect, different values of ΔN_c lead to different patterns when the surface coverage is represented as a function of time. The variation with the surface coverage is detailed below for the following simulation. A very large number of discs of radius 1 are sequentially supplied to a plane of surface 60×120 until 1230 discs are deposited using the localized and mobile adsorption algorithm. This number corresponds to the maximal coverage of 0.537, close to the jamming limit of 0.547 [4]. Different rates of disc supplies were implemented. It is important to note that the kinetic coefficient could only be determined for coverage of about 800–900 discs. This upper limit resulted from the fact that a one-unit time increment corresponds to 100,000 attempts at the end of the simulation.

Figures 3 and 4 portray the kinetic coefficient $K(N_{cs})$ of localized and mobile adsorption processes, respectively, as a function of the fractional surface coverage θ_{cs} for values of ΔN_{cs} equal to 5, 20, 50, and 100. The unique curve obtained over a large range of surface coverage displays the characteristic variation of the kinetic coefficient in the two processes. In the mobile adsorption simulation, the first 500 discs are immediately adsorbed. Therefore, infinite values of the kinetic coefficient are calculated using Eq. (13) [18].

III. NUMERICAL SIMULATION OF CHROMATOGRAPHY

A simple model is used to derive the solute partition between solid and liquid phases during flow of the polymer solution through the permeable medium. Contrarily to the assumption of the Langmuir-type process, the polymer partition is assumed to be only determined by the success or failure of adsorption attempts, taking into account that the interfacial adsorption energy is usually sufficiently high to allow the solute macromolecules to be extracted from the bulk. Polymer

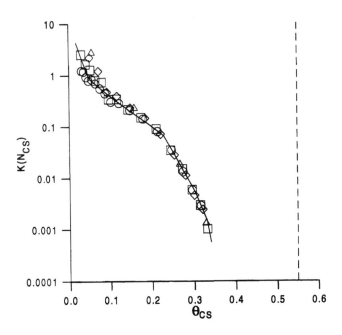

FIG. 3 Representation of the kinetic coefficient $K(N_{cs})$ as a function of the fractional surface coverage θ_{cs} in adsorption processes developing with features of the localized adsorption model for different rates of disc supply: ($\bigcirc$) 5; ($\square$) 20; ($\diamondsuit$) 50; and ($\triangle$) 100.

adsorption may develop quite irreversibly so that desorption of adsorbed neutral macromolecules is not expected to be rapidly induced even by dilution of the system [19]. Adsorbed macromolecules may slowly exchange with solute molecules of similar molecular weight (without changing the total amount of adsorption), this interfacial exchange being however blocked when the pure solvent faces the adsorbed polymer layer [20,21]. As far as very dilute polymer solutions are usually eluted in chromatography, interfacial exchange between adsorbed and solute discs is not taken into account in this numerical solution. Thus, deviation from the theoretical chromatograms is attributed to rapid interfacial exchange processes.

A. Probability of Adsorption in Chromatography

The following simulation is designed to determine the probability of adsorption of discs on a unique plane. Contrary to the situation described in Sec. II.3, discs do not attempt to adsorb twice on the same plane and discs failing to adsorb therefore disappear from the system. Our model considers the iterative adsorption

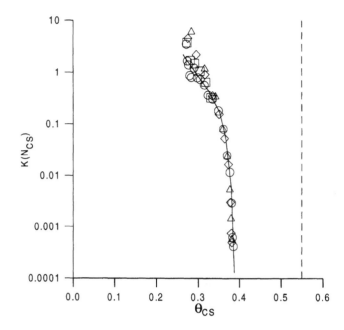

FIG. 4 Representation of the kinetic coefficient $K(N_{cs})$ as a function of the fractional surface coverage θ_{cs} in adsorption processes developing with features of the mobile adsorption model for different rates of disc supply: (○) 5; (□) 20; (◇) 50; and (△) 100.

from a mixture of 250 discs of radii r_1 and 250 of radius r_2 on a unique plane. The time is incremented by one unit when these 500 randomly selected discs have tried to adsorb on a plane of area 250 × 250, the adsorption probability $P(\theta_{cs})$ being defined by:

$$P(\theta_{cs}) = \frac{\Delta N_{cs}}{250} \qquad (14)$$

where θ_{cs} represents the portion of the covered surface area and ΔN_{cs} the number of discs of given radius adsorbed per unit time.

Results are given in Fig. 5 for the iterative adsorption from three mixtures of discs of radii 1 and 2, 1 and 3, and 1 and 7, in all three cases randomly selected from 500 discs. The three upper curves correspond to adsorption of discs of radius 1 and the common lower curves to adsorption of the larger discs, with the total degree of surface coverage θ_{cs} being represented on the abscissa.

Figures 1 to 4 clearly show that the progress of the surface coverage leads to a strong retardation of the random deposition of chains and discs on an adsorbing surface. Figure 5 shows that the retardation effect is very selective and, therefore,

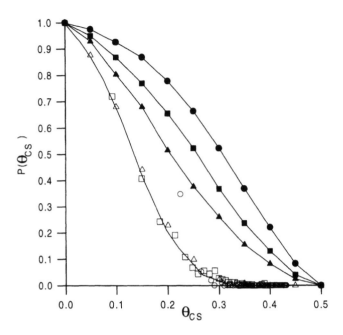

FIG. 5 Adsorption probability $P(\theta_{cs})$ as a function of the fractional surface coverage θ_{cs} for disc adsorption from mixtures of discs of radii 1 and 2 ($\triangle$), 1 and 3 ($\square$), 1 and 7 ($\bigcirc$). Open symbols correspond to the large discs and filled symbols to the small disc of radius 1.

adsorption chromatography is expected to induce a good separation of constituents of different sizes.

B. Size Effects in Adsorption Chromatography

In the numerical simulation study, the chromatographic column was schematized by 100 successive adsorbing square planes of individual area 62,500. At each run, small and large discs are randomly selected from a total number of 500. Since the discs attempt to adsorb only once on the same plate, the elutant composition is fixed only for adsorption on the first plate and the composition at the second plate results from failures in adsorption on the first. This situation is reproduced at all successive plates. At each run, following injection of 500 discs into the column, the model enabled determination for each plate of the total number $N_{cs}(r_1 + r_2)$ of adsorbed discs, the number $N_{cs}(r_1)$ of adsorbed discs of radius r_1, the number $N_{cs}(r_2)$ of adsorbed discs of radius r_2 and the relative area θ_{cs} occupied by the adsorbed discs. The number of runs was selected to obtain a constant surface coverage independent of the composition of the mixture and the disc radii. In all situations the radius r_1 of the small disc was chosen equal to 1. The radius r_2 of the large disc was chosen to be 2, 3, or 7 to investigate the influence of this parameter on the disc packing profile [6].

1. Localized Adsorption

As previously indicated, localized adsorption is simulated using the RSA algorithm. Discs of different radii are randomly chosen and deposited onto the plane at randomly selected coordinates. Adsorption occurs at this position if the area enclosing the coordinates is free of discs and if the surrounding free area is large enough to allow the new disc to be adsorbed; in contrast, adsorption does not occur if the selected position is already occupied.

Mixtures of 250 discs of radius 1 and 250 discs of radius 2 (500 injections), 3 (250 injections), or 7 (50 injections) were introduced into the column. Figure 6 represents the relative coverage as a function of plate number i. Figures 7 and 8 present the adsorption of discs of small r_1 and large r_2 radii, respectively. As previously noted, the disc supply is stopped when the same total covered surface area inside the column is obtained.

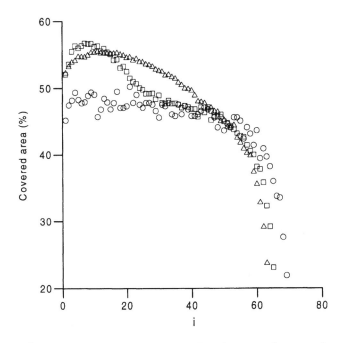

FIG. 6 Localized adsorption process. Relative covered area as a function of plate number i for different values of the radius of the large disc: ($\triangle$) 2; ($\square$) 3; and 7 ($\bigcirc$). The relative concentration of the two discs being supplied at plate 1 is 0.5.

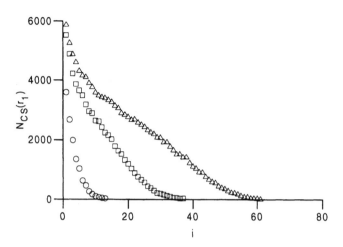

FIG. 7 Localized adsorption process. Number of discs of radius 1 adsorbed as a function of the plate number i for different values of the radius of the large discs: ($\triangle$) 2; ($\square$) 3; and 7 ($\bigcirc$). The relative concentration of the discs of radius 1 and the large discs being supplied at plate 1 is 0.5.

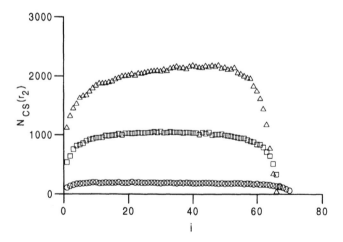

FIG. 8 Localized adsorption process. Number of large discs adsorbed as a function of the plate number i for different values of their radius: ($\triangle$) 2; ($\square$) 3; and 7 ($\bigcirc$). The relative concentration of the large discs and discs of radius 1 being supplied at plate 1 is 0.5.

For the mixture of discs of radii 1 and 7, the relative coverage (Fig. 6) is similar to the situation where only one species of disc is injected and the distribution of small discs through the column (Fig. 7) indicates that the area covered by the excess of small discs exactly compensates the lack of large discs on the 10 initial plates (Fig. 8). Dispersion of the points representing the adsorption per plate results from the low number of injections (50 injections). However, extrapolation to maximum coverage gives a value close to 0.49, lower than the jamming limit (0.547), which indicates that the surface area is far from being completely covered when the number of runs is limited.

For the mixture of discs of radii 1 and 3 the chromatogram presents an initial peak around $i = 10$ where the maximum area covered is close to 57%. Inspection of Fig. 7 leads to the conclusion that the peak is due to overadsorption of small discs while the pseudoplateau between plates 35 and 45 (Fig. 6) is only due to adsorption of large discs.

For the mixture of discs of radii 1 and 2, the chromatogram does not show a well-defined peak and the separation is of poor quality as can be concluded from the distribution of small and large discs, as reported in Figs. 7 and 8.

2. Mobile Adsorption

(a)   Parameters of the In-Plane Diffusion.   In mobile adsorption (MA) an adsorbing disc may be attracted by free portions of the surface area or repelled by already covered areas. The RSA algorithm was therefore adapted to take into account these interactions between the disc and the plane, which may modify the trajectory of the disc prior to adsorption and thus increase the adsorption probability. When a randomly selected position is already occupied, the disc will search for a free area in the close vicinity. However, as the close vicinity indicates that these interactions cannot greatly modify the position of adsorption relative to the initial randomly fixed coordinates, the maximum length of the walk is an important parameter. Adsorption occurs if the disc finds a free area, but if no free area is encountered during the random walk the adsorption attempt fails.

In order to determine the characteristics of the random walk, we consider the two following situations: a disc r_1 encounters an area occupied by a disc r_2, or a disc r_2 encounters an area occupied by a disc r_1. The unidirectional diffusion path allowing both discs to escape from the occupied position is given by:

$$(r_1 + r_2)^2 = 2(v/6)L^2 \qquad (15)$$

where L is the length of an elementary jump and v the number of jumps. For a given value of v the term L, which is expressed in Eq. (16) as a function only of the disc radius, expresses the probability of finding an unoccupied area. No allowance is made for the fact that discs of small radius may explore a greater area than those of large radius due to their higher relative Brownian mobility in solution. We point out that only interfacial characteristics are taken into account.

Therefore, an adsorbed disc of large radius r_2 is considered to more effectively screen the attractive interaction between an adsorbing disc r_1 and the plane than an adsorbed disc r_1 for an adsorbing disc r_2. This manner of defining the size-dependent probability of adsorption is introduced by correlating the jump length L with the radius r_a of the disc attempting to adsorb:

$$L = \frac{r_a f}{\lambda} \tag{16}$$

The relative "mobility" L of the species attempting to adsorb, which is scaled by the exponent f in Eq. (16), controls the in-plane mobility of the disc being adsorbed. Positive values of f are essential in this model of in-plane diffusion. The corresponding physically unrealistic assumption of an increase of the diffusive mobility of the disc with larger radius r_a is equivalent to the more realistic assumption of a decrease in the in-plane mobility of an adsorbed disc of increasing radius. This implies that a large adsorbed disc moves too slowly in the adsorption plane to facilitate by its own displacement the adsorption of small discs. Similarly, an increased mobility of the large discs in solution implies an increased mobility of the small discs in the adsorbed state. A small adsorbed disc is able to move rapidly in the plane and hence more readily allows successful adsorption of large discs. The exponent f in Eq. (16) thus scales the energy of the solute–adsorbent interactions: $f = 1$ corresponds to an in-plane mobility of the adsorbed discs inversely proportional to the square root of their area and $f = 2$ indicates that this characteristic is proportional to the area. From a physical point of view, $f = 1$ may describe the interactions of loose solutes like synthetic macromolecules, whereas $f = 2$ may be applicable to denser species like proteins.

The third parameter is the number of jumps, ν. In order to set ν, we now consider in two-dimensional space the random walk enabling an adsorbing disc to find a free area when initially it meets an already adsorbed disc. Three situations may be envisaged:

1. When a disc r_1 attempts to adsorb in an area r_2, Eqs. (15) and (16) give the following Eq. (17):

$$\nu_1 = 6\lambda^2 \left(\frac{r_1 + r_2}{r_1 f} \right)^2 \tag{17}$$

2. When a disc r_2 attempts to adsorb in an area r_1, one obtains:

$$\nu_2 = 6\lambda^2 \left(\frac{r_1 + r_2}{r_2 f} \right)^2 . \tag{18}$$

3. When the radii r_a of the adsorbing and adsorbed discs are equal, the relation is:

$$v_a = 24\lambda^2 \left[r_a^{2(1-f)} \right] \tag{19}$$

If we consider the adsorption of two types of disc, one of radius $r_1 = 1$ and a second of radius $r_2 = 2, 3,$ or 7 respectively, the calculated values of v_2 and v_a as a function f with $\lambda = 10$ lead to the curves shown in Fig. 9. Maximum efficiency of the in-plane diffusion to facilitate adsorption implies use of the corresponding values of v_1, v_2, and v_a, whereas smaller values of the jump frequency may lead to unsuccessful in-plane random walks. The jump frequency thus determines the adsorption probability according to Eqs. (5)–(7). Moreover, as assumed in the model, v also controls the mobility of the adsorbed species in the mobile adsorption process. The physically interesting limits of variation of v are given by Eq. (20):

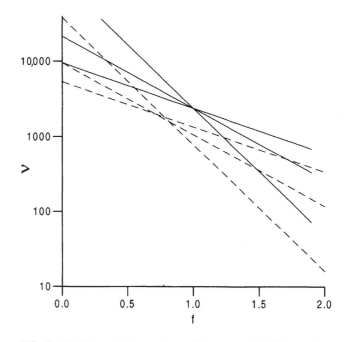

FIG. 9 Variation of the number v of jumps required for maximum efficiency of the in-plane diffusion as a function of the parameter f [Eq. (16)]. The solid and dashed lines represent v_a [Eq. (19)] and v_2 [Eq. (18)], respectively, and in the two cases, the negative slope of the variation increases with the size of the discs ($r_2 = 2, 3,$ or 7) attempting to adsorb on positions already occupied by discs of radius 1.

$$0 < \frac{v}{6\lambda^2} < (r_1 + r_2)^2 \tag{20}$$

Whereas the upper limit allows all particles to escape from all occupied areas, the lower limit corresponds to the LA process. Figure 10 represents the different domains of variation of $v/6\lambda^2$ investigated in the present study, under conditions whereby one disc of radius r_1 is already adsorbed and d is the distance of the impact point of the disc of radius r_2 from the center of the adsorbed disc 1.

Considering for clarity the case $f = 1$, adsorption statistically succeeds in the following situations:

$d > r_1 + r_2$: even the LA process leads to adsorption ($v = 0$).

$r_1 < d < r_1 + r_2$: owing to initial overlap of the discs 1 and 2, adsorption requires the MA process and succeeds for values of v between 0 and $6\lambda^2$.

$d < r_1$: the impact point of the adsorbing disc 2 is located in the area of the adsorbed disc 1 and adsorption occurs for values of $v/6\lambda^2$ between 1 and $(r_1 + r_2)^2$.

The set of parameters r_a, f, v, and λ entirely characterizes the mobile adsorption process which is thus based on simple geometrical considerations. Therefore, it was interesting to determine the influence of the parameter $v/6\lambda^2$, which counterbalances the opposing effects of surface area exclusion and in-plane mobility, on the concentrations of the two types of disc 1 and 2 and their relative adsorption on the successive plates of the chromatographic column.

(b) Influence of the Number of Jumps. A mixture of 250 discs of radius 1 and 250 discs of radius 3 was injected 250 times into the column. A similar simulation was done with the injection of 50 discs of radii 7. The parameter of interest was the number of jumps v in relation to the distribution of small and large discs along the column and the corresponding covered areas on the initial plates. In Figs. 11 and 12 the number of small (top schema) and large discs (bottom schema) is reported as a function of plate number i for the successive v values of 0 (LA process), 50, 500 (for Figs. 11 and 12), and 5000 (for Fig.

FIG. 10 Different domains of the efficient number of jumps according to the initial separation between the coordinates of an adsorbed disc 1 and the randomly selected coordinates of a disc 2 at the first adsorption attempt.

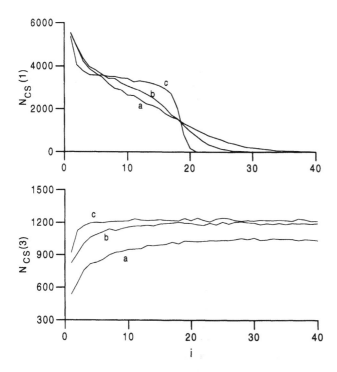

FIG. 11   Number of discs of radius 1 (top) and radius 3 (bottom) adsorbed as a function of plate number i for different values of the number of jumps ν: (a) 0; (b) 50; and (c) 500. The relative concentration of discs of radii 1 and 3 supplied at plate 1 is 0.5.

12), which correspond to an MA process characterized by in-plane diffusion paths of increasing length. One observes that the increase in disc mobility in the MA model allows a larger number of small discs to be adsorbed on the initial plates, whereas in the LA process about 40 and 15 plates are required for adsorption of the 62,500 and 12,500 small discs, respectively. In all situations, a tendency to exclusion of large discs from the initial plates offers an increased possibility for adsorption of the small ones. MA is more efficient than LA with regard to the surface area filling and the factor ν plays a major role in the disc compacting. Figures 13 and 14 show the relative area covered by adsorption when different jump frequencies are implemented. The increase in the disc mobility in the MA model induces a marked overadsorption of small discs in the first plates (20 and 5, respectively), whereas in the plateau, the coverage results only from the presence of large discs. Hence MA systematically leads to a greater degree of surface coverage than LA. Therefore, chromatographic separation of mixtures of small

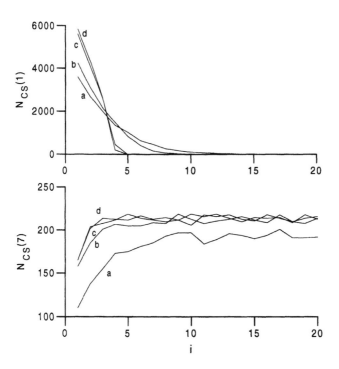

FIG. 12 Number of discs of radius 1 (top) and radius 7 (bottom) adsorbed as a function of plate number i for different values of the number of jumps v: (a) 0; (b) 50; (c) 500; and (d) 5000. The relative concentration of discs of radii 1 and 7 supplied at plate 1 is 0.5.

and large discs is expected to be strongly enhanced when the adsorption kinetics are modeled by the MA process.

C. Composition Effects in Adsorption Chromatography

As far as the composition C_{rel} of the mixture entering the column is fixed on the first plate throughout the experiment, this parameter strongly evolves with time, when the nonadsorbed discs become placed in front of the successive plates. Therefore, since the adsorption probability is modified from one plate to the following, we first present these effects by showing the covered areas as a function of the plate number i.

1. Localized Adsorption

Figures 15a–c and 16a and b show the covered area as a function of the plate number i when different mixtures of discs of radii 1 and 2, 1 and 3, and 1 and

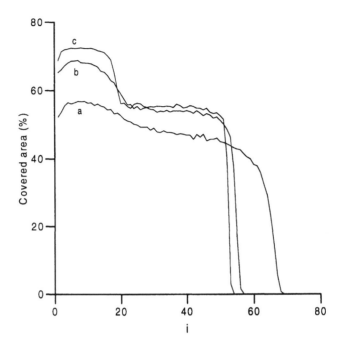

FIG. 13 Relative covered area as a function of the plate number i for adsorption of an equimolar mixture of discs of radii 1 and 3 for different values of the jump frequency v: (a) 0; (b) 50; and (c) 500.

7 are injected in the column to obtain a constant surface coverage inside the column. The composition C_{rel} is given by $N_c(r_1)/[N_c(r_2) + N_c(r_1)]$, where $N_c(r_i)$ is the number of discs of radius i, which is injected at each run.

The curves are functions of the disc radius and composition and may be categorized by their shape. In Fig. 15, the line corresponds to the injection of one type of disc. The covered area per disc increases with the relative concentration of discs of radius 1 as shown in Fig. 15 whereas it decreases in Fig. 16 above a threshold concentration.

Figure 15a–c shows results corresponding to the injection of discs of radii 1 and 2, 1 and 3, and 1 and 7, respectively. A maximum in surface coverage is observed in each situation. Before this point, the high concentration of small discs excludes too many large discs to allow maximal coverage, whereas after this point, a lack of small discs precludes maximal occupancy of the free surface area between large discs. When the small discs run short, the level of adsorption reaches that obtained for injection of a unique species. The peak corresponds to the best filling of the plateau by an ideal mixture of small and large discs.

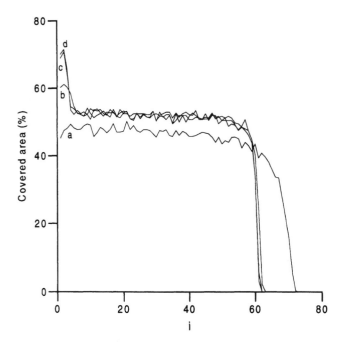

FIG. 14 Relative covered area as a function of the plate number i for adsorption of an equimolar mixture of discs of radii 1 and 7 for different values of the jump frequency ν: (a), 0; (b), 50; (c) 500; and (d) 5000.

Figure 16a and b shows results corresponding to the injection of discs of radii 1 and 2 and 1 and 3. The different curves present a maximum. It is shown that the situation of initial high concentration of small discs contributes to progressively exclude the large discs from the planes so that the previous maximal coverage per plane cannot be reached. The maximal degree of surface coverage within the column depends on the composition C_{rel} of the mixture being injected and on the size of the larger disc. In a given system, the highest value is obtained for a characteristic surface composition $C_{S,rel}$: for plates bearing adsorbed discs of radii 1 and 2, 1 and 3, or 1 and 7, the maximal surface coverage $\theta_{cS,max}$ is 0.565, 0.596, or 0.630. However, these degrees of surface coverage cannot be attained through direct injection at the given plate of a mixture of the required composition $C_{S,rel}$. The final surface distribution results from successes and failures in adsorption on the preceding plates, which progressively change the composition of the elutant encountering the plate of maximal coverage and thus modify the random access to the surface initially imposed at plate 1 by the composition C_{rel}.

Figure 17 shows the separation efficiency in the plates characterized by the

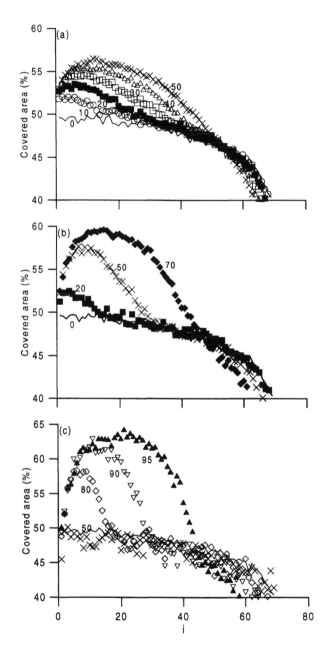

FIG. 15 Localized adsorption process. Relative coverage by adsorbed discs of radii 1 and 2 (a), 1 and 3 (b), 1 and 7 (c) as a function of the plate number i for different values of the relative concentration (%, as indicated on the curve) of the mixture supplied at plate 1.

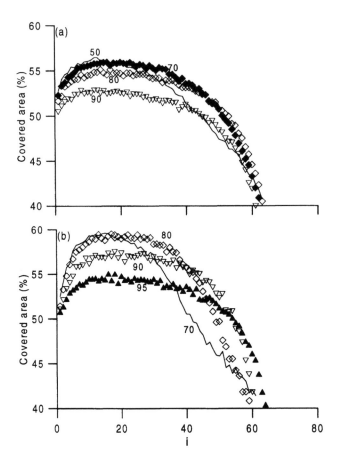

FIG. 16 Localized adsorption process. Relative coverage by adsorbed discs of radii 1 and 2 (a), 1 and 3 (b) as a function of the plate number i for different values of the relative concentration (%, as indicated on the curve) of the mixture supplied at plate 1.

maximal coverage in terms of $C_{S,rel}$ as a function of C_{rel}. Obviously, the separation efficiency is greater before since surface area exclusion of large discs is greater and smaller after this point as a result of the decay of the concentration of small discs. The selectivity of the chromatographic column, which can be defined by the ratio $C_{S,rel}/C_{rel}$ increases with the radius of the larger disc. Thus, for a symmetrical system ($C_{rel} = 0.5$) containing discs of radii 1 and 2, 1 and 3, and 1 and 7, the selectivity is 1.13, 1.57, or 1.86, respectively, and the efficiency of separation increases with the difference between the disc radii. The selectivity of the column strongly decreases with rising C_{rel}. This may be explained by the fact that the

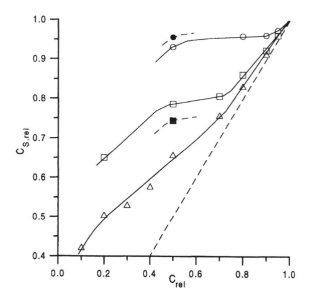

FIG. 17   Relative concentration of adsorbed discs of radii 1 and 2 ($\triangle$), 1 and 3 ($\square$), 1 and 7 ($\bigcirc$) at the plate characterized by the maximal coverage as a function of the relative concentration of the discs being supplied at plate $i = 1$. Open symbols refer to the localized adsorption model and filled symbols refer to the mobile adsorption model. The dashed line indicates the absence of selectivity.

small discs are always able to invade the free areas existing between adsorbed discs, whereas the larger ones may be unable to adsorb on some free areas. A nearly symmetrical system may therefore be separated with greater efficiency.

These considerations lead to a result of particular interest for optimal separation into the two fractions: the degree of surface coverage is lower before and after the plate of maximal coverage due to departure from the ideal surface composition. Hence, the best separate recovery of small and large discs is obtained when the chromatographic column is split into three sections a, b, and c, as shown in Fig. 18, after injection of a mixture of discs of radii 1 and 3 at the same concentration:

The first section (a) going from plate 1 to the plate of maximal coverage. It essentially contains the small discs,

The second section (b) going from the plate of maximal coverage to the plate where the curve joins the curve of reference corresponding to the injection of a unique species. It contains a mixture of small and large discs.

The last section (c), which only contains large discs.

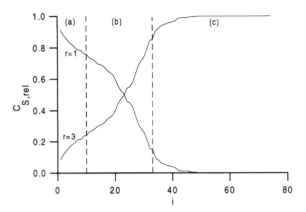

FIG. 18 Localized adsorption model. Relative concentration of discs of radii 1 and 3 as a function of the plate number *i* obtained after supply of an equimolar mixture at plate 1. (a) Corresponds to the region of overadsorption of discs of radius 1; (b) corresponds to a region of inefficient separation; and (c) corresponds to the region where only discs of radius 3 are recovered.

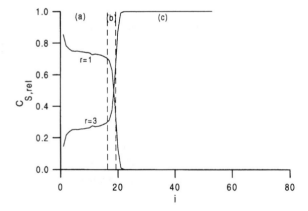

FIG. 19 Mobile adsorption process. Relative concentration of discs of radii 1 and 3 as a function of the plate number *i* obtained after supply of an equimolar mixture at plate 1. (a) Corresponds to the region of overadsorption of discs of radius 1; (b) corresponds to a region of inefficient separation; and (c) corresponds to the region where only discs of radius 3 are recovered.

2. Mobile Adsorption

No systematic investigations have been done using the mobile adsorption model. For comparison with the localized adsorption model, Fig. 17 shows the value of $C_{S,rel}$ for injection of the systems 1–3 and 1–7 at the initial concentration C_{rel} equal to 0.5 (black symbols). Figure 19 shows the variation of $C_{S,rel}$ with i for the system 1–3. As previously, cutting of the column into three sections leads to mainly obtain small discs in (a), a mixture of the two species in (b), and only large discs in (c). The mobile adsorption process essentially limits the zone (b) of mixture of the two species to a very small number of plates.

IV. EXPERIMENTAL SYSTEMS

A. The Adsorbents

1. Adsorption Experiments

Nonporous glass beads of industrial origin (Verre et Industrie) of 34 µm average diameter and 7.8×10^{-2} m^2/g specific surface area were used as raw material. The beads were washed with hot hydrochloric acid to extract or exchange with hydrogen ions all surface complexing ionic species and thoroughly washed free of acid with water to finally recover fully hydrated silica beads only bearing surface silanol (-SiOH) groups. Excess water was evaporated under reduced pressure at 40°C to maintain the fully hydrated silica surface [22].

Latex of sperical shape and narrow size distribution was kindly provided by the Laboratoire de Chimie et des Procédés de Polymérisation (Lyon, France). The latex particles were polymerized under emulsifier-free conditions using potassium persulfate as free radical initiator. The latex particles were hydrolyzed in the presence of Pyrex glass beads at 90°C during a week. This hydrolysis converts all surface groups in carboxylic acid groups [23,24]. Average diameters were the following: 840 nm from quasi-elastic light scattering and 860 nm from electronic microscopy determination.

2. Chromatography

Chromatographic separation experiments were carried out using Whatman glass microfiber filters GF/B whose retention is limited to particles above 1 µm. Each filter had a diameter of 5 mm and a thickness of 1.4 mm, the calculated mean diameter of the glass microfiber being close to 6 µm. The material was thoroughly washed with solvent before use.

B. The Polymer–Solvent Systems

Experiments have been carried out with polymers developing noncomplex polymer–adsorbent interactions. Adsorption isotherms and kinetics are given in this part in order to characterize the different systems.

To determine the amount of polymer adsorbed, the following procedure was implemented. The colloidal adsorbents were suspended in solvent and degassed under reduced pressure in order to generate a solid–liquid interface. The required mass of polymer solution was then added to the suspension. Mild controlled agitation was employed to homogenize the suspension during adsorption. The amount of adsorbed polymer was determined by comparing the initial specific radioactivity and that of the supernatant solution after adsorption. The isotherm represents the surface coverage N_s (mol/cm^2) as a function of the polymer concentration N (mol/ml) in solution at adsorption equilibrium.

To determine the rate of adsorption the following cell reactor (Fig. 20) and methodology were implemented in order to slow down the surface coverage during the initial step where the surface area coverage is small and to accelerate the process during the final step where the coverage tends to completion. The cell consisted of a cylindrical chrome or glass reservoir of volume V equal to 50 mL with a rotating magnetic bar D fixed on the bottom. The outlet aperture was fitted with a nonadsorbing filter E to avoid loss of beads during injection of the solution with an automatically driven syringe A. The adsorbent and a small amount of solvent were introduced into the cell reactor and the suspension was degassed. The cell was then filled with solvent and the suspension was homogenized by stirring with a magnetic bar D. At time zero, the radiolabeled polymer solution at concentration C_0 (cpm/mL) was injected at controlled rate J_v (mL/min) into the reactor. It was shown that the diffusion of macromolecules in the stagnant layer around the adsorbent can be neglected [25]. The chief advantage is the possibility of a quasi-continuous recording of the radioactivity of the efflu-

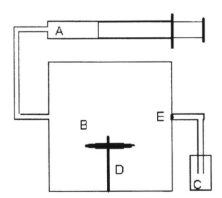

FIG. 20   Schematic representation of the experimental device employed for determination of the adsorption kinetics. (A) Automatically driven syringe for controlled polymer supply. (B) Reactor containing the suspension. (C) Effluent collector. (D) Magnetic bar. (E) Filter.

ent taken at C as a function of time, by which a high accuracy in the determination of the adsorption amount can be reached. The effluent was collected for a time Δt (min), and the successive samples n were analyzed for radioactivity. The increase in adsorption ΔA_s for an increment of time Δt was obtained using Eq. (21):

$$[\Delta A_s]_{n+1} = J_v C_0 \Delta t - A_{n+1} - \frac{V}{2J_v \Delta t}\{A_{n+2} - A_n\} \tag{21}$$

where A_n represents the total radioactivity of the nth sample. At time $t = 0$, $n = 0$. The increase ΔN_s (mol/cm^2) is related to ΔA_s through Eq. (22):

$$\Delta N_S = \frac{\Delta A_s}{SR_{spec}M_w} \tag{22}$$

where S, R_s, and M_w are the specific area of the adsorbent, the specific radioactivity, and the average molecular weight of the polymer, respectively. The value n corresponding to the end of the fast adsorption process was obtained by comparing the experimental variation of A_n as a function of n with the expected variation given by Eq. (23):

$$A_n = C_0 J_v \Delta t \left\{ 1 - \exp\left[\frac{J_v \Delta t}{V}(n_f - n)\right]\right\} + A_f \exp\left[\frac{J_v \Delta t}{V}(n_f - n)\right] \tag{23}$$

where A_f and n_f are the sample radioactivity and the sample number corresponding to the end of adsorption. Equation (23) means that the exponential behavior of A_n, which is determined in the absence of adsorption, is recovered when the adsorption has reached completion. The experimental kinetic coefficient is expressed by Eq. (24), according to Eq. (13):

$$\frac{\Delta N_s}{\Delta t} = K(N_s) \times N \tag{24}$$

1. The Diblock Copolymer of Styrene and Vinyl-2-pyridine

Copolymers were prepared by anionic polymerization of styrene with functional end-groups in tetrahydrofuran, followed by in situ polymerization of the vinyl-2-pyridine component [26]. Average molecular weights of polystyrene and poly-vinyl-2-pyridine blocks were determined by light scattering in tetrahydrofuran solution. Radiolabeling was carried out by trace quaternization of the pyridine moiety with $^{14}CH_3I$ in tetrahydrofuran leading to 0.1–0.3% pyridine iodide groups. In the text, the copolymer is referenced by [p-q] where p and q are the polymerization index of the styrene and vinyl-2-pyridine blocks, respectively.

Toluene was purified by distillation over sodium wire and maintained over molecular sieves to prevent the adsorption of moisture.

Toluene constitutes a selective solvent for the copolymer because it is a good solvent for the styrene block and a nonsolvent for the pyridine block. The copolymer solubilized in toluene below the critical micelle concentration (cmc) may have a segregated structure where the polyvinyl block is protected from precipitation by the polystyrene chain. Adsorption kinetics were determined from solution below and above the cmc, which was determined to be equal to 10^{-10} mol/mL.

Results of static light scattering measurements carried out on aged solutions indicated that the mean degree of micellization was of the order of 10 [27]. The Zimm plot is characteristic of a macromolecular system in poor solvent, as evidenced by the negative value of the second virial coefficient ($A_2 = -2 \pm 1 \times 10^{-4}$ g/mL) [28]. Moreover, the shape of the angular variation of the scattered intensity is characteristic of interacting spheres.

Since high adsorption levels were determined on hydrated silica whereas very low levels were determined on silanized silica, diblock copolymer adsorption on silica from toluene solution was attributed to silanol–pyridine interaction in nonpolar medium (Lewis acid–base interaction) [29]. Figures 21 and 22 show the kinetic coefficient $K(N_s)$ of adsorption on nonporous glass beads as a function of θ_s according to Eq. (24) for the copolymer [314–771] below and above the cmc, respectively. The fractional surface coverage θ_s was calculated by taking into account that the surface is fully covered with molecules when 2.7×10^{-12} mol/cm^2 are adsorbed (after 1 month) [30]. The fact that unique curves are obtained with changing the rate of supply of isolated and organized macromolecules indicates that surface area exclusion constitutes the rate-limiting phenomenon. Moreover, the adsorption kinetics during the fast process appears to be the same for isolated and organized macromolecules.

2. The Homopolymer Poly-4-vinylpyridine

Polyvinyl-4-pyridine was synthesized in ethylene glycol using azodiisobutyronitrile as initiator. After purification by successive precipitation in alkaline water and dissolution runs in ethanol, the polymer was fractionated in methanol-toluene mixtures. The polymerization index x and the radius of gyration R_G were determined from static light scattering measurements performed in 0.01 NaCl ethanolic solution and the following relation was obtained between x and R_G:

$$R_G(\text{nm}) = 0.94x^{0.45} \tag{25}$$

The polyvinyl-4-pyridine is soluble in water in acidic medium, and at pH 3.0 the degree of protonation is equal to 0.475 [31].

In order to determine polymer concentrations in very dilute solutions with a high degree of precision, radiolabeled polyvinyl-4-pyridine was obtained by quaternization of a few pyridine groups. In its final form, the polymer was composed of 99.9% pyridine groups and 0.1% quaternized pyridinium groups.

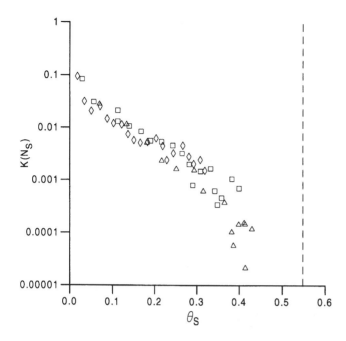

FIG. 21 Adsorption of block copolymer polystyrene-polyvinyl-2-pyridine [314-771] on silica from a solution of concentration equal to 10^{-10} mol/mL (cmc). Representation of the kinetic coefficient $K(N_s)$ (s^{-1}) as a function of the relative coverage θ_s, for different rates of polymer supply (mol/min cm^2): ($\diamond$) 2.8 $\times$ 10^{-14}; ($\square$) 4.95 $\times$ 10^{-14}; ($\triangle$) 12.9 $\times$ 10^{-14}.

Adsorption studies were carried out using polystyrene latex particles as adsorbent. Only electrostatic interactions between the pyridinium groups and the surface carboxylic acid groups are expected to occur. Interfacial reconformation of adsorbed polymer was evidenced by changing the rate of polymer supply in adsorption experiments. At the moment of initial contact with the surface, the polymer preserves its solution conformation. Under conditions of slow transfer of polymer to the sorbent surface, a slow reconformation of adsorbed macromolecules sets in, leading to a polymer flattening. This structure is unstable and overadsorption proceeds slowly to establish a thermodynamic equilibrium. Under conditions of fast transfer of polymer to the surface, the adsorbent is quite instantaneously covered with polymer characterized by their solution conformation. This interfacial structure was also unstable and thermodynamic equilibrium was attained via a desorption process. Equations (26) and (27) were found to describe

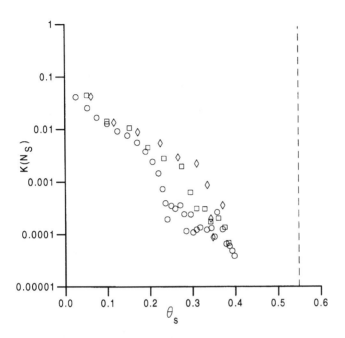

FIG. 22   Adsorption of block copolymer polystyrene-polyvinyl-2-pyridine [314-771] on silica from a solution of concentration equal to 3.5×10^{-10} mol/mL ($3.5 \times$ cmc). Representation of the kinetic coefficient $K(N_s)$ (s^{-1}) as a function of the relative coverage θ_s, for different rates of solution injection (mol/min): ($\diamondsuit$) 2×10^{-11}; ($\square$, $\bigcirc$) 10^{-10} (two experiments).

the slow relaxation process during delayed surface area coverage and the fast process during desorption of the exceeding macromolecules [31]:

$$\frac{\sigma - \sigma_{eq}}{\sigma_m - \sigma_{eq}} = \exp\left(\frac{-t}{\tau_{ads}}\right) \qquad (26)$$

$$\frac{\sigma - \sigma_{ads}}{\sigma_m - \sigma_{ads}} = \exp\left(\frac{-t}{\tau_{des}}\right) \qquad (27)$$

where σ_m is the area at the initial time of contact with the surface (equal to the cross-sectional area of the polymer coil), σ is the area at time t, σ_{eq} and σ_{ads} are the surface occupied at adsorption equilibrium by the isolated flat macromolecule and by the macromolecule at completion of the surface coverage, respectively. τ_{ads} and τ_{des} were determined to be equal to 2200 and 11 min, respectively. Since a unique equilibrium interfacial structure is established under slow and fast adsorption conditions, the adsorption is expected to be reversible. This characteristic was determined by determination of the electrophoretic mobility of the colloid–polymer complex. When the colloid suspension is washed out of nonadsorbed polymer by replacement of the supernatant polymer solution with pure water, the electrophoretic mobility of the bare colloid is recovered after a given time [32].

To represent the kinetic coefficient $K(N_s)$ as a function of the relative surface coverage θ_s, it is assumed that the maximal coverage of 9×10^{-14} mol/cm^2 corresponding to the fast adsorption regime prior to desorption corresponds to a relative coverage of about 0.4 for which the kinetic factor can be experimentally determined. It is noteworthy that for the model used in numerical simulation the 1231 discs that can adsorb on the plane strictly corresponds to the number of macromolecules able to adsorb on a unique latex particle and that the kinetics of the adsorption process is obtained for only 75% of the jamming limit. This means that the range of fast surface filling constitutes the experimentally measurable adsorption and that completion of coverage brings into play kinetic coefficients much lower than 10^{-4} s^{-1}, which are not accessible to our technique. On the other hand, it is obvious that higher coverage cannot be obtained as a result of the polymer interfacial reconformation. With this assumption, the experimental variation of $K(N_s)$ with θ_s shown in Fig. 23 agrees with that determined by the mobile adsorption model (see Fig. 4). Clearly, since no polymer is detected in the liquid phase for θ_s values below 0.2, adsorption of the injected macromolecules is instantaneous. The existence of a unique variation of $K(N_s)$ for different injection rates of polymer indicates that surface area exclusion constitutes the unique limiting factor in this range of surface coverage.

Since interfacial reconformation tends to slowly flatten the adsorbed molecule, it is clear that during the time devoted to a typical chromatography experiment, adsorbed polyvinyl-4-pyridine may not be considered as retaining its solution conformation. Nevertheless, the adsorbed polymer does not remain adsorbed at a given place but continuously moves on the surface.

C. Relevance of the Adsorbing Disc Model to Adsorbed Polymer Systems

1. The Diblock Copolymer Polystyrene-polyvinyl-2-pyridine

The diblock copolymer [p-q] is characterized by its solution β and surface β_s size asymmetry ratio defined as follows [33]:

$$\beta = \frac{p^{0.60}}{q^{0.33}}; \quad \beta_s = \frac{p^{0.60}}{q^{0.5}} \tag{28}$$

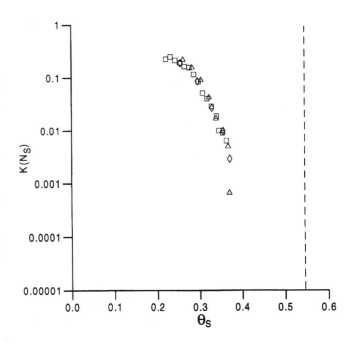

FIG. 23 Adsorption of polyvinyl-4-pyridine on carboxylate polystyrene latex in water at pH 3.0. Representation of the kinetic coefficient $K(N_s)$ (s^{-1}) as a function of the relative surface coverage θ_s for different rates of polymer supply (mol/min cm^2): ($\square$) 1.24 $\times$ 10^{-15}; ($\triangle$) 2.45 $\times$ 10^{-15}; ($\diamondsuit$) 4.82 $\times$ 10^{-15}.

Table 1 provides molecular characteristics of the copolymers used in the chromatography experiments. When β_s is great, the large bulk of the soluble block impedes the formation of the continuous interfacial cake comprising the nonsoluble blocks and small islands were expected to appear on the surface even at full surface coverage. When β_s is equal to 1 or less, isolated spherical cakes appear

TABLE 1 Molecular Characteristics of the Copolymer Polystyrene-polyvinyl-2-pyridine

[p-q]	β	β_s
[211–57]	6.5	3.3
[314–771]	3.4	1.1
[179–412]	3.0	1.1
[177–863]	2.3	0.7
[177–1581]	1.9	0.5

at small coverage, which give rise to a continuous film at full coverage. In the last situation, since the cross-sectional area corresponding to the volume of the soluble block is close to or smaller than the surface area of the nonsoluble block, no retardation effect is expected to result from the presence of the soluble block. The image of isolated adsorbed polymers characterized by $\beta_s = 1$ is given in Fig. 24. The surface deposition of the cake may thus be modeled by the disc adsorption scheme.

2. The Homopolymer Polyvinyl-4-pyridine

The number N_s of polymers giving rise to a monolayer was found to be close to the number N_s^* derived from the critical concentration of nonoverlapping spheres in two dimensions when the intrinsic viscosity [η] and radius of gyration R_G were determined in nonpolar medium [31]. This derivation was found to be valid in the case of the random sequential adsorption model when [η] was determined in 0.5 M NaCl solution, where long-range electrical forces are screened. It is well known that the reduced viscosity of polyelectrolyte strongly increases with dilution as a result of the effect of increasing electrostatic repulsive forces. When polymer characteristics determined in water or dilute electrolyte solution are employed to derive the adsorption of polymer of different average molecular weight, nonrealistic values were obtained as a result of an overvalued surface area exclusion effect. Conversely, since the effect of long-range forces disappears in concentrated electrolyte and polymer solution, we are brought to the conclusion that these forces may also be ignored at the interface. Our model of the adsorbed polyelectrolyte shown in Fig. 25 assumed that the area occupied by the adsorbed macromolecule is composed of (1) the central zone of high chain segment concentration, which constitutes the surface area being effectively excluded, and (2) the outer zone of smaller segment density, which does not exert such effects toward the central zone of solution macromolecules. If one further assumes that the volume of the central zone is close to that derived from the intrinsic viscosity determined in nonpolar medium, this image supports the adsorption model of uncharged discs of radius equal to the radius of the central zone.

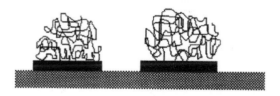

FIG. 24 Representation of the conformation of block copolymers adsorbed at the solid–liquid interface from a selective solvent.

FIG. 25   Representation of the conformation of a polyelectrolyte adsorbed at the solid–liquid interface. The rectangle indicates the limit of the zone corresponding to the excluded surface area.

V. SURFACE AREA EXCLUSION CHROMATOGRAPHY

Obviously, the efficiency of the chromatographic separation based on surface area exclusion effects may be improved when the rate of polymer transfer from the flowing phase (elutant) to the adsorbent (the stationary phase) is fast and thus experimental systems developing high surface to volume ratios may be advantageous. On the other hand, since high shear rates under fast injection may be detrimental to an efficient fractionation based on the present localized and mobile models of disc adsorption, we devised a very elementary column to show the limits and the possibilities of these idealized chromatographic separation methods.

A. Experimental Device and Methodology

1. The Chromatographic Column

The chromatographic column that is schematically represented in Fig. 26 is composed of a calibrated glass syringe (Tacussel, Lyon) and the corresponding pistons (C), which were modified to be fitted with input and output apertures of 0.5 mm, allowing the solvent or polymer solution to be injected at a controlled rate with the aid of an automatically driven syringe at one point and recovered at the other one. Two 2-mm-thick "nonadsorbing" Teflon discs (E) of large porosity were clamped close to the input piston inside the column to immediately establish a homogeneous distribution of the injected solution through the first and successive Whatman glass microfiber filters (C). The small amount of copolymer and polyelectrolyte adsorbed on these two "nonadsorbing" Teflon discs is reported in all chromatograms to show the absence of polymer retention and/or fractionation by adsorption. The column was filled with 56 filters and the height of the adsorbent in the column was reduced from 8 to 4.5 cm after stacking. The glass fiber volume thus represented 10% of the void volume of the column, which was equal to 0.77 mL.

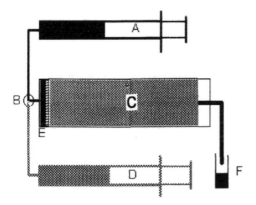

FIG. 26 Schematic representation of the experimental device employed for determination of the adsorption profiles in chromatography. (A) Automatically driven syringe for controlled polymer supply. (B) Automatic valve for swapping solution for solvent. (C) Stacked glass fiber filters (the stationary phase). (D) Automatically driven syringe for controlled solvent supply. (E) Nonadsorbing Teflon filters. (F) Elutant collector.

2. The Chromatographic Separation

To establish reproducible adsorption characteristics on the solid–liquid interface, the column was carefully saturated with use of a syringe (D) and eluted with toluene in the case of experiments performed with diblock copolymers and acidic water at pH 3.0 in the case of experiments performed with polyvinyl-4-pyridine. The polymer solution was injected at a controlled rate employing the syringe (A) enabling the polymer to be adsorbed on the successive filters. When the fixed volume of solution was injected, the injection was stopped and the elution was continued with solvent by running of the valve (B). In some instances, the solution in the void volume was pushed out by injection of air. The effluent was collected at the outlet with a sample collector (F).

At the end of the experiment, the filters were successively taken out of the column with care, individually deposited into glass vials, and counted for radioactivity content using a Tricarb spectrometer (Packard). The surface area available to polymer adsorption was estimated to be 1 dm^2/filter taking into account adsorption measurements of polyvinyl-4-pyridine of molecular weight 3.6×10^5 on polystyrene latex particles. Actually, no major difference in the adsorption behavior was expected to occur on silica and latex particles because at the pH level of the experiments silica and polystyrene latex particles are characterized by a surface charge density of the order of 2.8 $\mu C/cm^2$. Quite similar interactions are expected to be established between the pyridinium groups of the polymer and the dissociated silanol or carboxylic acid groups [34]. Therefore, the radioactivity (cpm) of each filter was converted to the number of adsorbed polyelectrolyte and the chromatogram obtained by plotting the adsorbed amount (mol/dm^2 or µg/dm^2) as a function of the filter number i. For simplicity, the number of moles adsorbed per filter unit of area equal to 1 dm^2 is always defined by N_s. For elution with a mixture of two polymers of different molecular weight, only one of the two polymer samples was radiolabeled and the experiment was duplicated to determine the adsorption amounting to each polyelectrolyte. The parameters used in the experiments are the same as previously defined, namely, the polyelectrolyte concentration C_0 (cpmL/mL) or N_0 (mol/mL) in solution, the injection rate J_v (mL/min), and the rate of polyelectrolyte supply to the first filter J_vC_0 (mol/min).

B. Chromatographic Separation of Unimers and Micelles from Micellar Solutions of the Diblock Copolymer Polystyrene-polyvinyl-2-pyridine

All experiments were carried out in a room thermostated at 25°C by injecting aged copolymer solutions (5–8 days) for which reproducible Zimm plots have been obtained.

1. Influence of the Irreversibility of the Copolymer Adsorption

To determine the irreversible nature of the adsorption, a small amount of copolymer is injected and the elution is continued with different volumes of solvent (10 and 20 mL). At the same time, the effluent is collected and analyzed for radioactivity content. The first test concerns the irreversibility of the adsorption of isolated copolymer molecules to definitely establish the validity of the localized adsorption model.

(a) The Copolymer [314–771]. The bulk and surface size asymmetry ratios are 3.4 and 1.1, respectively. This means that the possible protecting effect of the solubilizing block against localized adsorption is relatively low. Figure 27a shows the radioactivity (cpm) as a function of the filter number i. A unique curve is obtained for the amount adsorbed per filter when 10 or 20 mL of toluene is injected immediately after elution with the copolymer solution. This result indicates that polymers are not exchanged on the surface by solvent molecules and do not move in the interface. Figure 27b shows the radioactivity of the effluent (0.6 mL) as a function of time. An increase in radioactivity is observed, which stops when the injection of the copolymer solution is replaced by injection of solvent. Clearly, a small amount of polymer does not adsorb although the second half of the column remains fully free of adsorbed polymer. The ratio R of the maximal radioactivity per filter and in the effluent is found to be close to 20.

(b) The Copolymer [211–57]. The bulk and surface size asymmetry ratios are 6.5 and 3.3, respectively. This means that the tendency to adsorb is much lower for this copolymer as a result of the high solubilizing action of the large polysty-

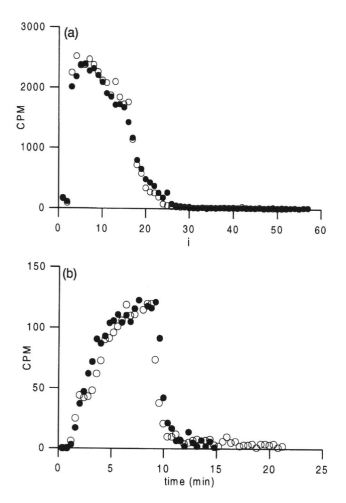

FIG. 27 (a) Radioactivity counted (cpm) per filter as a function of the filter number i after injection of the copolymer [314–771] and elution with 10 (●) and 20 mL (○) of toluene. (b) Radioactivity (cpm) counted per 0.6 mL of elutant as a function of time (min).

rene block. One may wonder if the injection of different volumes of solvent further provides unique variations of the radioactivity as a function of the filter number i and time. This information is given in Fig. 28a and b. Obviously, the presence of a small adsorbing block ensures the conditions of a localized adsorption. Nevertheless, R is found to be close to 6, indicating that relatively more macromolecules flow through the column without interacting with the adsorbent than was determined for the copolymer [314–771].

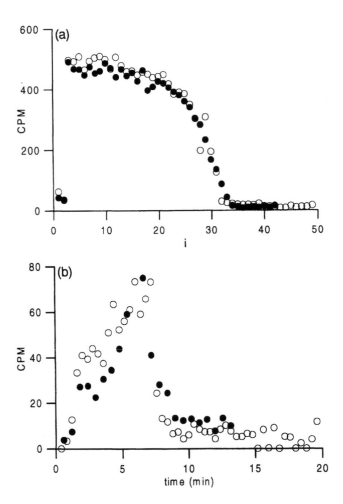

FIG. 28   (a) Radioactivity counted (cpm) per filter as a function of the filter number i after injection of the copolymer [211–57] and elution with 10 (●) and 20 mL (○) of toluene. (b) Radioactivity (cpm) counted per mL of elutant as a function of time (min).

These combined results demonstrate that the copolymer molecules irreversibly adsorb on the successive filters and that a very small number of macromolecules adopted solution conformations that impede any adsorption. When the polystyrene block prevents interaction between pyridine and silica, from this effect appears to be more marked with copolymers characterized by large β and β_s values [35].

The amount of solvent used as elutant does not modify the copolymer distribution inside the column. Therefore, we estimate that injection of solutions of different concentration provides a unique adsorption pattern. Conclusively, when different patterns are determined, other effects have to be invoked.

2. Influence of the Rate of Polymer Supply

The modification of the concentration C_0 of the copolymer solution providing modified rates of polymer supply $J_v C_0$ may induce different adsorption profiles in the column. One may imagine that under conditions of fast elution, the adsorption on a given filter i is delayed in time when the macromolecule are transferred from filter i to the following without attempting to adsorb. With this assumption, the adsorption peak is expected to be higher when the rate of polymer supply is slow. Moreover, for micellar solutions, when the micelles have to relax prior to adsorption [36], the adsorption level is expected to be higher on the first filters when the rate of supply is slow, and the molecules are expected to be transferred to the last filters when the rate of supply is high. To avoid such effects, the elution rate is held constant in the following experiments, and the effect of the rate of polymer supply $J_v C_0$ is determined by only varying the concentration C_0. This has been done for injection of the copolymer [314–771] and Fig. 29 shows the adsorption profiles obtained under the conditions defined in Table 2.

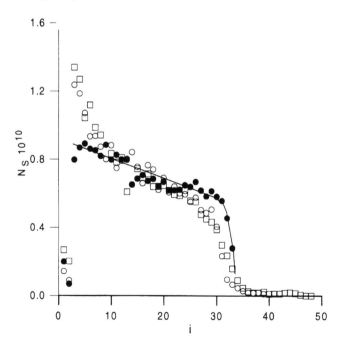

FIG. 29 Copolymer [314–771]. Effect on the adsorption profile of the rate of polymer supply $J_v C_0$ determined by varying the concentration C_0: (●) 1.15×10^{-10} mol/mL; (○) 1.01×10^{-9} mol/mL; (□) 9.5×10^{-9} mol/mL. The complementary experimental parameters are given in Table 2.

TABLE 2   Parameters of Experiments Reported in Fig. 29

$10^9 \times C_0$ (mol/mL)	J_v (mL/min)	$10^9 \times J_v C_0$ (mol/min)	$10^9 \times N_s$ (mol)
0.115	1.5	0.17	2.16
1.01	1.5	1.52	2.15
9.5	1.5	14.2	2.06

Clearly, the major effect of varying the copolymer concentration from the cmc to $100 \times$ cmc is to modify the adsorption profile in the column because similar curves are not obtained at low and high rates of polymer supply although similar adsorptions are obtained at the end of the experiment. The full line corresponds to the injection of isolated macromolecules. For injection of the micellar solution, an adsorption peak is determined at the column inlet and comparatively lower molecules are adsorbed at the end of the adsorption zone. This clearly shows that micellar solutions initially contain unimers (isolated molecules) and micelles (organized molecules) that behave differently on adsorption.

3. Influence of the Injection Rate

To definitely eliminate relaxation effects and delay in adsorption resulting from changes in the elution rate, solutions of different copolymer [p-q] were eluted at the same concentration but at different rates as indicated in Table 3. Obviously, different shapes were obtained for the different systems but no definite effect

TABLE 3   Parameters of Experiments Reported in Fig. 30

Code [p-q]	$10^9 \times C_0$ (mol/mL)	J_v (mL/min)	$t(s)$ injection	$10^9 \times J_v C_0$ mol/min	$10^9 \times N_s$ (mol)	Figure number
[314–771]	9.5	1.5	11	14.2	2.06	(a)
[314–771]	9.5	0.15	105	1.42	2.08	(a)
[179–412]	2.00	1.5	55	3.0	2.30	(b)
[179–412]	1.95	0.15	495	0.29	2.12	(b)
[177–863]	1.88	1.5	50	2.8	2.29	(c)
[177–863]	1.88	0.15	470	0.28	2.18	(c)
[177–1581]	1.14	1.5	88	1.71	1.98	(d)
[177–1581]	1.18	0.15	813	0.177	2.05	(d)
[211–57]	2.04	1.5	338	3.06	15.1	(e)
[211–57]	2.04	0.3	1950	0.612	15.0	(e)

would be attributed to a decrease (by a factor 10) of the injection rate. Therefore, when these experimental criteria are employed in chromatography, modifications in the shape of the adsorption profile have to be attributed to changes in the structure and/or composition of the copolymer solutions. The most simple situation represented in Fig. 30e corresponds to the copolymer [211–57] for which no micelles could be detected by light scattering in solution of concentrations

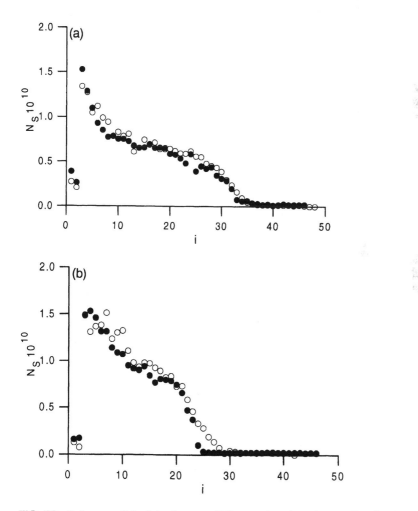

FIG. 30 Influence of the injection rate. Effect on the adsorption profile after injection of the different copolymers: (a) [314–771]; (b) [179–412]; (c) [177–863]; (d) [177–1581]; and (e) [211–57]. The complementary experimental parameters are given in Table 3.

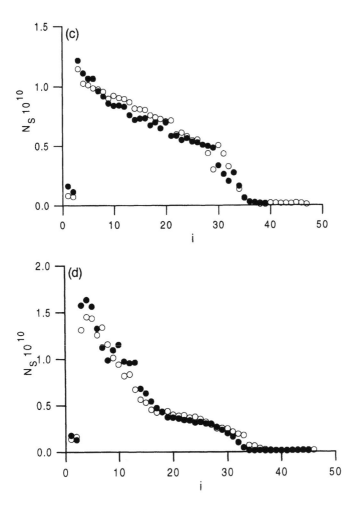

FIG. 30 Continued

between 3.6×10^{-8} and 3.6×10^{-7} mol/mL. The chromatogram profile corresponds to the progressive filling of the successive filters by components of identical characteristics. Figure 30a and b corresponds to copolymers of similar size asymmetry ratios. Figure 30c and d corresponds to systems whereby the surface coverage is only slightly impeded by the solubilized polystyrene block. For the micellar solution, the presence of peaks at the column inlet is expected to result from the relative excess of unimers and default of micelles as determined for localized adsorption processes (see Figs. 6–8).

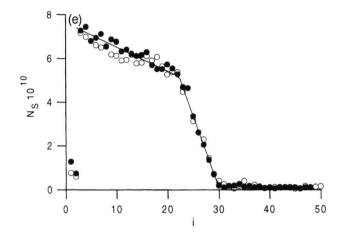

FIG. 30 Continued

4. Influence of the Interfacial Reconformation of the Polyvinyl-2-pyridine Block

Reconformation of adsorbed homopolymers has been previously determined as a process that can strongly influence the surface coverage [37–40]. Similar interfacial phenomena were expected to address the adsorbed chain of block copolymers (belonging to isolated and organized molecules) and may modify the separation efficiency. This should particularly apply to copolymers of size asymmetry ratios β_s smaller than 1 such as the copolymer [117–1581] [33] (Fig. 31). Experiments were carried out with the parameters reported in Table 4 including a very short injection time of 9 s.

A monotonous decay is determined at the cmc with a maximal coverage of 0.9 $\times 10^{-10}$ mol on the first adsorbing filter. The decay appears to be more complex at 10 and 100 $\times$ cmc, where initial peak and plateau adsorptions were observed. Overadsorption at the inlet of the column should be attributed to preferential adsorption of unimers, whereas values of the final plateau should be attributed only to adsorption of micelles. Therefore, adsorbed unimers are expected to occupy a smaller area than adsorbed micelles, since the number of molecules adsorbed in the plateau region is greater than the values that may be determined from plateaus observed at 10 and 100 $\times$ cmc. Moreover, in the experiment where the solution was injected during 88 s (and longer, as indicated in Fig. 30d), lower micelles are adsorbed than in the experiment where the solution was injected during 9 s; fast aging seems to flatten the interfacial conformation of adsorbed micelles. However, a more pertinent information of these results is not possible because no correlation is established between the actual areas being occupied by

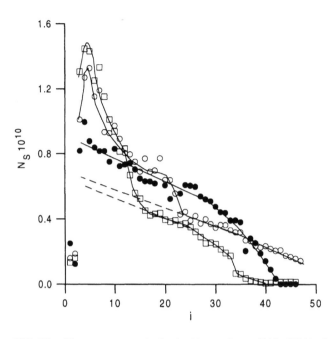

FIG. 31 Chromatograms obtained with copolymer [117–1581] after injection of the different concentrations of 0.10, 1.14, and 10.8 $\times$ 10^{-9} mol/mL for 16 min, 82 and 9 s, respectively. The complementary experimental parameters are given in Table 4.

isolated and organized molecules. Progress in the interpretation of the adsorption profiles could only be obtained when this correlation has been established.

5. Correlation Between Experimental and Simulated Adsorption Profiles in Chromatography

Actually, experimental and simulated adsorption profiles can be correlated for organized systems for which exists a proportionality between the area occupied by the micelle and the number of unimers forming the micelle, and, finally, be-

TABLE 4 Parameters of Experiments Reported in Fig. 31

$10^9 \times C_0$ (mol/mL)	J_v (mL/min)	$10^9 \times J_v C_0$ (mol/min)	$10^9 \times N_s$ (mol)	t(injection)
0.10	1.5	0.153	2.21	16 min
1.14	1.5	1.71	1.98	82 s
10.8	1.5	16.2	2.45	9 s

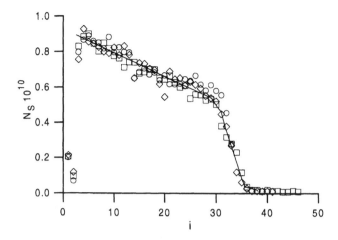

FIG. 32 Copolymer [314–771]. Chromatogram obtained for injection of solutions of concentration equal to the cmc. The complementary experimental parameters are given in Table 5.

tween the area covered by the micelles and the radioactivity counted on the filter. This has been verified for the copolymer [314–771], which is characterized by a β_s value of 1.1. Figure 32 shows the chromatograms for three experiments where a total amount of $2.11-2.17 \times 10^{-9}$ mol was injected in the column. The corresponding parameters are given in Table 5. The three experiments indicate that a maximal coverage of 0.9×10^{-10} mol is obtained at the column inlet. The monotonous decay of the coverage well characterizes the adsorption of a unique species. Figure 32 should be compared to Fig. 33, which represents the chromatogram corresponding to the same total coverage obtained after injection of a solution at the concentration of 10^{-9} mol/mL ($10 \times$ cmc). The parameters of the experiments are given in Table 6. Taking into account the analogy existing between the adsorption profiles of Figs. 6–8 obtained from simulation of the injec-

TABLE 5 Parameters of Experiments Reported in Fig. 32

$10^{10} \times C_0$ (mol/mL)	J_v (mL/min)	$10^{10} \times J_v C_0$ (mol/min)	$10^9 \times N_s$ (mol)	$t(s)$ (injection)
1.15	1.5	1.72	2.16	850
1.15	1.5	1.72	2.11	850
1.12	0.6	0.67	2.17	2220

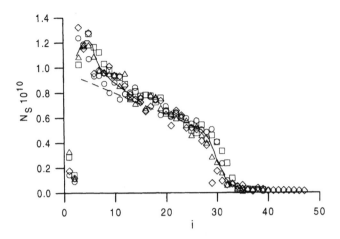

FIG. 33   Copolymer [314–771]. Chromatogram obtained for injection of solutions of concentration equal to 10 × cmc. The complementary experimental parameters are given in Table 6.

tion of a mixture of discs of radii 1 and 3, and that represented in Fig. 33, we may conclude that the initial peak corresponds to the overadsorption of unimers and the plateau region, where the unimers are excluded, to micelles. In Fig. 33, extrapolation of the plateau to the inlet of the column sets a line that reproduces the coverage determined in Fig. 32. This indicates that adsorption at a constant full coverage by unimers or micelles corresponds to the interfacial deposition of the same number N_s of macromolecules. Taking into account that the determination by light scattering provided an average degree of micellization equal to 10, it can be concluded that a micelle composed of 10 molecules occupies the area corresponding to 10 molecules. This may be expressed differently by assuming that the radii of the adsorbed unimer and micelle are equal to 1 and 3.16, respectively, so that the numerical simulation of the chromatography using discs of radii

TABLE 6   Parameters of Experiments Reported in Fig. 33

$10^9 \times C_0$ (mol/mL)	J_v (mL/min)	$10^9 \times J_v C_0$ (mol/min)	$10^9 \times N_s$ (mol)	t(s) (injection)
1.01	1.5	1.52	2.15	86
1.01	0.15	0.152	2.3	990
1.09	0.6	0.65	2.14	237
1.09	0.6	0.65	2.28	240

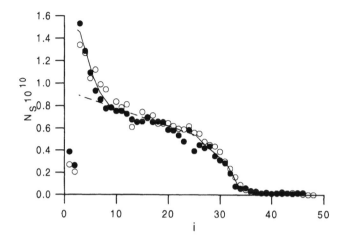

FIG. 34 Copolymer [314–771]. Chromatogram obtained for injection of solutions of concentration equal to 95 × cmc. The complementary experimental parameters are given in Table 3.

1 and 3 may be used as a model to interpret quantitatively the chromatographic separation of unimers and micelles for the symmetrical copolymers.

The relative concentration N_{rel} of unimers for the copolymer solution of concentration equal to 10 × cmc is 0.52, so that reference to the situation of C_{rel} = 0.5 is justified. The deviation from the adsorption behavior of that of monosized discs is in fact observed in the first half of the chromatogram in simulation and experiments.

Figure 34 shows the chromatogram corresponding to the injection of a solution of 9.5×10^{-9} mol/mL (95 × cmc) for which N_{rel} is 0.09. The parameters of the experiments are given in Table 3. The new information relative to experiments already reported in Fig. 30a is the following: the dashed line always schematizes the adsorption of micelles, whereas the peak may be compared to the adsorption profile reported in simulation (Fig. 15b for C_{rel} = 20%), which displays the typical deviation in the very initial section of the column and the clearly defined plateau region.

C. Chromatographic Separation of the Protonated Polyvinyl-4-pyridine

The protonated form of polyvinyl-4-pyridine may be viewed as being composed of chain segments of two types: 0.475 chain segments are vinyl-4-pyridinium groups (which may be dissociated or ion-paired with counterions) whereas 0.525

chain segments are pure vinyl-4-pyridine segments. Nothing is known about the distribution of these different groups in the coil volume, but due to the nonsolubility in water of the neutral segment it is expected that charged and neutral segments are nonhomogeneously distributed in the coil volume and along the chain. With this assumption, it is expected that the relative local concentration of all the groups strongly depends on the density of the chain segments and, as a result, of the molecular weight too. Fractionation as a function of molecular weight, charge density, or hydrodynamic volume is thus extremely complex and the use of a given technique does not provide polyelectrolytes of well-defined characteristics.

As indicated in Sec. IV.B.2, the adsorption kinetics can be described by the mobile adsorption process which assumes the adsorbed polymer to be mobile on the surface and desorbed when the equilibration solution is replaced by the solvent. Actually, the rate of the surface-to-solution transfer of the polyelectrolyte at constant composition of the system has not been investigated. Moreover, nothing is known about the interfacial exchange rate between polyelectrolytes of small and high molecular weight.

Determination of the characteristics of polyelectrolytes sampled at the outlet of a chromatographic column is very difficult and in this chapter we present some results recently obtained in the separation by molecular weight of an equimolar mixture of two fractionated polyelectrolytes of molecular weight equal to 1,057,000 (sample F1, $x = 10,067$) and 78,000 (sample F2, $x = 743$) [34].

1. Adsorption Profiles After Elution with a Solution Containing Only One Polyelectrolyte

Due to relaxation phenomena of characteristic times τ_{ads} and τ_{des} defined by Eqs. (26) and (27), the adsorption on the column was studied under the two conditions of incomplete and maximal coverage. In the first situation, the injection time was short, whereas in the second situation the injection time was of the order of τ_{des} in order to determine the influence of the fast interfacial relaxation in the adsorption levels. In the two cases, the adsorption profiles were determined before and after elution of the column by water at pH 3.0 in order to optimize the parameter of an efficient separation.

(a) Adsorption at Low Coverage. The sample F1 at the concentration of 1.05 $\times 10^{-11}$ mol/mL is injected for 2.5 min at 1.5 mL/min. The column was analyzed for content first. The experiment was repeated twice with solvent injection for 6 and 9 min after polymer injection. Figure 35 shows the adsorption N_s as a function of the filter number i, which appears to be only slightly affected by elution with solvent. Clearly, the molecules adsorbed at the inlet of the column appear not to be affected by the column washing, whereas the molecules adsorbed for a lower period appear to be slightly desorbed by the solvent. Actually, the fact that the lines a–c start at the same level shows that all of the polymer being injected was

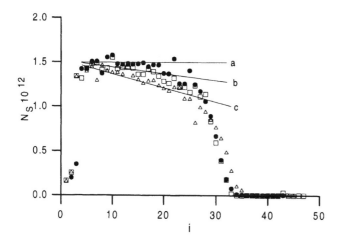

FIG. 35 Polyvinyl-4-pyridine at pH 3.0. Chromatogram obtained after injection for 2.5 min at 1.5 mL/min of the sample F1 ($x = 10,067$) at the concentration of 1.05×10^{-11} mol/mL (a), and after elution of the column with water at pH 3 for 6 min [line (b)] and 9 min [line (c)].

adsorbed before washing, in agreement with the mobile adsorption process, which assumed the adsorption to be fast and complete on free planes (the initial state). This confirms that at low coverage interfacial reconformation does not induce polymer desorption. Nevertheless, since lines b and c progressively deviate from a, this deviation may indicate that polymer molecules being adsorbed for smaller times are only weakly held on the surface.

Determination of the radioactivity content of the effluent during elution with the polyelectrolyte solution indicated that no polyelectrolyte was found in the effluent. This is converse to the situation of the diblock copolymers, where a small portion of the copolymer flows through the column without being adsorbed [see Secs. V.B.1 (a) and (b)].

(b) Adsorption at High Coverage. A second set of experiments with a more concentrated solution (1.01×10^{-10} mol/mL) injected for 14 min at 0.15 mL/min provided the results reported in Fig. 36. This experiment was also repeated twice where the solvent was injected for 5 and 10 min after polymer injection. Aging of the interfacial layer induces a decrease of the surface coverage, which may be attributed to a slight flattening of the adsorbed polymer. The straight line indicates coverage at the theoretical time zero (no relaxation) to be close to 3.3 $\times 10^{-12}$ mol/dm^2. The existence of plateaus for the adsorption profiles indicates that the washing process does not induce the desorption of the adsorbed molecules but only replaces the polymer solution in the void by the pure solvent, the polymer

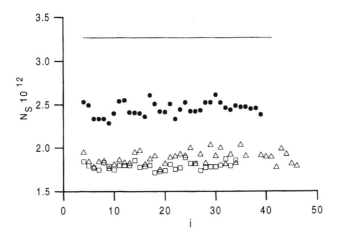

FIG. 36   Polyvinyl-4-pyridine at pH 3.0. Chromatogram obtained after elution with sample F1 ($C_0 = 1.01 \times 10^{-10}$ mol/mL; $J_v = 0.15$ mL/min; $t = 13.5$ min; $J_v C_0 = 1.51 \times 10^{-11}$ mol/min): (●), no solvent injection; after (△) 10 mL and (□) 15 mL solvent injection. The line gives the surface coverage prior to polyelectrolyte reconformation.

desorption being only the result of the fast interfacial relaxation. Moreover, the polyelectrolyte layer appears to be stabilized at the "end" of the relaxation process.

(c)   Excluded Area for Samples F1 and F2.   Figure 37a and b shows the adsorption chromatograms obtained for experiments carried out using the parameters given in Table 7. After elution, the column is rapidly eluted with solvent to displace the excess polymer from the void. The amount of polymer adsorbed is expressed in μg/dm² in Fig. 37a and in mol/dm² in Fig. 37b. Figure 38 shows the chromatogram obtained after saturation of the column under conditions described in Table 8.

The difference in the adsorption amounts of 1 and 1.6 μg/dm² observed in Fig. 37a may result from the variation of the density of polymer chain segments already existing in solution. This similar establishment of monolayers confirms that the full surface area of the glass fiber filter is available for adsorption of both samples. Actually, the experimental adsorption agrees with the value that is calculated taking into account the molecular dimensions of the two polyelectrolytes [31]. From Figs. 37 and 38 we conclude that polyelectrolyte F1 effectively occupies an interfacial area nine-fold larger than sample F2. Therefore, to quantitatively interpret the chromatograms obtained for mixtures of F1 and F2, we refer to results of numerical simulation of the mobile adsorption processes obtained for discs of radii 1 and 3.

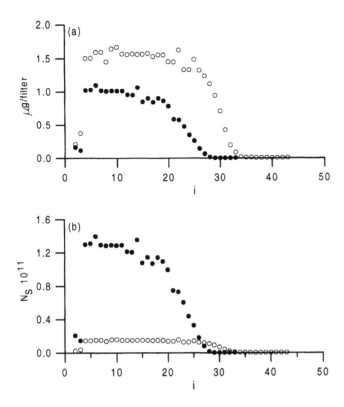

FIG. 37 Polyvinyl-4-pyridine at pH 3.0 Chromatograms obtained after elution with samples F1 (○) and F2 (●) inducing a relatively low surface coverage. Representation of the adsorbed amount per filter ($\mu g/dm^2$) (a) and (mol/dm^2) (b) as a function of the filter number i. The complementary experimental parameters are given in Table 7.

TABLE 7 Parameters of Experiments Reported in Fig. 37a and b

Sample	$10^{11} \times C_0$ (mol/mL)	J_v (mL/min)	$10^{11} \times J_v C_0$ (mol/min)	t (min) (injection)
F1	1.05	1.5	1.58	2.5
F2	1.00	1.5	1.50	14

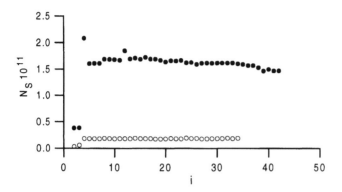

FIG. 38 Polyvinyl-4-pyridine at pH 3.0. Chromatograms obtained after elution with samples F1 (○) and F2 (●) inducing a relatively high surface coverage. Representation of the adsorbed amount per filter (mol/dm²) as a function of the filter number i. The complementary experimental parameters are given in Table 8.

2. Adsorption Profiles After Elution with a Solution Containing an Equimolar Mixture of Samples F1 and F2

This experiment was duplicated and equimolar mixtures contained at once a radiolabeled sample of one polyelectrolyte and a nonlabeled sample of the other. Injection of mixtures at a total concentration of 2×10^{-10} mol/mL at 0.15 mL/min for 15 min and rapid washing with solvent resulted in the histograms shown in Fig. 39. The solid line represents the previous situation of injection of F1. The amount of polymer injected into the column corresponded to a full coverage of about 130 filters. This number should be compared to the 54 plates covered in simulation by small and large discs (Fig. 13). Since curve (c) in Fig. 13 reveals coadsorption of small and large discs on 17 plates, we investigated the type of coverage on the equivalent 40 microfiber filters, where simulation predicted coadsorption of fractions F1 and F2.

The results of Fig. 39 may be interpreted on the basis of the mobile adsorption

TABLE 8 Parameters of Experiments Reported in Fig. 38

Sample	$10^{10} \times C_0$ (mol/mL)	J_v (mL/min)	$10^{11} \times J_v C_0$ (mol/min)	t (min) (injection)
F1	1.01	0.15	1.51	14.5
F2	1.48	1.5	2.22	5

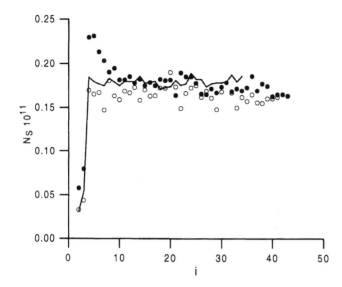

FIG. 39   Polyvinyl-4-pyridine at pH 3.0. Chromatograms after elution with an equimolar mixture of samples F1 (○) and F2 (●). Representation of the number of moles adsorbed per filter (area = 1 dm^2) as a function of the filter number i. The parameter of the elution for each sample are as follows: $C_0 = 1.05 \times 10^{-10}$ mol/mL; $J_v = 0.15$ mL/min; $t = 15$ min; $J_v C_0 = 1.5 \times 10^{-11}$ mol/min. The line represents the situation of injection of F1 as given in Fig. 38.

model because adsorption of 0.17×10^{-11} mol/dm^2 effectively corresponds to the maximum adsorption determined from experiments reported in Fig. 36. On the first filters, a slightly denser occupation by the polyelectrolyte F2 can be noted, as expected from Fig. 11 (top), curve (c). The plateaus indicate equal adsorption of polyelectrolytes F1 and F2 whereas, according to our simulation, the number of moles F2 was expected to be threefold greater than that of F1. This implied that the solution continuously flowing through the column is more enriched in F2 than determined from simulation and, on the average, always identical to the mixture being injected into plate 1. Therefore, the surface coverage and the number of large discs being adsorbed on plate 1 for injection of an equimolar mixture should be recovered on the successive plates because the injected mixture remained equimolar on the average. Figure 11 (bottom), curve (c) shows that the number of discs of radius 3 on plate 1 is only equal to 75% of the number corresponding to the plateau value, so that we expect that the maximal coverage by polyelectrolyte F1 should not exceed $0.75 \times 0.17 \times 10^{-11}$, i.e., 0.13 $\times 10^{-11}$ mol/dm^2.

The unexpected adsorption of 0.17×10^{-11} mol/dm^2 experimentally determined may be interpreted as follows: our schema reveals a situation of selective adsorption of large discs that only sets in when the total area allotted to a set of adjoining small discs is large enough to allow adsorption of one large disc. Therefore, at any moment one large disc replaces the corresponding set of small discs, which again attempt to adsorb on the subsequent plates. Therefore, for selective reversible adsorption, the situation will be similar on each plate and the histogram of the size exclusion chromatography will be flat. Similarly, one polyelectrolyte F1 may replace several adsorbed polyelectrolytes F2 when the positions of F2 are distributed in such a way that the areas belonging to F2 and the interstitial areas are large enough. Conversely, when the area portions are too small to allow mobile adsorption of F1, the interfacial exchange is impeded. Maximal instantaneous replacement of small polyelectrolytes by large ones should result from the in-plane mobility of the adsorbed polyelectrolytes.

This result lead us to propose an alternative interpretation for the desorption observed in Fig. 35, when solvent is used as elutant. If we assume that a fractionated sample may be characterized by a given mass polydispersity, when this sample is continuously supplied at the inlet of the column, adsorbed polyelectrolytes of relatively lower molecular weight are progressively displaced toward the column outlet to the benefit of polymers of larger molecular weight. Thus, we may conclude that a continuous elution of the column by solvent progressively desorbs polyelectrolytes of lower molecular weight and that an efficient fractionation will result from the continuous elution with solvent.

VI. CONCLUSION

Concerning polymeric systems, the chromatographic technique most frequently used for fractionation by molecular weight is gel permeation chromatography. The fractionation is based on a sieve effect that excludes the large macromolecules from the small pores, thus inducing an elution retardation proportional to their hydrodynamic volume. In hydrodynamic chromatography, depletion at the wall of the largest macromolecules leads to retardation of small macromolecules. Field flow fractionation depends on the action of a field (centrifugal, electrical, thermal, or hydraulic) applied perpendicular to the channel flow, which gives rise to partitioning of the components into regions of different flow velocity. In all of these techniques, the macromolecules are assumed not to interact with the solid phase or the wall [41].

Usual two-dimensional gel chromatography or chromatographic fractionation may serve to separate macromolecules having a low energy of interaction with the gel phase or beads, such that macromolecules may be displaced by solvent under given conditions of temperature and solvent quality. Orthogonal chromatography combining two sizes of exclusion chromatography also relies on this principle. However, some limitation of the fractionation efficiency may result from partial adsorption and partition effects [42].

Separation of macromolecules having a strong interaction with the solid phase constituted a very difficult challenge. This preliminary work aimed at determination of the separation efficiency of surface area exclusion chromatography when small and large constituents compete in irreversible adsorption, assuming that the components reversibly or irreversibly adsorb on the stationary phase. Our intention was to determine the conditions whereby random deposition of the different species may induce a fractionation within the column. Different systems may benefit from such fractionation processes, essentially those combining hydrophobic and electrical interactions with hydrogen bonding. Proteins may behave in this manner, as may synthetic macromolecules which remain adsorbed even when solvent is injected into the column [43].

Adsorption chromatography seems to be suited for analysis of such molecules as surfactants and proteins, which may coexist in solution under different forms of molecular organization. Actually, since relaxation processes are able to affect the structural composition of the flowing phase in the usual chromatographic methods, such processes appear to not modify the solution composition and molecular organization in adsorption chromatography.

REFERENCES

1. M. Muthukumar and A. Baumgärtner, Macromolecules 22:1937 (1989).
2. Y. Lipatov and L. M. Sergeeva, Adsorption of Polymers, (John Wiley and Sons, New York, 1970.
3. J. Feder, J. Theor. Biol. 87:237 (1980).
4. E. L. Hinrichsen, J. Feder, and T. Jossang, J. Stat. Phys. 44:793 (1986).
5. B. Senger, P. Schaaf, J. C. Voegel, A. Johner, A. Schmitt, and J. Talbot, J. Chem. Phys. 97:3813 (1992).
6. A. Elaissari, G. Chauveteau, C. Huguenard, and E. Pefferkorn, J. Colloid Interface Sci. 173:221 (1995).
7. R. H. Fowler and E. A. Guggenheim, Statistical Thermodynamics, Cambridge University Press, New York, 1939.
8. W. Mayer and E. R. Tompkins. J. Am. Chem. Soc. 69:2866 (1947).
9. E. Glueckauf, Trans. Faraday Soc. 51:34 (1955).
10. E. Glueckauf, Ion Exchange and Its Application, Society of Chemical Industry, London, 1955, p. 34.
11. P. J. Flory, Principles of Polymer Chemistry, Cornell University Press, Ithaca, NY, 1953, Chapter 12.
12. A. Silberberg, J. Chem. Phys. 48:2835 (1968).
13. E. Pefferkorn, A. Haouam, and R. Varoqui, Macromolecules 21:2111 (1988).
14. D. Poland, Macromolecules 24:3361 (1991).
15. Z. Adamczyk, B. Siwek, M. Zembala, and P. Belouschek, Adv. Colloid Interface Sci. 48:151 (1994).

16. G. Y. Onoda and E. G. Liniger, Phys. Rev. A *33*:715 (1986).
17. J. Feder, J. Theor. Biol. *87*:237 (1980).
18. A. Elaissari, A. Haouam, C. Huguenard, and E. Pefferkorn, J. Colloid Interface Sci. *149*:68 (1992).
19. W. H. Grant, L. E. Smith, and R. R. Stromberg, Faraday Discuss. Chem. *59*:209 (1975).
20. E. Pefferkorn, A. Haouam, and R. Varoqui, Macromolecules *222*:2677 (1989).
21. E. Pefferkorn, A. Carroy, and R. Varoqui, J. Polym. Sci.: Polym. Phys. Ed. *23*:1997 (1985).
22. R. K. Iler, *The Chemistry of Silica*, John Wiley and sons New York, 1979.
23. A. A. Kamel, Ph.D. dissertation, Lehigh University, 1981, pp. 95–110.
24. A. A. Kamel, M. S. El-Aasser, and J. W. Vanderhoff, J. Dispers. Sci. Technol. 2: 213 (1981).
25. E. Pefferkorn, A. Carroy, and R. Varoqui, Macromolecules *18*:2252 (1985).
26. M. Fontanille and P. Sigwalt, C. Roy. Acad. Sci. *251*:2947 (1960).
27. C. Huguenard, A. Elaissari, and E. Pefferkorn, Macromolecules *27*:5277 (1994).
28. B. H. Zimm, J. Chem. Phys. *16*:1093 (1948).
29. E. Pefferkorn, Q. Tran, and R. Varoqui, J. Polym. Sci.: Polym. Chem. Ed. *19*:27 (1981).
30. C. Huguenard, R. Varoqui, and E. Pefferkorn, Macromolecules *24*:2226 (1991).
31. E. Pefferkorn and A. Elaissari, J. Colloid Interface Sci. *138*:187 (1990).
32. J. Widmaier, personal communication.
33. C. M. Marques, J. F. Joanny, and L. Leibler, Macromolecules *21*:1051 (1988).
34. C. Huguenard, J. Widmaier, A. Elaissari, and E. Pefferkorn, Macromolecules *30*: 1434 (1997).
35. A. Johner and J. F. Joanny, Macromolecules *23*:5299 (1990).
36. A. Halperin and S. Alexander, Macromolecules *22*:2403 (1989).
37. E. Pefferkorn, A-C. Jean-Chronberg, and R. Varoqui, C. Roy. Acad. Sci. Paris, *308, Serie II*: 1203 (1989).
38. E. Pefferkorn, A-C. Jean-Chronberg, and R. Varoqui, Macromolecules *23*:1735 (1990).
39. E. Pefferkorn and A. Elaissari, J. Colloid Interface Sci. *138*:187 (1990).
40. A. Elaissari and E. Pefferkorn, J. Colloid Interface Sci. *143*:85 (1991).
41. M. J. R. Cantow, *Polymer Fractionation*, Academic Press, New York, 1967.
42. H. G. Barth and J. W. Mays, *Modern Methods of Polymer Characterization*, John Wiley and Sons, New York, 1991.
43. J. D. Andrade, *Surface and Interfacial Aspects of Biomedical Polymers: Protein Adsorption*, Plenum Press, New York, 1985.

10
Separation of Polymer Blends by Interaction Chromatography

HARALD PASCH   Department of Polymer Analysis, German Plastics Institute, Darmstadt, Germany

I.   Introduction   387

II. Liquid Chromatography of Polymer Blends   389
     A. Thermodynamic fundamentals   389
     B. Size exclusion chromatography   392
     C. Gradient HPLC   397
     D. Chromatography under limiting conditions of solubility   400

III. Separation of Polymer Blends by Liquid Chromatography at the
     Critical Point of Adsorption   403
     A. General relations   403
     B. Separation of binary blends of homopolymers   406
     C. Separation of blends of homo-and copolymers   415

IV. Analysis of Polymer Blends by Coupling Liquid Chromatography
     and Selective Detectors   419
     A. Combination of LCCC and a viscometer detector   420
     B. Coupled liquid chromatography and FTIR spectroscopy   423

     References   431

I. INTRODUCTION

Polymer blends are mixtures of two or more polymer components of different chemical composition. They are of increasing commercial importance for a number of applications, particularly in the construction and automotive industries. The advantage of polymer blends is the useful combination of the properties of the components without the creation of chemically new polymers. This approach in many cases is more feasible than developing new tailor-made polymer structures. Two of many prominent blend systems are rubber-modified polystyrene, known as high-impact polystyrene, and mixtures of butyl rubber and styrene-butadiene copolymers for the production of car tires.

The chemical composition of polymer blends may be rather complex because as blend components homopolymers as well as copolymers may be used. Accordingly, binary polymer blends may be composed of two homopolymers, a homopolymer and a copolymer, or two copolymers. In addition, very frequently compatibilizers are used in technical blends due to the fact that most homopolymers are immiscible.

In view of the complexity of polymer blends, the following protocol can be formulated for polymer analysis:

Determination of the molar mass distribution (MMD) of the blend
Identification of the blend components and quantitative determination of the blend composition
Determination of the MMD of the blend components
Determination of the chemical composition of the blend components (in the case of copolymers)

The identification and quantitative determination of blend components is complicated. Depending on the specific chemical structure a variety of different analytical methods must be used. Spectroscopic methods such as Fourier transform infrared spectroscopy (FTIR) and nuclear magnetic resonance (NMR) can help to identify blend components. However, in many cases they are unable to answer the question of whether two chemical structures are combined to yield a copolymer or a blend or both. For example, in analyzing a rubber mixture one can identify styrene and butadiene as the monomer units. However, using FTIR or NMR one is unable to say if the sample is a mixture of polystyrene (PS) and polybutadiene (PB), or a copolymer of styrene and butadiene, or a blend of a styrene-butadiene copolymer and PB. For the last case, even the copolymer composition cannot be determined just by running an FTIR or NMR spectrum.

In most cases, for the precise determination of the blend composition including the composition and MMD of the components a separation step is required. Only after obtaining fractions comprising the different blend components can an analysis with regard to chemical composition and MMD be conducted. This chapter discusses different options to use liquid chromatography for the separation of polymer blends. It shows that optimum results can be obtained when selective separation techniques are combined with selective detectors.

II. LIQUID CHROMATOGRAPHY OF POLYMER BLENDS

A. Thermodynamic Fundamentals

Chromatographic separation of macromolecules by size, chemical composition, or architecture relates to the selective distribution of the macromolecules between the mobile and the stationary phase of a given chromatographic system. This distribution process occurs multiple times within one chromatographic run, and depending on the strength and type of the interactions, more or less resolved component peaks are obtained. The separation process is described by

$$V_R = V_i + VK_d \tag{1}$$

where V_R is the retention volume of the solute, V_i is the interstitial volume of the column, V is the volume of the packing, i.e., the "stationary" volume, and K_d is the distribution coefficient, which is equal to the ratio of the analyte concentration in the stationary phase and in the mobile phase. Note that V can comprise the pore volume V_p, surface area V_a, or volume of chemically bonded "stationary phase" on the packing V_{stat}, depending on the separation mode and type of packing.

K_d is a function of the change in Gibbs' free energy ΔG related to the analyte partitioning between the mobile and the stationary phases [1].

$$\Delta G = \Delta H - T\Delta S = -RT \ln K_d \tag{2}$$

$$K_d = \exp (\Delta S/R - \Delta H/RT) \tag{3}$$

The change in Gibbs' free energy may be due to different effects:

1. Inside the pore, which has limited dimensions, the macromolecule cannot occupy all possible conformations and therefore the conformational entropy ΔS decreases.
2. When penetrating the pores, the macromolecule may interact with the pore walls resulting in a change in enthalpy ΔH.

Depending on the chromatographic system and the chemical structure of the macromolecule, only entropic or enthalpic interactions, or both, may be operating. Therefore, in the general case the distribution coefficient may be expressed as:

$$K_d = K_{SEC}K_{LAC} \tag{4}$$

where K_{SEC} is based on entropic interactions, whereas K_{LAC} characterizes the enthalpic interactions. Depending on the magnitude of entropic and enthalpic effects, the size exclusion mode (SEC, ΔS-driven) or the liquid adsorption mode (LAC, ΔH-driven) will be predominant.

In size exclusion chromatography, separation is accomplished with respect to the hydrodynamic volume of the macromolecules. The stationary phase is a swollen gel with a characteristic pore size distribution, and depending on the size of the macromolecules a larger or lesser fraction of the pores is accessible to the macromolecules.

In *ideal SEC*, separation is exclusively directed by conformational changes of the macromolecules and ΔH by definition is zero. Thus:

$$K_d = K_{SEC} = \exp(\Delta S/R) \tag{5}$$

Since the conformational entropy decreases ($\Delta S < 0$), the distribution coefficient of ideal SEC is $K_{SEC} < 1$. The maximum value, $K_{SEC} = 1$, is related to zero change in conformational entropy, i.e., to a situation where all of the pore volume is accessible to the macromolecules (separation threshold). At $K_{SEC} = 0$, the analyte molecules are too large to penetrate the pores (exclusion limit). Accordingly, the separation range is $0 < K_{SEC} < 1$.

The retention volume for ideal SEC is

$$V_R = V_i + V_p K_d = V_i + V_p K_{SEC} \tag{6}$$

If enthalpic effects, due to electrostatic interactions between macromolecules and the pore walls, have to be taken into account, the distribution coefficient K_d of *real SEC* is as follows:

$$K_d = \exp(\Delta S/R - \Delta H/RT) = \exp(\Delta S/R)\exp(-\Delta H/RT) = K_{SEC}K_{LAC} \tag{7}$$

In this case, the retention volume is a function of K_{SEC} and K_{LAC}. If electrostatic interactions occur at the outer surface of the stationary phase as well, an additional term $V_{stat}K_{LAC}$ has to be accounted for.

In adsorption chromatography (LAC), where separation is directed by adsorptive interactions between the macromolecules and the stationary phase, an ideal and a real case may be defined as well. In *ideal LAC* conformational changes are assumed to be zero ($\Delta S = 0$) and the distribution coefficient is exclusively determined by enthalpic effects.

$$K_d = K_{LAC} = \exp(-\Delta H/RT) \tag{8}$$

Depending on the pore size of the stationary phase two possible cases have to be discussed:

1. For narrow-pore stationary phases, separation occurs exclusively at the outer surface. The pores are not accessible to the macromolecules ($K_{SEC} = 0$). Accordingly, the retention volume is a function of the interstitial volume and the volume of the stationary phase (V_{stat}):

$$V_R = V_i + V_{stat}K_{LAC} \tag{9}$$

2. If the solute can freely penetrate the pore volume of the stationary phase ($K_{SEC} = 1$), the pore volume adds to the interstitial volume.

$$V_R = V_i + V_p + V_{stat}K_{LAC} \tag{10}$$

In *real LAC* only a fraction of the pores of the packing is accessible and therefore entropic interactions must be assumed. Accordingly, the distribution coefficient is a function of ΔH and ΔS (compare to real SEC). The retention volume now is a function of enthalpic interactions at the surface of the packing, entropic effects owing to the limited dimensions of the pores, and possible electrostatic interactions inside the pores. Therefore, the expression for V_R in real LAC is formally similar to that in real SEC:

$$V_R = V_i + V_p(K_{SEC}K_{LAC}) + V_{stat}K_{LAC} \tag{11}$$

As the enthalpic interactions are based on a multiple attachment mechanism, it is clear that the retention volume increases with increasing molar mass [2].

Real SEC and real LAC are often mixed mode chromatographic methods with predominance of entropic or enthalpic interactions. With chemically heterogeneous polymers, effects are even more dramatic because exclusion and adsorption act differently on molecules of different composition.

In a more general sense, the size exclusion mode of liquid chromatography relates to a separation regime whereby entropic interactions are predominant, i.e., $T \Delta S > \Delta H$. In the reverse case, $\Delta H > T \Delta S$, separation is mainly directed by enthalpic interactions. As both separation modes in the general case are affected by the macromolecule size and the pore size, a certain energy of interaction ε may be introduced, characterizing the specific interactions of the monomer unit of the macromolecule and the stationary phase. ε is a function of the chemical composition of the monomer unit, the composition of the mobile phase of the chromatographic system, the characteristics of the stationary phase, and the temperature. The theory of adsorption at porous adsorbents predicts the existence of a finite critical energy of adsorption ε_c, where the macromolecule starts to adsorb at the stationary phase. Thus, at $\varepsilon > \varepsilon_c$ the macromolecule is adsorbed, whereas at $\varepsilon < \varepsilon_c$ the macromolecule remains unabsorbed. At $\varepsilon = \varepsilon_c$ the transition from the unabsorbed to the adsorbed state takes place, corresponding to a transition from SEC to adsorption. This transition is termed "critical point of adsorption" and relates to a situation where the adsorption forces are exactly compensated by entropy losses [3,4].

$$T \Delta S = \Delta H \tag{12}$$

$$\Delta G = 0 \tag{13}$$

Accordingly, at the critical point of adsorption the Gibbs free energy is constant and the distribution coefficient is $K_d = 1$, irrespective of the molar mass of the

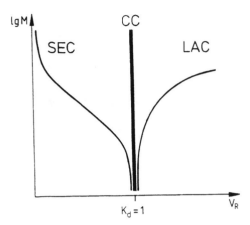

FIG. 1 Chromatographic behavior molar mass versus retention volume in the three modes of liquid chromatography: size exclusion (SEC), critical mode (CC), and liquid adsorption (LAC).

macromolecules and the pore size of the stationary phase. The molar mass versus retention volume behaviour in the *critical mode of liquid chromatography* is schematically represented in Fig. 1 in comparison to SEC and LAC. The critical point of adsorption relates to a very narrow range between the size exclusion and adsorption modes of liquid chromatography, a region very sensitive to temperature and mobile phase composition.

The transition from one to another chromatographic separation mode by changing the temperature or the composition of the mobile phase for the first time was reported by Tennikov et al. [5] and Belenkii et al. [6,7]. They showed that a sudden change in elution behavior may occur by small variations in the solvent strength. Thus, just by changing the eluent composition gradually, a transition from the SEC to the LAC mode and vice versa may be achieved. The point of transition from SEC to LAC is the critical point of adsorption and chromatographic separations at this point are termed *liquid chromatography at the critical point of adsorption* (LCCC).

B. Size Exclusion Chromatography

SEC is the most commonly used method for the determination of MMD of polymers. It separates macromolecules with respect to their hydrodynamic volume, and, using an appropriate calibration, the hydrodynamic volume of homopolymer molecules can be directly correlated with chain length. However, when analyzing heterogeneous systems such as polymer blends, one must be very careful. The dimensional distribution of macromolecules can in general be unambiguously correlated with MMD only within one heterogeneity type. For samples consisting of molecules of different chemical composition, the distribution obtained represents an average of dimensional distributions of molecules having a different composition and, therefore, cannot be attributed to a certain type of macromolecule.

The inadequacy of using SEC without further precaution for the determination of MMD of polymer blends can be explained with reference to Fig. 2 [8]. For a linear homopolymer distributed only in molar mass, fractionation by SEC results in one molar mass being present in each retention volume. The polymer at each retention volume is monodisperse. If a blend of two linear homopolymers is fractionated, then two different molar masses can be present in one retention volume. If one of the blend components is a copolymer, then a multitude of different combinations of molar mass, composition, and sequence length can be

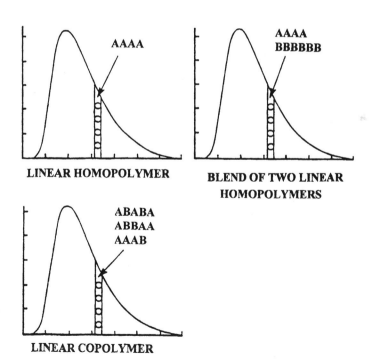

FIG. 2 SEC fractionation showing composition at a given retention volume. (From Ref. 8.)

combined to give the same hydrodynamic volume. In this case, fractionation with respect to molecular size is completely ineffective in assisting the analysis of composition or MMD.

For demonstration, the SEC behavior of different polymethacrylates is given in Fig. 3. On silica gel as the stationary phase and methyl ethyl ketone as the eluent all polymethacrylates elute in the conventional SEC mode. The calibration curves of elution volume versus molar mass for polymethyl methacrylate (PMMA), poly-*t*-butyl methacrylate (PtBMA), poly-*n*-butyl methacrylate (PnBMA), and polydecyl methacrylate (PDMA) reflect the inability of the system to separate different polymethacrylates of similar molar mass. Samples of molar mass of about 100,000 g/mol would elute at elution volumes of 3.08 mL (PnBMA), 3.21 mL (PtBMA), 3.32 mL (PDMA), and 3.43 mL (PMMA). Since the peak width of a narrow disperse sample is about 0.5 mL, even PnBMA and

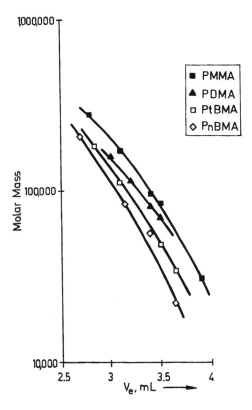

FIG. 3   SEC calibration curves of molar mass versus elution volume for different poly-methacrylates. Stationary phase: LiChrospher 300 + 1000 Å, eluent: methyl ethyl ketone.

PMMA could not be separated into two individual peaks. In terms of molar mass differences the following situation is apparent for the separation of a mixture of PMMA having a molar mass of 200,000 g/mol and other polymethacrylates into two elution peaks (distance of the elution peak maxima shall be an elution volume of 0.5 mL), it is necessary that the molar mass of the second component be less than 32,000 g/mol and 65,000 g/mol for PnBMA and PDMA, respectively.

A more feasible approach to the analysis of polymer blends by SEC is the combination of separation by size with multiple detection. SEC with dual detection is known to provide information on the composition at any point of the elution curve [9–13]. This requires two detectors with sensitivities to the components that are sufficiently different, or a detector that can measure two variables simultaneously, such as a diode array photometer or an infrared detector [14,15]. For practical reasons, this technique has been applied mostly to UV-absorbing polymers because the use of an IR flow-through detector cell is rather limited due to the absorption of the eluent [16].

The common combination of the refractive index (RI) with the UV detector can, however, only be applied if at least one of the monomers absorbs a suitable wavelength and if the UV spectra of both components are sufficiently different. Successful applications of this setup are the analysis of mixtures of PS with PMMA, PB, PVC, or PtBMA. The RI detector provides the total elution profile, whereas the UV detector yields the elution profile of PS. Substracting the latter from the former, the elution profile of the nonabsorbing component can be generated.

For polymer systems without UV activity the combination of the RI detector with a density detector can be used. The working principle of the density detector is based on the mechanical oscillator method. Since this detector yields a signal for every polymer, provided that its density is different from the density of the mobile phase, this detector can be regarded as universal [17–19]. The separation of mixtures of PS and PB by SEC with dual-density RI detection is presented in Figs. 4 and 5. In a first set of experiments, the response factors of both polymers in the two detectors have to be determined. Then, from the intensity of each slice of the elution curves in both detectors, the mass distribution of both polymers along the elution volume axis can be calculated.

As can be seen in Fig. 4, a separation into the component peaks is obtained due to the fact that the molar masses of PS and PB are sufficiently different. For both components the individual elution profiles can be determined and using corresponding calibration curves for PS and PB the individual MMDs can be calculated. The same information can be extracted from an experiment, where the molar masses of the components are similar and SEC separation does not work (see Fig. 5). Again the individual mass distributions are obtained and the MMDs for PS and PB can be determined.

The limitation of SEC with dual detection is that only binary blends of homo-

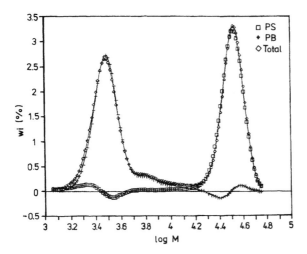

FIG. 4 Mass distribution and separated distributions of the components of a mixture of PS 50,000 and PB 3000 from SEC with D-RI detection. Stationary phase: styragel; mobile phase: chloroform. (From Ref. 20.)

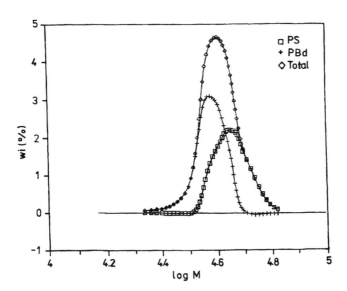

FIG. 5 Mass distribution and separated distributions of the components of a mixture of PS 50,000 and PB 31,400 from SEC with D-RI detection. Stationary phase: styragel; Mobile phase: chloroform. (From Ref. 20.)

polymers can be investigated successfully. In the case of ternary blends, more than two detectors must be used or one of the detectors must be able to detect two components simultaneously. For blends of homopolymers and copolymers this approach cannot be used because the copolymer itself is chemically heterogeneous.

C. Gradient HPLC

Apart from using entropic interactions to promote separation as in SEC, enthalpic interactions or a combination of entropic and enthalpic interactions can be used for separating polymer blends. In particular, enthalpic interactions of the solute molecules and the stationary phase may be used for the separation of copolymers and polymer blends with respect to chemical composition. For stationary phases of a certain polarity, very specific precipitation/redissolution processes are able to promote such separations. Using solvent mixtures as the mobile phase, the precipitation/redissolution equilibria may be adjusted to a specific retention behavior.

Polymer samples that are rather homogeneous in chemical nature and chain length can be eluted isocratically, meaning that the composition of the mobile phase is constant throughout the chromatographic run. If, however, the sample consists of species of different chemical composition, isocratic elution cannot be applied because the different species exhibit retention properties that vary too much. For example, in a given chromatographic system one component could be irreversibly adsorbed due to very strong enthalpic interactions, whereas another component could elute without being retained. By changing the composition of the mobile phase during the chromatographic run, the solubility and the enthalpic interactions of the sample components may be changed. Thus, using gradient elution techniques, the polymer sample may be fractionated with respect to composition over a very broad range of different compositions. Gradient elution in the normal phase mode requires the use of a polar stationary phase together with a gradient whose polarity increases in the course of the run. In the reversed phase mode of gradient elution a nonpolar stationary phase is combined with a gradient whose polarity decreases in the course of the run.

Very frequently, precipitation processes are intentionally used for the separation according to chemical composition. In this case the initial mobile phase is a nonsolvent for one component, which precipitates on the top of the column, while the other component is eluted. By increasing the solvent strength, the initially precipitated component is also recovered. If the polarity of the stationary phase has no effect on elution, adsorption of the redissolved macromolecules on the stationary phase can be excluded, and separation is solely governed by solubility. An overview on different techniques is given in a number of reviews [1,21–23].

Various modifications of gradient HPLC of polymers have been described in the literature, which differ mainly in the nature of gradient employed. In many cases it is not possible to judge whether a certain separation occurs solely by adsorption or whether entropic interactions are involved. Also, the extent of precipitation involved cannot be completely predicted. The composition of the mobile phase at which a given sample is dissolved can be determined by turbidimetric titration [1,24–26].

Most of the work in gradient HPLC of polymers was conducted with respect to chemical composition analysis of copolymers. One of the first separations of random copolymers was carried out by Teramachi et al. [27]. Mixtures of poly(styrene-co-methyl acrylate)s were separated by composition on silica columns through a carbon tetrachloride/methyl acetate gradient. When increasing the content of methyl acetate in the eluent, retention increased with increasing methyl acrylate content in the copolymer. Similar separations could be achieved on other columns as well, including polar bonded-phase columns (diol, nitrile, amino columns) [1]. Other applications have been published by Glöckner, Mori, Schunk, and others, showing the usefulness of gradient HPLC [28–43].

Compared to random copolymers, a rather limited amount of work has been devoted to mixtures of homopolymers. The separation of mixtures of poly(meth)-acrylates by gradient HPLC has been successfully conducted by Mourey [44]. The effect of the alkoxy group on the retention behavior of the poly(meth)acrylates provided the chance of separating different (meth)acrylate homopolymers through normal phase gradient elution. Figure 6A shows the separation of polymethacrylate esters on silica gel using a toluene–methyl ethyl ketone gradient. As was expected from the polarity of the alkoxy group, polybutyl methacrylate was less retained than polymethyl methacrylate. The same sequence holds for polyacrylates; see Fig. 6B. From the chromatograms the composition of the samples may be quantitatively determined. As in this case separation is accomplished with respect to chemical composition and not molar mass, a second chromatographic method must be used for the determination of the MMD of the chromatographic fractions.

The fractionation of polymer blends by multiple solvent gradient elution was discussed by Jansen et al. [45]. Following the precipitation of the blends in a nonsolvent, fractionation was achieved according to solubility by multiple solvent elution with increasing solvent power. Reproducible baseline resolved fractionation of some commercial blends was obtained, including blends of polyphenylene oxide/polystyrene (PPO/PS) and ethylene-propylene-diene copolymer/polystyrene (EPDM/PS) (see Fig. 7). In the first case a mobile phase of isooctane-THF-chloroform was used, whereas in the second case methanol-toluene formed the mobile phase.

Precipitation chromatography was used by Mori for the identification of polymer components in mixtures [46]. He injected PS, PVC, polyvinyl acetate, and

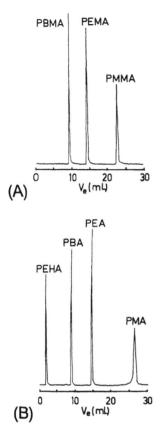

FIG. 6 Separation of polymethacrylates (A) and polyacrylates (B) by gradient HPLC. Stationary phase: silica gel, mobile phase: toluene–methyl ethyl ketone. (From Ref. 44.)

various polymethacrylates onto a silica column and eluted with different solvents, including THF, chloroform, ethyl acetate, methanol, etc. Depending on the solubility of a specific component, this component was eluted or retained in the column with a specific solvent. By testing different solvents, different polymer components were eluted and could be identified by their solubility. An extensive study on the correlation of retention times, cloud points, and solubility parameters has been published recently by Staal [47]. He investigated the chromatographic behavior of PS, PMMA, and PB in different eluents and discussed the mechanism of separation and possibilities of predicting the elution behavior.

The characterization of polymer mixtures by temperature-rising gradient HPLC was published by Lee and Chang [48]. Using a C18 reversed phase and

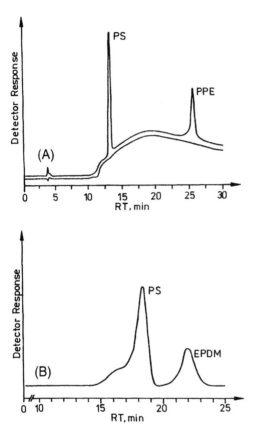

FIG. 7   Gradient HPLC separation of PPO/PS (A) and EPDM/PS blends (B). (From Ref. 45.)

an eluent of methylene chloride–acetonitrile 57:43% (v/v) the authors were able to separate a mixture of 5 PMMA and 11 PS calibration standards, see Fig. 8 (Table 1). Keeping the composition of the mobile phase constant, the temperature was increased from 5°C to 45 °C within the chromatographic run. The PMMA samples eluted before the PS samples in the order of decreasing molar masses, whereas PS eluted in the order of increasing molar masses. Obviously, both size exclusion and adsorption mechanisms are responsible for this type of separation.

D.   Chromatography Under Limiting Conditions of Solubility

This type of liquid chromatography of polymers refers to a situation where the macromolecules move with the solvent zone and elute at the "limit" of their

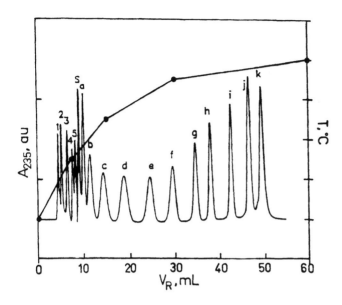

FIG. 8 Temperature gradient HPLC analysis of a set of 11 PS and 5 PMMA standards. Stationary phase: Nucleosil-C18 100 + 500 + 1000 Å; mobile phase: CH_2Cl_2-ACN 57: 43% (v/v). Temperature program as indicated.

solubility. The limiting conditions of solubility are achieved when the polymer sample is dissolved in a thermodynamically good solvent and injected in a mobile phase that is a weak nonsolvent for the polymer. In this case, homopolymers with different molar masses elute from the chromatographic column at the same retention volume, which is roughly equal to the volume of liquid in the column. Accordingly, under limiting conditions of solubility there is no separation ac-

TABLE 1 PMMA and PS Standards Separated by Temperature Gradient HPLC

Code	Polymer	M_w (g/mol)	Code	Polymer	M_w (g/mol)
1	PMMA	1,500,000	d	PS	22,000
2	PMMA	501,000	e	PS	37,300
3	PMMA	77,500	f	PS	68,000
4	PMMA	8,500	g	PS	114,000
5	PMMA	2,000	h	PS	208,000
a	PS	1,700	i	PS	502,000
b	PS	5,100	j	PS	1,090,000
c	PS	11,600	k	PS	2,890,000

cording to molar mass, and the method can be applied to separations based on other properties of the polymer sample [49–51].

The following mechanism is suggested by Berek et al. to cause the limiting condition phenomena [52]: At low levels of nonsolvent, the calibration curves shift slightly to lower retention volumes due to the suppression of adsorption and a reduced effective pore size of the column packing. At higher quantities of nonsolvent, in the vicinity of the θ-composition, the thermodynamic quality of the solvent is strongly reduced. If such a mixture is used as the eluent and the injected polymer is dissolved in a good solvent, the macromolecules move together with the zone of their initial solvent. If macromolecules move faster due to exclusion processes, they encounter the nonsolvent and precipitate. As the injection zone (good solvent) reaches the precipitated macromolecules, they redissolve and move with the injection zone. This "microgradient" process of

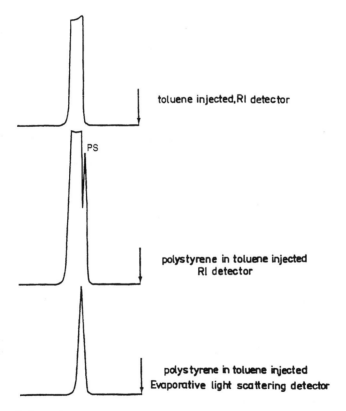

FIG. 9 Elution of polystyrene and solvent in a limiting condition experiment using RI and ELSD detection. (From Ref. 54.)

precipitation/redissolution occurs multiple times with the polymer eluting just in the front part of the solvent zone (Fig. 9). As a consequence, the macromolecules move with a velocity similar to that of the solvent zone and elute at the limit of their solubility. Because the polymer elutes very close to the solvent peak, an evaporative light scattering detector (ELSD) must be used to "see" the polymer peak.

So far, limiting conditions have been reported for PS, PMMA, and polyvinyl acetate [53,54]. Using this technique, blends of PS and PMMA have been separated on silica gel as the stationary phase and THF-n-hexane 82:18% (v/v) as the limiting condition eluent. In this case, PS elutes under SEC conditions with its MMD correctly estimated, whereas PMMA elutes at limiting conditions. Further, it has been shown that under limiting conditions of solubility the tacticity of macromolecules can be estimated [55].

III. SEPARATION OF POLYMER BLENDS BY LIQUID CHROMATOGRAPHY AT THE CRITICAL POINT OF ADSORPTION

A. General Relations

As was already pointed out in Sec. II.A, LCCC refers to a situation where the entropy losses exactly compensate the adsorptive interactions in the chromatographic system. At the critical point of adsorption, the Gibbs free energy is constant and the distribution coefficient K_d is 1, irrespective of the molar mass of the macromolecules. Under these conditions, all homopolymer molecules of the same chemical structure elute at the same retention volume and separation is accomplished with respect to other types of molecular heterogeneity, including chemical composition, functionality, and molecular architecture [56–58]. LCCC has been successfully used for the determination of the functionality type distribution of telechelics and macromonomers [59–65], for the analysis of block copolymers [66–71], and for the separation of macrocyclic polymers [72].

The general behavior of a binary blend in different chromatographic modes is summarized in Fig. 10 [73]. When separating with respect to chain length, the retention behavior for both blend components is very similar in the size exclusion and the adsorption modes, and the calibration curves log molar mass versus retention time suggest that in these cases normally one retention time corresponds to two molar masses (one molar mass on each calibration curve). Thus, for comparable molar masses of the components an overlapping of the elution zones is obtained. Accordingly, a sufficient separation of the components using SEC or adsorption chromatography may be achieved only when their molar masses are quiet different.

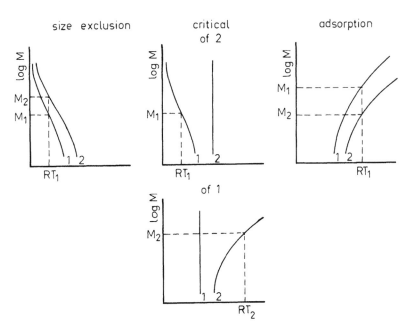

FIG. 10   Behavior of the calibration curves log molar mass versus retention time of a binary polymer blend in the three chromatographic modes. (From Ref. 73.)

A completely different behavior of the blend components is obtained when chromatographic conditions are used, corresponding to the critical mode of one of the components. In this case, the elution zones are separated from each other over the entire molar mass range, and separation is achieved even for components of similar molar mass. This type of separation holds much promise for the development of general separation schemes of polymer blends. With chromatographic conditions corresponding to the critical point of component 2, separation should be possible in any case. In addition, since component 1 is eluted in the SEC mode, its molar mass distribution can be determined. Changing the chromatographic system to the critical point of component 1, one can analyze component 2 with respect to its MMD.

From a thermodynamic point of view, the total change in the Gibbs free energy of a binary blend $A_n + B_m$ can be regarded as the sum of the contributions of the components A_n and B_m.

$$\Delta G_{AB} = \Sigma \ (n_A \Delta G_A + n_B \Delta G_B) \tag{14}$$

As specific interactions between components A and B are usually negligible in the chromatographic system, an interaction parameter χ_{AB} must not be introduced in Eq. (14). Using chromatographic conditions, corresponding to the critical point of homopolymer A, component A_n will be eluted at $K_d = 1$ irrespective of molar mass, whereas component B_m will be separated according to its molar mass. The total change in the Gibbs free energy is solely due to component B_m.

$$\Delta G_A = 0 \tag{15a}$$

$$\Delta G_{AB} = n_B \Delta G_B \tag{15b}$$

Vice versa, at the critical point of homopolymer B, B_m will be eluted at $K_d = 1$.

$$\Delta G_B = 0 \tag{16a}$$

$$\Delta G_{AB} = n_A \Delta G_A \tag{16b}$$

Depending on the polarity of components A_n and B_m in the binary blend A_n + B_m, and the polarity of the stationary phase, different chromatographic situations can be encountered, (Fig. 11). For example, at the critical point of homopolymer A, homopolymer B may be separated either in the SEC mode (1) or the adsorption mode (2). The same is true for the critical point of B, where A may be eluted according to (3) or (4). Preferable, of course, are the cases (1) and (3), whereas in the cases (2) and (4) for high molar mass polymers irreversible adsorption may be encountered. Let us now consider that the polarity of A_n is higher than

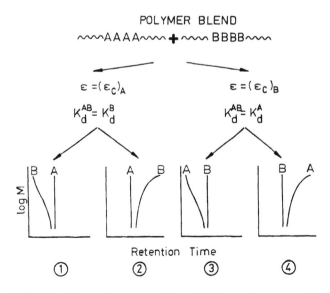

FIG. 11   Schematic representation of different chromatographic situations in LCCC of binary polymer blends. For explanation of 1–4, see text.

the polarity of B_m. In this case, chromatographic behavior according to (1) is achieved, when silica gel is used as the stationary phase (silica gel separates in the order of *increasing* polarity). On the other hand, separation according to (3) is obtained on a reversed phase, such as RP-8 or RP-18 (separation in the order of *decreasing* polarity).

B. Separation of Binary Blends of Homopolymers

Blends of PS and PMMA are very common model systems. Depending on the preparation procedure they form immiscible blends of different morphology. Since the hydrodynamic volumes of PS and PMMA are rather similar, it is not possible to separate blends of them by SEC when the molar masses of the components are close to each other. Therefore, critical chromatography shall be used to separate PS-PMMA blends. Following the discussion on elution behavior as a function of column polarity (see Fig. 11), different stationary phases must be selected for establishing the critical points of PS and PMMA, respectively. Since PMMA is the more polar component, a polar (silica gel) column is chosen for establishing its critical point. PS is then eluted in the SEC mode. For establishing the critical point of PS, however, a reversed stationary phase must be used. PMMA is eluted in the SEC mode under these conditions.

The behavior of PMMA of different molar masses on silica gel in eluents comprising methyl ethyl ketone (MEK) and cyclohexane is given in Fig. 12A. At concentrations of MEK >73% by volume the SEC mode is operating, whereas at concentrations <73% by volume of MEK, adsorption takes place. The critical point of PMMA is obtained at an eluent composition of MEK-cyclohexane 73: 27% (v/v). At this point, all PMMA samples are eluted at the same retention time irrespective of their molar mass. Depending on the size of the macromolecules under investigation, similar to conventional SEC the pore size of the stationary phase has to be adjusted to the desired molar mass range. Thus, for higher molar mass samples the investigations must be carried out on column sets with larger pores (see Fig. 12B for a two-column set of silica gel with pores sizes of 300 and 1000 Å). For these columns the critical point of PMMA corresponds to a mobile phase composition of MEK-cyclohexane 70:30% (v/v).

For establishing the critical point of PS, a reversed phase Nucleosil RP-18 is used (Fig. 13). The figure indicates that at concentrations of (THF) >88% by volume, separation is predominantly driven by entropic effects and the SEC mode is operating. In contrast, the LAC mode with predominantly enthalpic interactions is operating at concentrations of THF <87% by volume in the eluent. The critical point of adsorption of PS is obtained at an eluent composition of THF–water 88.8:11.2% (v/v). At this point, all PS samples regardless of their molar mass elute at one retention time. This, by definition, indicates, that the PS polymer

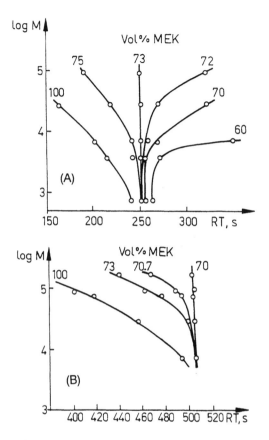

FIG. 12 Critical diagrams molar mass versus retention time of PMMA. Stationary phase: Nucleosil Si-100 (A) or LiChrospher Si-300 + Si − 1000 (B). Mobile phase: MEK–cyclohexane. (From Ref. 67.)

chain is "chromatographically invisible," i.e., does not contribute to retention. Accordingly, using these chromatographic conditions, blends of PMMA and PS can be analyzed with respect to the PMMA component.

For higher molar mass samples the investigations must be carried out on column sets with larger pores; see Fig. 13B for a two-column set of RP-18 300 + 1000 Å. For this column set, critical conditions were found to be operating at an eluent composition of THF–water 88.1 : 11.9% by volume.

In a first set of experiments, PS-PMMA blends are separated under chromatographic conditions, corresponding to the critical point of PS. By the use of

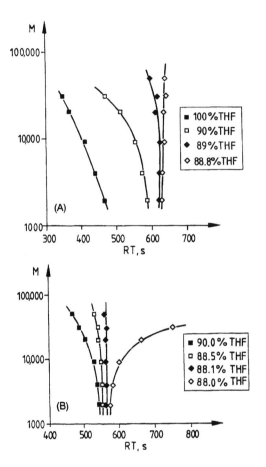

FIG. 13 Critical diagrams molar mass versus retention time of PS. Stationary phase: Nucleosil RP-18 100 Å (A) or RP-18 300 Å + 1000 Å (B); mobile phase: THF-water.

Nucleosil RP-18 with an average pore size of 100 Å, the blends are completely separated into their components (Fig. 14), although the blends under investigation are composed of PS and PMMA of similar molar masses.

The separation of the PS-PMMA blends can also be carried out under conditions corresponding to the critical point of PMMA (Fig. 15). With silica gel Si-100 as the stationary phase and MEK-cyclohexane as the eluent, different chromatographic modes can be established. In pure MEK, the size exclusion mode is operating for both components. Under these conditions PS and PMMA may be separated only if their molar masses are different. For low molar mass samples

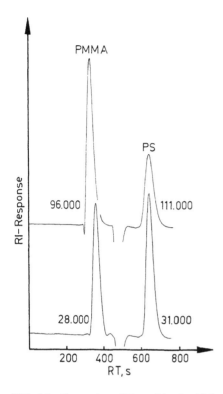

FIG. 14 Separation of binary blends of PS and PMMA at the critical point of PS. Stationary phase: Nucleosil RP-18 100 Å; mobile phase: THF-water 88.8:11.2% (v/v).

(PS, PMMA ~ 30,000 g/mol), the two components of the blend can be identified but separation is poor. For higher molar masses (PS, PMMA ~ 150,000 g/mol); however, one symmetrical elution peak, similar to the elution profile of a homopolymer, is obtained. When cyclohexane is added to MEK, the elution behavior of PMMA changes dramatically, whereas for PS it remains nearly constant. At the critical point of PMMA, which corresponds to a mobile phase composition of MEK-cyclohexane 73:27% (v/v) (see Fig. 12A), a complete separation of the elution zones of PMMA and PS is achieved. Regardless of the molar mass, all PMMA fractions are eluted at the same retention time, whereas for the PS fractions a size exclusion mode is operating, and retention time decreases with increasing molar mass.

The quantitative analysis of the PS-PMMA blends is somewhat straightforward. The blend composition, i.e., the amounts of the components PS and PMMA, is determined via corresponding calibration curves peak area versus con-

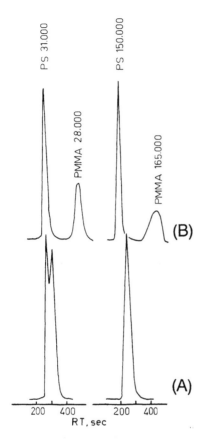

FIG. 15 Chromatograms of PS-PMMA blends in the SEC (A) and in the critical mode (B). Stationary phase: Nucleosil Si-100; mobile phase: MEK-cyclohexane 100:0 (A) or 73:27% (v/v) (B).

centration. This determination can be conducted under critical conditions of PMMA, as well as at critical conditions of PS. The determination of the MMD of the blend components is carried out using conventional SEC calibration procedures. Separating the blend under critical conditions of PS, one can elute and quantify the PMMA in the SEC mode. When the blend is separated under critical conditions of PMMA, the PS is eluted in the SEC mode and its MMD is calculated via a conventional PS calibration curve.

It is known that using a refractive index detector in SEC, very frequently a negative peak is observed at the end of the chromatogram. This "solvent peak" or "system peak" is due to preferential adsorption, residual water, or impurities in the mobile phase. In SEC the solvent peak is well separated from the polymer peaks, and interference does not occur. In LCCC, however, where at the critical point of adsorption the polymer molecules elute close to the solvent peak, an overlapping of the polymer peaks and the solvent peak may occur. This can be avoided when instead of a refractive index detector an on-line viscometer is used; see Fig. 16 for the separation of blends of PMMA and poly-n-butyl methacrylate (PnBMA) [74]. Since a slight change in the composition of the mobile phase due to preferential adsorption does not contribute to a change in viscosity, a solvent peak is not obtained. Accordingly, the elution peaks of the blend components can be detected without interference. It must be taken into account, however, that the viscosity detector, unlike the RI detector, does not yield a concentration signal. The viscometer response is a function of the concentration and the intrinsic viscosity ($c[\eta]$), and unless both components have the same $[\eta]$, the viscometer output cannot be used for the determination of concentrations.

For the determination of the component concentrations by a differential refractometer a different approach may be helpful [75]. It is known from the principles

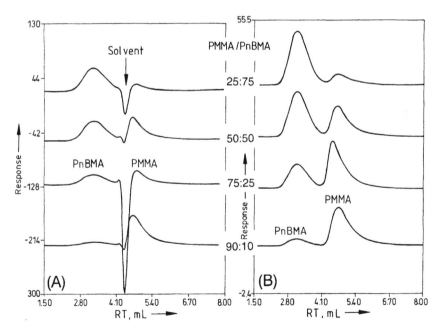

FIG. 16 Chromatograms of PnBMA-PMMA blends of different composition at critical conditions of PMMA. Stationary phase: Nucleosil Si-300 + Si-1000; mobile phase MEK-cyclohexane 72:28% (v/v). Detector: differential refractometer (A) or viscometer (B). (From Ref. 74.)

of critical chromatography that the chromatographic behavior of a polymer in the transition range from SEC to LAC is very sensitive to mobile phase composition. A slight change in the eluent composition may cause a shift from critical conditions to the adsorption mode. Since the position of the solvent peak is rather insensitive to eluent composition, this fact can be used to separate the PMMA elution peak from the solvent peak. Figure 17 demonstrates the changes in the chromatogram when changing the mobile phase composition from MEK-cyclohexane 72:28% to 68:32% (v/v), corresponding to a slight adsorption mode. In this case, the PMMA elution peak is well separated from the solvent peak and quantification can be carried out. The PnBMA elutes in the SEC mode and the solvent peak appears between the elution peaks of the components. Another option is to use an evaporative light scattering detector (ELSD) instead of a refractometer. Since in the ELSD the solvent is evaporated prior to detection, a solvent peak cannot appear in the chromatogram [76–79].

The broad applicability of the LCCC approach to the separation of binary blends is documented in Fig. 18. Operating at the critical point of adsorption of PMMA, all polymethacrylates of lower polarity, including polyethyl methacrylate (PEMA), poly-*n*-butyl methacrylate (PnBMA), poly-*t*-butyl methacrylate (PtBMA), and polydecyl methacrylate (PDMA), elute in the ideal SEC mode.

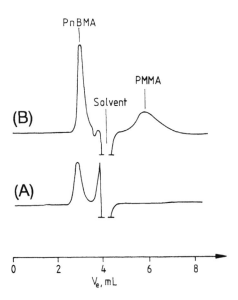

FIG. 17 Chromatograms of a PnBMA-PMMA blend at different compositions of the mobile phase. Stationary phase: LiChrospher Si-300 + Si-1000; mobile phase: MEK-cyclohexane 72:28% (A) or 68:32% (B). (From Ref. 75.)

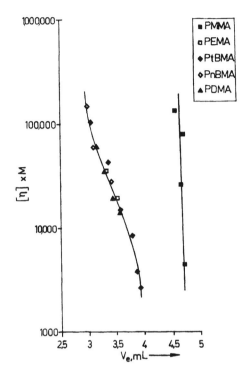

FIG. 18 Universal calibration curve for different polymethacrylates. For chromatographic conditions, see Fig. 16.

Accordingly, a common universal calibration curve is obtained for these polymethacrylates that can be used for molar mass calculations. In any case, the elution zone of PMMA is completely separated from the elution zones of the other polymethacrylates, providing for optimum separation of corresponding binary blends. Partial overlapping of the elution peaks is then exclusively due to axial dispersion.

As has been shown, the key experiment for the successful separation of binary blends is the determination of critical conditions for one of the blend components. Similar to the critical point of PMMA, the critical point of PtBMA can be established on silica gel LiChrospher Si-300 + Si-1000. The critical point of PtBMA corresponds to a mobile phase composition of MEK-cyclohexane 18.8:81.2% (v/v).

The separation of blends of PS and PtBMA using chromatographic conditions, corresponding to the critical point of PtBMA, is shown in Fig. 19. As was expected, in all cases, regardless of the molar masses of the components, a complete

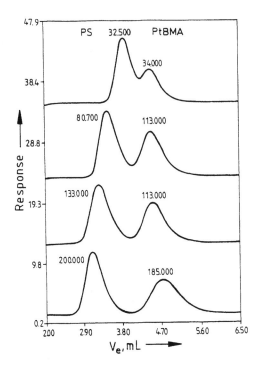

FIG. 19   Separation of blends of PS and PtBMA at the critical point of PtBMA. Stationary phase: LiChrospher Si-300 + Si-1000; mobile phase: MEK-cyclohexane 18.8:81.2% (v/v). Detector: viscometer.

separation is obtained. Since PS is the less polar component in the blend, it elutes first from the column in the SEC mode. For the low molar mass blend, comprising PS 32,500 and PtBMA 34,000, the peaks are not baseline-separated. A better resolution can be obtained by changing the separation range of the stationary phase. A useful combination for this molar mass range would be Nucleosil Si-100 + Si-300.

The ability of liquid chromatography at the critical point of adsorption to separate polymer blend components of minimum structural differences is demonstrated in Fig. 20. Even blends of such very similar polymers like PtBMA and PnBMA can be separated without problems.

With the same resolving power the separation of PtBMA-PnBMA blends can be conducted on a reversed phase system operating at the critical point of PnBMA (Fig. 21). In this case, critical conditions correspond to a mobile phase of THF-acetonitrile 53.1:46.9% (v/v).

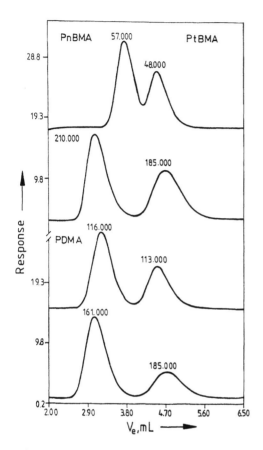

FIG. 20 Separation of blends of PnBMA and PtBMA at the critical point of PtBMA. For chromatographic conditions, see Fig. 19.

To summarize, the mobile phase compositions corresponding to the critical points of PS and a number of polymethacrylates are given in Table 2. As can be expected for a reversed phase system (RP-18), the amount of good solvent in the mobile phase (THF) necessary to elute the polymer from the column increases with increasing hydrophobicity of the polymer. For the normal phase silica system, the amount of good solvent (MEK) increases with increasing polarity of the polymer.

C. Separation of Blends of Homo- and Copolymers

Similar to the separation of blends of homopolymers, blends of homo- and copolymers can be separated by LCCC provided that the polarities of the blend compo-

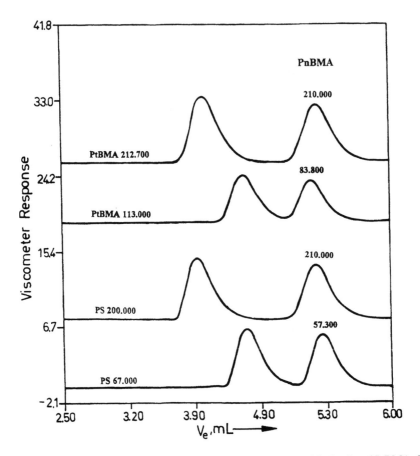

FIG. 21 Separation of blends of PnBMA and PtBMA at the critical point of PtBMA. Stationary phase: Nucleosil RP-18 300 + 1000 Å; mobile phase: THF-ACN 53.1:46.9% (v/v).

TABLE 2 Mobile Phase Compositions Corresponding to Critical Conditions for PS and Polymethacrylates

Polymer	THF-ACN, RP-18[a]	MEK-Cyclohexane, Si[b]
PMMA	—	72:28
PS	49.4:50.6	—
PtBMA	49.6:50.4	18.8:81.2
PnBMA	53.1:46.9	14.3:85.7
PDMA	78.5:21.5	3.34:96.66

[a] Nucleosil RP-18, 300 Å + 1000 Å, 250 × 4 mm ID.
[b] LiChrospher Si, 300 Å + 1000 Å, 200 × 4 mm ID.

nents are different. Since the critical conditions always relate to a specific homopolymer, the best way to carry out such separations is to operate at the critical point of the homopolymer. As has been shown previously, blends of PMMA with poly(styrene-co-methyl methacrylate) and poly(styrene-co-acrylonitrile), respectively, can be separated at the critical point of PMMA [73].

The separation of technical blends of homopolymers and poly(n-butyl methacrylate-co-methyl methacrylate), refered to as P(nBMA-co-MMA), will be discussed in the following section. The samples under investigation are blends of the copolymer and PMMA having the following composition:

Sample	Composition (%) PMMA/copolymer	M_w (g/mol)	M_n (g/mol)
1	25/75	78,700	44,700
2	50/50	79,300	45,700
3	75/25	81,000	46,900

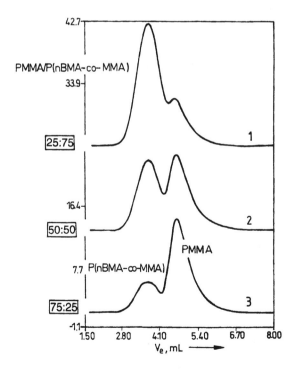

FIG. 22   Separation of polymer blends of PMMA and P(nBMA-co-MMA). For chromatographic conditions, see Fig. 16.

In the previous section it has been demonstrated that due to their different polarity PMMA and PnBMA can be separated by LCCC. If now the polarity of the copolymer P(nBMA-co-MMA) is compared to the polarities of the homopolymers, then the copolymer can be assumed to be more polar than PnBMA but less polar than PMMA. Of course, the polarity of the copolymer depends on the composition; the more MMA is incorporated into the copolymer the higher is the polarity.

If a polar stationary phase is selected for the separations, it can be assumed that the PMMA homopolymer is more strongly retained than the P(nBMA-co-MMA). Optimum separation should occur when the mobile phase composition corresponds to the critical point of PMMA. Figure 22 shows the separation of

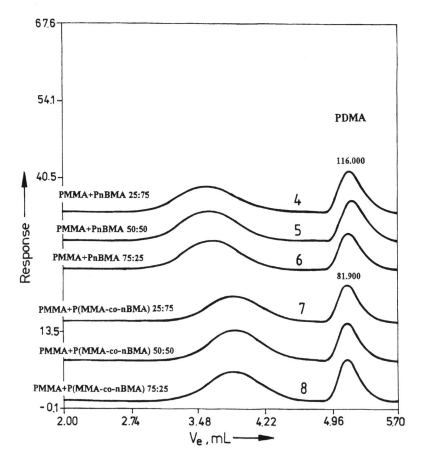

FIG. 23 Separation of PDMA blends with homo- and copolymers. Stationary phase: Nucleosil RP-18; mobile phase: THF-ACN 87.5:21.5% (v/v).

TABLE 3 Average Molar Masses of the Blend Components Determined by LCCC

| | Nominal | | Experimental | | |
| | | | | | |
Sample	c (mg/mL)	M_w (g/mol)	V_e (mL)	$[\eta]$ (dL/g)	M_w (g/mol)
4. PDMA	2.545	116,000	5.355	0.307	116,700
Blend A (25:75)	2.154	104,300	3.79	0.514	109.700[a]
5. PDMA	2.525	116,000	5.39	0.306	116,000
Blend A (50:50)	2.415	97,400	3.825	0.512	102,800[a]
6. PDMA	2.654	81,900	5.355	0.246	79,000
Blend A (75:25)	2.294	107,300	3.84	0.516	94,600[a]
7. PDMA	2.601	81,900	5.13	0.245	78,400
Blend B (25:75)	2.128	78,700	3.90	0.474	82,200[a]
8. PDMA	2.797	81,900	5.125	0.251	81,800
Blend B (75:25)	2.346	81,000	3.885	0.496	84,700[a]

Blend A: PMMA + PnBMA; Blend B: PMMA + P(nBMA-co-MMA).
[a] PMMA calibration.

the copolymer blends under these conditions. Since the polydispersity of the blend components is somewhat high, broad elution peaks for the copolymer and PMMA are obtained and some overlapping of the elution zones cannot be avoided. Nevertheless, separate peaks for the blend components are obtained and, in the present case, the peak intensities reflect the relative component concentrations.

When a copolymer is modified with a less polar homopolymer component, a reversed phase system can be used for separation. Again, operation at the critical point of the homopolymer is prefered. The separation of blends with PDMA as the homopolymer is presented in Fig. 23. The other blend components are technical blends of PnBMA + PMMA and PMMA + P(nBMA-co-MMA).

As can be seen, PDMA is properly separated from the more polar blend components which are eluted in the SEC mode. Using a conventional PMMA calibration curve, these components may be quantified with respect to molar mass (Table 3). The molar mass of PDMA can be calculated from the intrinsic viscosity via the corresponding Mark-Houwink relationship. In the present case, the concentration of the components are known from the blend preparation procedure.

IV. ANALYSIS OF POLYMER BLENDS BY COUPLING LIQUID CHROMATOGRAPHY AND SELECTIVE DETECTORS

In addition to sophisticated separation techniques, the use of selective detectors can help to disclose the composition of a polymer blend. In liquid chromatography of polymers different types of detectors are used, including concentration-sensitive and molar mass–sensitive detectors. These detectors can be universal, measuring a bulk property of the eluate, or selective, measuring a specific property of the solute. Among the universal detectors the most frequently used are the differential refractometer and the evaporative light scattering detector. As for selective detectors, the UV and IR detectors are most common. In particular, FTIR can be of exorbitant value in terms of selective structural information, as will be shown later. Molar mass–sensitive detectors, e.g., differential viscometers and light scattering detectors, are frequently used in SEC because they yield the molar mass of each fraction of a polymer peak. Since the response of such detectors depends on both concentration and molar mass, they have to be combined with a concentration-sensitive detector. An overview of different detection systems in liquid chromatography of polymers is given in Refs. [80–87].

A. Combination of LCCC and a Viscometer Detector

So far the differential viscometer has been treated as a means to avoid solvent peaks in the chromatogram (see Secs. III.B and III.C). However, since the detector signal bears molar mass information, it can be used for obtaining selective information on the components of polymer blends.

The viscosity of a polymer solution is related to the molar mass M via the Mark-Houwink-Sakurada equation:

$$[\eta] = KM^a \tag{17}$$

where $[\eta]$ is the intrinsic viscosity, and K and a are coefficients that are characteristic for a specific polymer, solvent, and temperature.

Viscosity measurement in SEC can be performed by measuring the pressure drop P across a capillary, which is proportional to the viscosity η of the flowing liquid (the viscosity of the pure mobile phase is denoted as η_0). The intrinsic viscosity is defined as the limiting value of the ratio of specific viscosity η_{sp} and concentration c for $c \rightarrow 0$:

$$[\eta] = \lim (\eta - \eta_0)/\eta_0 c = \lim \eta_{sp}/c \tag{18}$$

When a polymer solution passes the capillary, the pressure drop is increased by ΔP. In viscosity detection, one has to determine the viscosity η of the sample solution and the viscosity η_0 of the pure mobile phase. The specific viscosity $\eta_{sp} = \Delta\eta/\eta$ is obtained from $\Delta P/P$. As the concentrations in SEC are typically very low, $[\eta]$ can be approximated by η_{sp}/c.

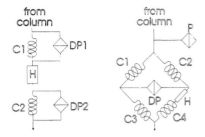

FIG. 24 Schematic representation of differential viscometers.

A schematic representation of a differential viscometer is given in Fig. 24. In the two-capillary design the capillaries C1 and C2 are connected in series, and each is connected to a differential pressure transducer (DP1 and DP2), with a sufficiently large holdup reservoir (H) in between. With this design, one measures the sample viscosity η from the pressure drop across the first capillary, and the solvent viscosity η_0 from the pressure drop across the second capillary. Another design is a differential viscometer, in which four capillaries are arranged similar to a Wheatstone bridge. In the "bridge" design, a holdup reservoir in front of the reference capillary (C4) makes sure that only pure mobile phase flows through the reference capillary, when the peak passes the sample capillary (C3). This design offers considerable advantages: The detector measures actually the pressure difference ΔP at the differential pressure transducer (DP) between the inlets of the sample capillary and the reference capillary, which have a common outlet, and the overall pressure P at the inlet of the bridge. The specific viscosity η_{sp} is thus obtained from $\Delta P/P$. In order to calculate $[\eta]$ from η_{sp}, the concentration at each point of the chromatogram must be determined. Therefore, the viscometer must be coupled to a concentration detector.

As was already discussed in sec. III, LCCC offers the opportunity to separate the components of polymer blends. Chromatographic conditions in this case correspond to the critical point of component A, whereas component B is eluted in the SEC mode. When using a concentration detector, component B can be analyzed with respect to molar mass, whereas component A at the critical point of adsorption does not provide separation according to molar mass.

The molar mass versus elution volume behavior of a typical binary blend is shown in Fig. 25A. PtBMA exhibits conventional SEC behavior, whereas for PMMA under critical conditions the elution volume is independent of molar mass.

On the other hand, both blend components exhibit normal Mark-Houwink behavior, i.e., regardless of their chromatographic behavior the typical $[\eta]-M$ relationship is always operating (see Fig. 25B). Accordingly, from $[\eta]$ the molar

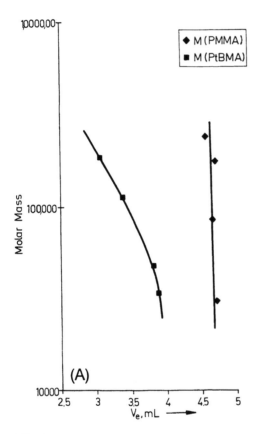

FIG. 25 Molar mass versus elution volume diagram (A) and Mark-Houwink diagram (B) for PMMA and PtBMA under critical conditions for PMMA. Stationary phase: LiChrospher Si-300 + Si-1000; mobile phase: MEK-cyclohexane 72:28% (v/v).

mass of both components can be calculated directly from the Mark-Houwink plot.

The separation of binary blends of PMMA and PtBMA is presented in Fig. 26. Operating at critical conditions for PMMA, PtBMA elutes in the SEC mode (cf. Fig. 25A). The peak areas of the viscometer traces for both components correspond to $[\eta]c$. By determining c with an ELSD detector, the intrinsic viscosities $[\eta]_{PMMA}$ and $[\eta]_{PtBMA}$ can be calculated from the ratio of the viscometer and the ELSD signals. Using the corresponding Mark-Houwink functions (Fig. 25B), from the intrinsic viscosities the molar masses M_{PMMA} and M_{PtBMA} can be calculated (Table 4).

Thus, by combining the selectivity of LCCC with a viscometer and a concen-

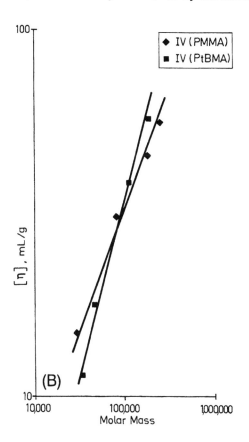

(B)

tration detector, the following information of a binary blend can be obtained in one experiment:

Concentration of component A in the blend
Concentration of component B in the blend
Molar mass of component A in the blend
Molar mass of component B in the blend

B. Coupled Liquid Chromatography and FTIR Spectroscopy

When analyzing a commercial polymer blend, very frequently the first step must be the determination of the gross composition. Only when the chemical structures of the blend components are known, sophisticated separation techniques such as LCCC can be adapted to a specific analysis.

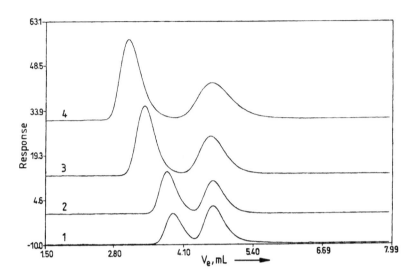

FIG. 26 Separation of binary blends of PMMA and PtBMA under critical conditions for PMMA. For chromatographic conditions, see Fig. 25.

The most frequently used techniques for a "flash" analysis are infrared spectroscopy and SEC. Infrared spectroscopy provides information on the chemical substructures present in the sample, whereas SEC gives a first indication of the molar mass range of the components. Information on both molar mass and composition is obtained when SEC or a comparable chromatographic method is com-

TABLE 4 Average Molar Masses of the Components of PMMA-PtBMA Blends, Determined by LCCC and Viscometric Detection

| Sample | Nominal | | Experimental | | | |
	c (mg/mL)	M_p (g/mol)	V_e (mL)	$[\eta]$ (mL/g)	M (g/mol)	PtBMA/PMMA (wt%)
9. PtBMA	1.946	34,000	4.05	11.3	33,700	45/55
PMMA	2.233	30,500	4.81	14.8	31,100	
10. PtBMA	1.756	48,000	3.93	17.3	46,400	48/52
PMMA	1.990	30,500	4.81	14.8	30,100	
11. PtBMA	1.498	113,000	3.49	37.8	114,600	50/50
PMMA	1.548	85,100	4.83	28.5	85,000	
12. PtBMA	1.254	185,000	3.18	58.0	188,200	48/52
PMMA	1.332	175,000	4.88	40.6	146,400	

bined with an IR detector. In the past, numerous workers have tried to use IR detection of the SEC column effluent in liquid flow cells. The problems encountered relate to obtaining sufficient signal-to-noise (S/N) ratio even with FTIR instruments, flow-through cells with minimum pathlengths, and mobile phases with sufficient spectral windows. Attempts to use FTIR detection with liquid flow-through cells and high-performance columns have not been very successful due to the requirement of considerably less sample concentration for efficient separation.

A rather broad applicability of FTIR as a detector in liquid chromatography can be achieved when the mobile phase is removed from the sample prior to detection. In this case, the sample fractions are measured in pure state without interference from solvents. Experimental interfaces to eliminate volatile mobile phases from HPLC effluents have been tried with some success [88–90] but the breakthrough toward a powerful FTIR detector was achieved only by Gagel and Biemann, who formed an aerosol from the effluent and sprayed it on a rotating aluminum mirror. The mirror was then deposited in an FTIR spectrometer and spectra were recorded at each position in the reflection mode [91–93].

Recently, Lab Connections Inc. introduced the LC-Transform, a direct HPLC–FTIR interface based on the invention of Gagel and Biemann and discussed first applications in polymer analysis [94–96]. The design concept of the interface is shown in Fig. 27. The system is composed of two independent modules: the sample collection module and the optics module. The effluent of the liquid chromatography column is split with a fraction (frequently 10% of the total effluent) going into the heated nebulizer nozzle located above a rotating sample collection disc. The nozzle rapidly evaporates the mobile phase while depositing a tightly focused track of the solute. When a chromatogram has been collected on the sample collector disc, the disc is transfered to the optics module in the FTIR

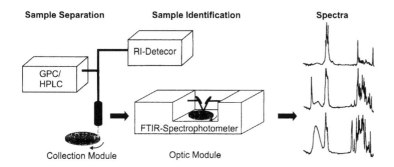

FIG. 27 Schematic representation of the principle of coupled liquid chromatography and FTIR spectroscopy.

for analysis of the deposited sample track. A control module defines the sample collection disc position and rotation rate in order to be compatible with the run time and peak resolution of the chromatographic separation. Data collection is readily accomplished with software packages presently used for GC-FTIR. The sample collection disc is made from germanium, which is optically transparent in the range 6000–450 cm⁻¹. The lower surface of the disc is covered with a reflecting aluminum layer.

As a result of the investigation, a complete FTIR spectrum for each position on the disc and, hence, for each sample fraction is obtained. This spectrum bears information on the chemical composition of each sample fraction. The set of all spectra can be arranged along the elution time axis and yields a three-dimensional plot in the coordinates elution time–FTIR frequency-absorbance.

One of the benefits of coupled SEC-FTIR is the ability to directly identify the individual components separated by chromatography. A typical SEC separation of a polymer blend is shown in Fig. 28.

Two separate elution peaks 1 and 2 are obtained, indicating that the blend contains at least two components of significantly different molar masses. However, a quantification of the components with respect to concentration and molar mass cannot be carried out as long as the chemical structure of the components is unknown.

The analysis of the chemical composition of the sample is conducted by coupled SEC-FTIR using the LC transform. After separating the sample with respect

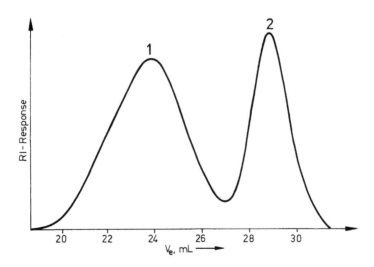

FIG. 28   SEC separation of a binary blend. Stationary phase: Ultrastyragel 2 × linear + 10⁵ Å; mobile phase: THF.

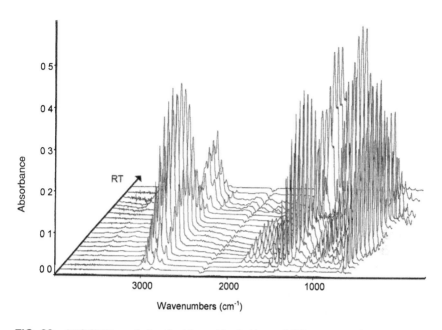

FIG. 29   SEC-FTIR analysis of a binary blend: "waterfall" representation.

to molecular size, the fractions are deposited on the germanium disc and FTIR
spectra are recorded continuously along the sample track. In total, a set of about
80 spectra is obtained, which is presented in a three-dimensional plot (Fig. 29).
The projection of the 3D plot on the retention time–IR frequency coordinate
system yields a two-dimensional representation, where the intensities of the absorption peaks are given by a color code. Such a "contour plot" readily provides
information on the chemical composition of each chromatographic fraction see
(Fig. 30).

It is obvious from Fig. 30 that the chromatographic peaks 1 and 2 have different chemical structures. By comparison with reference spectra that are accessible
from corresponding data bases, component 1 can be identified as polystyrene,
whereas component 2 is polyphenylene oxide. With this knowledge, appropriate
calibration curves can be used for quantifying the composition and the component
molar masses of the blend.

Coupled SEC-FTIR becomes an inevitable tool when blends comprising copolymers have to be analyzed. The contour plot in Fig. 31 reveals two components of different molar masses, component 2 being identified as polycarbonate
by its characteristic absorption peaks. Component 1, however, shows absorption
peaks for styrene units in addition to an intense peak at 2237 cm^{-1}. This peak

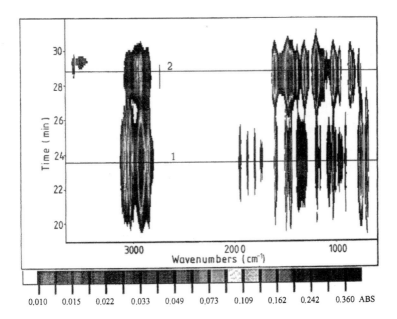

FIG. 30   SEC-FTIR analysis of a binary blend: ''contour plot'' representation.

is due to nitrile groups and indicates that component 1 is composed of styrene and acrylonitrile units. Accordingly, component 1 can be identified as a styrene-acrylonitrile copolymer (SAN), which is frequently used as a modifier of polycarbonate. The composition of the copolymer can easily be determined from its FTIR spectrum, comparing the relative intensities of the nitrile and the styrene absorption peaks.

Very frequently components of similar molar masses are used in polymer blends. In these cases resolution of SEC is not sufficient to resolve all component peaks (see Fig. 32 for a binary blend containing an additive). The elution peaks of the polymer components 1 and 2 overlap and, thus, the molar masses cannot be determined directly. Only the additive peak 3 at the low molar mass end of the chromatogram is well separated and can be quantified.

A first indication of the composition of the present sample can be obtained from the contour plot shown in Fig. 33. Component 3 shows typical absorption peaks of a phenyl benzotriazole and can be identified as a UV stabilizer of the Tinuvin type. Component 2 exhibits absorption peaks that are characteristic for nitrile groups (2237 cm^{-1}) and styrene units (760, 699 cm^{-1}), whereas component 1 shows a strong ester carbonyl peak around 1740 cm^{-1} and peaks of styrene

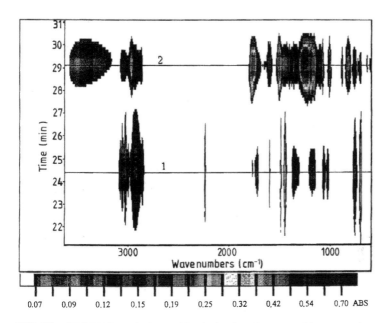

FIG. 31 SEC-FTIR analysis of a blend of polycarbonate and SAN copolymer. For chromatographic conditions, see Fig. 28.

units. In agreement with the peak pattern in Fig. 31, component 2 is identified as styrene-acrylonitrile copolymer.

Component 1 could be a mixture of polystyrene and PMMA or a styrene-methyl methacrylate copolymer. Since the FTIR spectra over the entire elution peak are uniform, it is more likely that component 1 is a copolymer.

One important feature of the SEC-FTIR software is that from the contour plot specific elugrams at one absorption frequency can be obtained. Taking the elugram at 2230 cm^{-1}, which is specific for the nitrile group, the elution peak of the SAN copolymer can be presented individually. For the presentation of component 1 the elugram at the carbonyl absorption frequency is drawn. Thus, via the "chemigram" presentation the elution peak of each component is obtained (Fig. 34).

The total concentration profile can be obtained from the chemigram at the frequency of the C-H valence vibrations (2800–3100 cm^{-1}). The specific chemigrams that are characteristic for one component each represent the elution profile of this component. Accordingly, the chemigrams can be used for the calculation of the molar masses of the components.

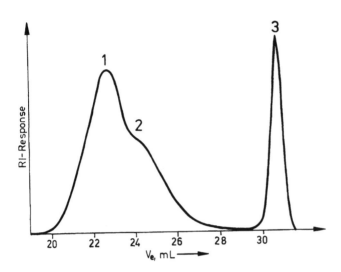

FIG. 32 SEC separation of a blend of two copolymers and an additive. For chromatographic conditions, see Fig. 28.

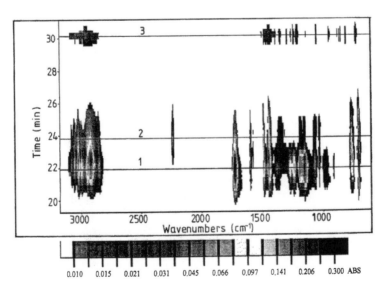

FIG. 33 Contour plot of the SEC-FTIR analysis of a blend of two copolymers and an additive.

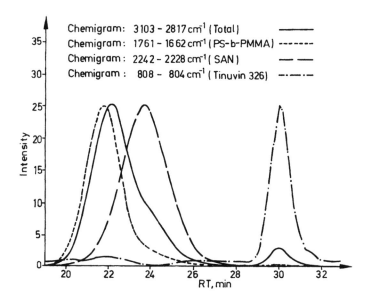

Chemigram: 3103 – 2817 cm⁻¹ (Total) ———
Chemigram: 1761 – 1662 cm⁻¹ (PS-b-PMMA) -------
Chemigram: 2242 – 2228 cm⁻¹ (SAN) — —
Chemigram: 808 – 804 cm⁻¹ (Tinuvin 326) —··—··—

FIG. 34 Chemigrams taken from the contour plot in Fig. 33.

ACKNOWLEDGMENT

Financial support of a part of this work from the Bundesminister für Wirtschaft through the Arbeitsgemeinschaft industrieller Forschungsvereinigungen e.V. (AiF) (grant no. 10947) is gratefully acknowledged.

REFERENCES

1. G. Glöckner, *Gradient HPLC of Copolymers and Chromatographic Cross-Fraction-ation*, Springer-Verlag, Berlin, 1991, Chapter 3.
2. G. Glöckner, *Polymer Characterization by Liquid Chromatography*, Elsevier, Amsterdam, 1987.
3. S. G. Entelis, V. V. Evreinov, and A. V. Gorshkov, Adv. Polym. Sci. *76*:129 (1986).
4. S. G. Entelis, V. V. Evreinov, and A. I. Kuzaev, *Reactive Oligomers*, Khimiya, Moscow, 1985.
5. M. B. Tennikov, P. P. Nefedov, M. A. Lazareva, and S. J. Frenkel, Vysokomol. Soedin. *A19*:657 (1977).
6. B. G. Belenkii, E. S. Gankina, M. B. Tennikov, and L. Z. Vilenchik, Dokl. Acad. Nauk USSR *231*:1147 (1976).
7. A. M. Skvortsov, B. G. Belenkii, E. S. Gankina, and M. B. Tennikov, Vysokomol. Soedin. *A20*:678 (1978).

8. S. T. Balke, in *Modern Methods of Polymer Characterization* (H. G. Barth and J. M. Mays, eds.), Wiley-Interscience, New York, 1991, Chapter 1.
9. W. W. Yau, J. J. Kirkland, and D. D. Bly, *Modern Size Exclusion Liquid Chromatography*, John Wiley and Sons, New York, 1979.
10. H. E. Adams, in *Gel Permeation Chromatography* (K. H. Altgelt and L. Segal, eds.), Marcel Dekker, New York, 1971.
11. A. Revillon, J. Liq. Chromatogr. *3*:1137 (1980).
12. S. Mori and T. Suzuki, J. Liq. Chromatogr. *4*:1685 (1981).
13. W. W. Yau, Chemtracts-Macromol. Chem. *1*:1 (1990).
14. E. Kohn and M. E. Chisum, in *Detection and Data Analysis in Size Exclusion Chromatography* (Th. Provder, ed.), ACS Symp. Ser. 352, ACS, Washington, D.C., 1987.
15. F. M. Mirabella, E. M. Barrall, and J. F. Johnson, J. Appl. Polym. Sci. *20*:959 (1976).
16. R. Bruessau, GIT Fachz. Lab. *31*:388 (1987).
17. B. Trathnigg, J. Liq. Chromatogr. *13*:1731 (1990).
18. B. Trathnigg, J. Chromatogr. *552*:505 (1991).
19. B. Trathnigg and Ch. Jorde. J. Chromatogr. *385*:17 (1987).
20. B. Trathnigg and X. Yan, Chromatographia *33*:467 (1992).
21. B. G. Belenkii, and L. Z. Vilenchik, *J. of Chromatogr. Library*, Vol. 25, Elsevier, Amsterdam, 1983.
22. S. Mori, Trends Polym. Chem. *R138* (1994).
23. C. G. Smith, P. B. Smith, and A. J. Pasztor, Anal. Chem. *65*:217R (1993).
24. G. Glöckner, Z. Phys. Chem. *229*:98 (1965).
25. G. Glöckner, Chromatographia *25*:854 (1988).
26. G. Glöckner, Chromatographia *23*:517 (1987).
27. S. Teramachi, A. Hasegawa, Y. Shima, M. Akatsuka, and M. Nakajima. Macromolecules *12*:992 (1979).
28. H. Sato, H. Takeuchi, and Y. Tanaka, Macromolecules *19*:2613 (1986).
29. R. W. Sparidans, H. A. Claessens, G. H. J. van Doremaele, and A. M. van Herk, J. Chromatogr. *508*:319 (1990).
30. S. Mori, J. Appl. Polym. Sci. *38*:95 (1989).
31. G. Glöckner and A. H. E. Müller, J. Appl. Polym. Sci. *38*:1761 (1989).
32. H. Sato, M. Sasaki, and K. Ogino. Polym. J. *21*:965 (1989).
33. G. Glöckner and D. Wolf, Chromatographia *34*:363 (1992).
34. M. Augenstein and M. A. Müller, Makromol. Chem. *191*:2151 (1990).
35. G. Glöckner, M. Stickler, and W. Wunderlich, J. Appl. Polym. Sci. *37*:3147 (1989).
36. S. Mori and M. Mouri, Anal. Chem. *61*:2171 (1989).
37. S. Mori, J. Chromatogr. *507*:473 (1990).
38. S. Mori, J. Chromatogr. *503*:411 (1990).
39. S. Mori, Anal. Chem. *62*:1902 (1990).
40. S. Mori, J. Chromatogr. *541*:375 (1991).
41. T. C. Schunk, J. Chromatogr. *661*:215 (1994).
42. T. C. Schunk and T. E. Long, Chromatogr. *692*:221 (1995).
43. T. C. Schunk, J. Chromatogr. *A656*: 591 (1993).
44. T. H. Mourey, J. Chromatogr. *357*:101 (1986).
45. J. A. J. Jansen, in: *Integr. Fundam. Polym. Sci. Technol. Proc. Int. Meet. Polym.*

Sci. Technol., Vol. 5 (P. J. Lemstra and L. A. Kleintjens, eds.), Elsevier, Amsterdam, 1991.

46. S. Mori, J. Liq. Chromatogr. *16*:1 (1993).
47. W. J. Staal, Ph.D. thesis, Technical University Eindhoven, The Netherlands, 1996.
48. H. C. Lee and T. Chang, Macromolecules *29*:7294 (1996).
49. D. Hunkeler, D. Macko, and T. Berek, Polym. Mat. Sci. Eng. *65*:101 (1991).
50. D. Hunkeler, T. Macko, and D. Berek, in: *Chromatography of Polymers: Characterization by SEC and FFF* (T. Provder, ed.), ACS Books, Washington, D.C., 1993.
51. M. Janco, T. Prudskova, and D. Berek. J. Appl. Polym. Sci. *55*:393 (1995).
52. D. Berek, Macromol. Symp. *110*:33 (1996).
53. D. Hunkeler, M. Janco, V. Guryakova, and D. Berek, ACS Adv. Chem. Ser. *247*: 13 (1995).
54. D. Hunkeler, M. Janco, and D. Berek, in: *Strategies in Size Exclusion Chromatography* (M. Potschka and P. L. Dubin, eds.), ACS Symp. Ser. 635, ACS, Washington, D.C., 1996.
55. D. Berek, M. Janco, T. Kitayama, and K. Hatada. Polym. Bull. *32*:629 (1994).
56. H. Pasch, H. Much, and G. Schulz, in: *Trends in Polymer Science, Vol. 3, Research Trends*, Trivandrum, India, 1993.
57. H. Pasch, Adv. Polym. Sci. *128*:1 (1997).
58. H. Pasch, H. Much, G. Schulz, and A. V. Gorshkov, LC-GC Int. *5*:38 (1992).
59. N. N. Filatova, A. V. Gorshkov, V. V. Evreinov, and S. G. Entelis, Vysokomol. Soedin. *A30*:953 (1988).
60. A. V. Gorshkov, S. S. Verenich, V. V. Evreinov, and S. G. Entelis, Chromatographia *26*:338 (1988).
61. A. V. Gorshkov, T. N. Prudskova, V. V. Guryakova, and V. V. Evreinov, Polym. Bull. *15*:465 (1986).
62. A. V. Gorshkov, H. Much, H. Becker, H. Pasch, V. V. Evreinov, and S. G. Entelis, J. Chromatogr. *523*:91 (1990).
63. H. Krüger, H. Pasch, H. Much, V. Gancheva, and R. Velichkova, Makromol. Chem. *193*:1975 (1992).
64. R.-P. Krüger, H. Much, and G. Schulz, Macromol. Symp. *110*:155 (1996).
65. H. Pasch and I. Zammert, J. Liquid Chromatogr. *17*:3091 (1994).
66. H. Pasch, C. Brinkmann, H. Much, and U. Just. J. Chromatogr. *623*:315 (1992).
67. H. Pasch, C. Brinkmann, and Y. Gallot, Polymer *34*:4099 (1993).
68. H. Pasch and M. Augenstein, Makromol. Chem. *194*:2533 (1993).
69. H. Pasch, M. Augenstein, and B. Trathnigg, Makromol. Chem. *195*:743 (1994).
70. H. Pasch, Y. Gallot, and B. Trathnigg, Polymer *34*:4986 (1993).
71. H. Pasch, GIT Fachz Lab. *37*:1068 (1993).
72. H. Pasch, A. Deffieux, I. Henze, M. Schappacher, and L. Rique-Lurbet, Macromolecules *29*:8776 (1996).
73. H. Pasch, Polymer *34*:4095 (1993).
74. H. Pasch and K. Rode, Macromol. Chem. Phys. *197*:2691 (1996).
75. H. Pasch, K. Rode, and N. Chaumien, Polymer *37*:4079 (1996).
76. M. Lafosse, L. Elfakir, L. Morin-Allory, and M. Dreux, J. High Res. Chromatogr. *15*:312 (1992).

77. K. Rissler, U. Fuchslueger, and H. J. Grether, J. Liq. Chromatogr. *17*:3109 (1994).

78. S. Brossard, M. Lafosse, and M. Dreux, J. Chromatogr. *591*:149 (1992).

79. M. Dreux, M. Lafosse, and L. Morin-Allory, LC-GC Int. *9*:148 (1996).

80. S. R. Abbott and J. Tusa, J. Liq. Chromatogr. *6*:77 (1983)

81. R. P. W. Scott, *Liquid Chromatography Detectors*, Elsevier, Amsterdam (1986).

82. E. S. Yeung, *Detectors for Liquid Chromatography*, John Wiley and Sons, New York, 1986.

83. K. H. Shafer, S. L. Pentoney, and P. R. Griffiths, J. High Res. Chromatogr., Chromatogr. Commun. *7*:707 (1984)

84. J. M. Willis, J. L. Dwyer, and L. Wheeler, *Proceedings International Symposium on Polymer Analysis and Characterizationn*, Crete, Greece (1993).

85. A. E. Hamielec, A. C. Ouano, and L. L. Nebenzahl, J. Liquid Chromatogr. *1*:527 (1978).

86. M. A. Haney, J. Appl. Polym. Sci. *30*:3037 (1985).

87. W. W. Yau, S. D. Abboutt, G. A. Smith, and M. Y. Keating, ACS Symp. Ser. *352*: 80 (1987).

88. R. M. Robertson, J. A. de Haseth, J. D. Kirk, and R. F. Browner, Appl. Spectrosc. *42*:1365 (1988).

89. P. R. Griffiths and C. M. Conroy, Adv. Chromatogr. *25*:105 (1986)

90. J. W. Hellgeth and L. T. Taylor, J. Chromatogr. Sci. *24*:519 (1986).

91. J. J. Gagel and K. Biemann, Anal. Chem. *58*:2184 (1986).

92. J. J. Gagel and K. Biemann, Anal. Chem. *59*:1266 (1987).

93. J. J. Gagel and K. Biemann, Microchim. Acta *11*:185 (1988).

94. L. M. Wheeler and J. N. Willis, Appl. Spectrosc. *47*:1128 (1993).

95. J. N. Willis, J. L. Dwyer, and L. M. Wheeler, Polym. Mat. Sci. *69*:120 (1993).

96. J. N. Willis, J. L. Dwyer, and M. X. Lui, Proceedings International GPC Symposium 1994, Lake Buena Vista, FL, USA (1995).

Index

Acceptor
  characteristics, 127
  -donor interactions, 61, 126, 127, 133, 136
  number, 52, 54, 67, 111, 128, 130, 154–156
  probe, 62
  reference, 131
  site, 54, 131, 139
Acid–base
  character, 44, 62, 64, 111
  concept, 54, 64, 108, 130
  contribution to adhesion, 126
  exchanges, 43
  number, 108,
  probes, 108
  properties, 48, 52, 61, 66, 76, 108, 113, 150
    evaluation, 155
  reaction enthalpy, 154
  scale, 63, 153, 154
  surface properties, 105, 110
Acid
  boric, 316
  dispersive effect, 53
  number, 53, 68, 130
  oxidation, 88

[Acid]
  perchloric, 195, 230
  polysilicic, 175
  styrene–methacrylic, 175
  treatment, 293, 355
Acidic
  constant, 130
  mobile phase, 178
  probe, 47, 54, 64
  protein retention, 292
Acidity
  of hydrogen bond, 155
  of substances, 52, 110, 111
Additivity
  of group contribution, 230, 231
  of radii, 23
  of surface energies, 130
Adhesion
  between fibers, 138, 140
  between surfaces, 126
  induced by grafting, 135
  polymer-matrix, 281
  strength, 95
  work of, 42, 56, 138
Adsorbate
  -adsorbent
    area, 25

[Adsorbate]
mobility, 330
boiling point, 17
cavity interaction, 30
cavity model, 9, 10
molecular structure, 35
nonspecific, 24
retention time definition, 5
size, 24
solid interaction, 35
zeolite interaction, 26
Adsorbed
film, speading pressure of, 57
molecule
amount of, 163
interfacial orientation of, 257
surface area of, 51, 66, 76, 158
polyelectrolytes
amount of, 318
in-plane mobility of, 384
repulsion between, 322
structure of, 320
polymer
hydrodynamic thickness of, 301
reconformation of, 359, 373, 379
structure, 313
species, surface blocking effect of, 334
water, role of, 153
Adsorbent
retention on polar, 229
structural information, 2
surface model, 8
Adsorbing disc model, 361
Adsorption
at porous adsorbents, 391
at stationary phases, 174, 181, 195, 199, 215–218
capacity of resins, 113
centers, 247
composition effect in, 348
constant at equilibrium, 231
critical point of, 391, 403
-desorption free energy, 58
end of, 357
energy distribution, 162–169
enthalpy of, 66, 126, 130, 150, 332

[Adsorption]
free energy of, 56, 61, 130, 150, 332
in micropores, 2
in slit-like pores, 3
irreversible, 149, 366, 405
isosteric heats of, 26, 163
isotherm
(BET), 44, 48, 290
(organic molecules), 48, 148, 161
(polyelectrolyte), 318
(probe), 51
kinetic coefficient of, 331, 336, 358, 361
layer thickness, 319–322
localized, 330, 333, 341, 349
memory effect in, 195
mobile, 330, 334, 343, 355
model, 363
nonspecific surface, 264
of chains, 330
of diblock copolymer, 358
of heterogeneous polymers, 391
of hydrocarbons, 58
of hydrophilic polymer, 274, 313
of linear hydrocarbons, 157
of micelles, 377
of polyelectrolyte, 314 -318, 322, 366, 378
of polymer, 331
of proteins, 285, 292, 312
of specific ions, 324
of two discs, 345
of viruses, 283
on a lattice, 332
preferential, 373, 384, 410
prevention, 189, 282, 283
probability of, 338, 339, 343, 344, 345
profile in chromatography, 371, 374, 378
properties of carbon black, 85–88,
random sequential model of, 330,
rate of polymer, 331, 356,
supply in, 359
reduction by coating, 276
separation efficiency in, 385
single surface approach, 18

[Adsorption]
site accessibility, 157, 333
size effect in, 340
standard entropy of, 129
surface, 128, 282
thermodynamic treatment of, 330
virial coefficient treatment of, 3
virial equation of state for, 5
Aerosil, 158
Aggregation, 182
Alkanes, 33, 59, 233, 289, 313
as nonspecific adsorbates, 24, 48, 61,
    127, 130, 136, 147
branched, 158
flexible, 157
insertion of, 157
retention volume of, 149, 151, 242
surface area of, 51, 76
vapor pressure of, 155
Alkanethiol, as stabilizer, 185
Aluminium oxide
surface energy of, 103
specific interaction with, 229
Amphoteric
character, 87
probes, 48, 54, 136
properties, 154
surface, 63, 87, 151
Analytic solution
Hansen's, 34
Anhydrous silylation, 274, 279, 294
Aromatic cycle
number of, 228
retention of, 251
Atomic van der Waals radii, 49
Average size of colloids, 179, 200

Band shift of elution chromatogram, 264
Bending energy
of molecules, 36
Bimodal distribution, 164, 166, 187
Biocompatible, 272
Biopolymers,
binding of, 264
molecular weight determination of,
    182

Blends, 387
binary, 395
characterization of, 393, 419
elaboration of, 42, 79
fractionation of, 398
interaction parameters in, 88
polymer, 127
ternary, 397
Boiling point, 48, 128
correlation with
B_{2S} data, 17
energy, 36
van der Waals attraction, 36
Brush
layer, 300, 305
-type network, 69, 286, 294

Calcium
carbonate, 48, 63
treatment of, 60
oxide, 48
surface energy of, 63,
Calibration
column, 14, 46, 181, 186, 189, 199,
    216, 222, 265, 270, 305
curves for polymer, 175, 392, 394,
    402
universal, 175, 228, 273, 413
Capillary
bridge design, 421
column, 46, 209
electrophoresis, 312, 316
model, 194
rise, 138
viscosity detection in, 420
zone electrophoresis, 218
Carbon
atom, number of, 17, 35, 130, 151,
    228, 230, 233, 242, 246, 254
black, 19, 152, 157
surface energy of, 88
fiber, 66, 75, 78, 82, 110, 111, 113,
    139
microporous, 23
type of atom, 242, 246, 250, 255
with flat surfaces, 30

Carrier gas, 2, 42, 127, 153, 168
  flow rate of, 47, 54, 148
  pressure of, 13
Cellophane
  dispersive energy of, 69
  film, 46, 51
Cellulose
  acetate fibers, 85
  as stationary phase, 179
  fibers, 68, 108, 128, 133
    chemically modified, 133, 134
  structure of, 92
  surface
    energy, 69, 72
    properties, 44
Chain length
  effect on column permeability, 304
  of polymer, 397, 403
Charge density
  and layer thickness, 322
  and polyelectrolyte adsorption, 318, 322, 365
  and potential, 185
Chemical composition
  of polymer blends, 387, 426
    and interaction energy, 391
  of surfaces, 43, 166
  samples of different, 397
  separation according to, 397
Chemical stability
  of polymer coating, 273
  of polymer grafting, 300, 308
Chemical treatments
  and surface heterogeneity, 162
  on fibers, 133, 136, 139
Chemically-bonded silica column, 283
Clay
  as a lamellar solid, 157
  characterization, 98
  mechanical properties of, 148
Coacervation, 314, 317
Coating, 88, 100, 135, 267, 273, 283, 317, 326
  instability of, 274,
  of paper, 61, 91

Collapse
  of electrical double layer, 191
  of grafted chain, 301
Colloidal silica, 174
  separation of, 176
Column
  calibration, 216
  conditioning, 168
  dead volume of, 132
  efficiency, 312, 316
  flow rate, 5, 149
  gas compression in, 162
  interstitial volume in, 5, 389
  overloading, 212
  packing, 46
  permeability, 301–304, 308
  plate number, 148, 176
  porosity, 301
  pressure drop in, 129
  resolution, 269, 272, 284
  schematic representation of, 340
  temperature, 14
Compatibility
  surface, 42, 77, 95
  interfacial, 77, 88
Compatibilizer, 388
Composite, 42, 77, 103
  acceptor, donor number of, 67
  adhesive force in, 278
  elaboration of, 108
  interaction parameter in, 113
  mechanical properties of, 82, 126
  papers, 132
    internal bond strength of, 138
    surface characterics of, 133
Compressibility
  correction factor for gas, 56, 149
Condensation
  approximation, 164
  in micropores, 24, 33
  of silanol groups, 279
  prevention of, 15
  reaction, 274
Configurational entropy, 322

Conformation
  of adsorbed polymer, 333, 361
  of grafted layer, 301, 308
  of macromolecules, 389
  of micelles, 373
  of probe, 51
Connectiveness index, 236
Connectivity index, 228
Contact angle measurement, 44, 72, 126,
  131, 138, 140, 146, 152
Copolymer blend, 393
Cork, 94, 113
Critical chromatography, 406, 412
Critical point of adsorption, 391, 403,
  406, 411, 414, 421
Cross-linking, 277
  agent, 267
  degree of, 266
Crystalline silica,
  structure of, 157
  surface energy of, 98

Debye length, 176, 182
Degree of micellization, 376
Degree of polymer coating, 267
Desorption, 45, 188, 269, 315, 331, 337,
  359, 360, 379, 384
  free energy of, 57, 58
  free entropy of, 61
  isotherm, 161
Destabilization
  colloid, 180
Detector, 12, 15, 45, 131, 147, 209, 317,
  395, 397, 403, 411, 425
  selective, 419, 420
Dipolar interactions, 152
Disc mobility, 347
Dispersion, 88, 105, 146
  axial, 413
  forces, 27, 30, 166
Dispersive
  characteristics, 140
  contribution, 126–133, 152
  energy, 26, 61, 69, 72, 74, 76, 82, 85,
    88, 92, 98, 104

[Dispersive]
  interactions, 126, 127, 147, 150,
    164
  properties, 68, 94, 108
Distribution
  coefficient, 389, 403
  Gaussian, 5
  nonequilibrium, 14
  of adsorbed discs, 343, 346, 350
  of adsorption energy, 162–166,
    169
  of chain segments, 333
  of chemical groups, 160, 378, 403
  of colloid in a capillary, 219
  of contacts, 136
  of electrons, 232, 247
  of energy, 198
  of molecular weight, 272, 278, 388,
    395
  of molecules, 2, 240, 244, 333, 389,
    393
  of potential, 198
  of size, 177, 181, 187, 190, 208, 212,
    222, 278, 312, 355, 368, 390
  selective, 217
Donor
  number, 52, 67, 111, 128, 131, 136,
    146, 154
  probe, 62, 66
  properties, 127
  site, 54
Double layer thickness, 199

Efficiency, 148, 251, 312, 316
  of fractionation, 385
  of purification, 72
  of separation, 212, 213, 314, 385
  of silanol, 267
Electrical double layer, 181, 189, 195,
  198, 213, 217
Electrolyte
  concentration effect of, 195, 198, 199,
    213, 283, 284, 393
  valence effect of, 198
Electrophoretic mobility, 322, 324,
  326

Electrostatic
forces, 30, 146, 363
induction, 26, 30
interactions, 27, 54, 153, 198, 247,
  292, 312, 359, 390
Elutant, 364, 368, 384, 398
Elution
characteristics, 126
mode, 392
multiple solvent, 398
nonsteric effect in, 273
of probes, 127, 229, 282
order, 285
point method, 161
profile, 219, 233, 378, 395, 409, 428
rate, 369
sequence, 232, 247, 312
size effect in, 284, 312
time, 426
volume, 127, 175, 264, 292, 315, 385,
  394
zone, 403
Energetic
heterogeneity, 246
parameter, 26
stabilization, 322
Energy
barrier, 257
dispersion, 30
dispersive, 44, 48, 52, 74, 77, 82, 92–
  95
electrostatic, 30
Gibbs free, 389, 404
interaction, 27, 30, 31
minimization, 23
of adsorption, 3, 34, 56, 61, 130, 150,
  151, 162, 231, 330, 332, 337, 391
  distribution of, 163–168, 198
of bending, 36
of desorption, 57
of electrons, 228
of gas–solid interaction, 3, 19, 22
of interaction, 7, 10, 22, 157, 232, 344,
  384
of molecules, 35
of self-association, 54

[Energy]
of solvation, 155
standard free, 129
surface area of high, 23, 24, 44, 76,
  138
surface, 42, 44, 56, 69, 72, 85, 102–
  105, 128, 130, 146, 148, 160
Enthalpy
of adsorption, 66, 67, 126, 129, 146,
  150, 332, 337
of desorption, 61
of formation, 53, 154
of interaction, 131, 389
of reaction, 54, 154
of vaporization, 36, 67, 156
Entropic
effect in interfacial tension, 138
interactions, 389, 391, 397
Entropy
chain, 330
configurational, 322, 389, 390
of adsorption, 129, 146
of desorption, 61
standard, 60
Environment
of carbon atoms, 233
of immobilized molecules, 313
of silanol groups, 160
Equation of Drago, 154
Equilibrium partition coefficient, 265
Exchange process, 338
Exclusion
ion, 178
limit, 390
of heterogeneous polymer, 391
size, 150, 156, 384
  chromatography, 174, 190–194, 389,
    392
  mode, 389, 391, 403, 408
  surface area, 346, 352, 358, 361
  chromatography, 364, 385

Fiber(s), 42, 147
acid–base properties of
  carbon, 110, 111
  cellulose, 108, 110

[Fiber(s)]
  glass, 111
  polyethylene, 111
  pulp, 108
  untreated carbon, 111
  untreated glass, 111
  untreated polyethylene, 111
  acid number of, 68
  adhesion between, 138
  adsorption isotherm on, 48
  aramid, 44
  based composite, 42
  carbon, 68, 77, 82
  graphitized, 82
  cellulose acetate, 85
  cellulosic, 68, 72
  surface tension of, 133
  chemical characteristics of, 43
  column, 46
  dispersive energy of, 76, 82
  glass, 82–84, 364
  heat of adsorption of, 133
  interaction parameter between matrix
    and, 67–68
  lignocellulosic, 55, 73, 74, 76, 126,
    131, 132
  micro-, 355
  paper, 128
  polarity of, 133
  polyamide, 84
  polyethylene, 85, 126, 131, 132, 138
  polyethylene terephthalate, 85
  specific interaction parameter of, 131
  sulfite pulp, 76
  surface area of, 44, 380
  surface modified, 78
  thermally treated, 82
  thermoplastic, 133
  wetting characteristics of, 138
Finite concentration, 69
  Henry's law at, 45
  inverse gas chromatography at, 160,
    168
Flow rate
  of carrier gas, 2, 4, 5, 12–17, 46, 55,
    69, 129, 148, 162

[Flow rate]
  of eluent, 175, 206, 212, 216, 272, 275
Fluorescence detection, 209
Form factor, 186
Fractionation, 178, 186, 187, 678, 384,
    393, 394
  field flow, 223
  of blends, 398
  under shear, 364
Frequency
  absorption, 429,
  jump, 336, 345
Fugacity, 3
Fumed silica, surface energy of, 103
Functionality
  polymer, 403
  surface chemical, 274, 281

Gaussian peak, non, 14, 15
Geometric area, 24, 33
Gold colloid, 187, 208, 209, 219, 224
Gradient elution, 397, 398
Grafted
  filler, 72
  silica, 98, 103
Graft polymerization, 293, 296, 299
Grafting agent, 43, 74, 82–84, 88, 98,
    103, 108, 158
  polymeric, 133–135, 284
Grinding
  of alumina, 98
  of muscovite, 158, 166
  of silica, 98
  prevention of, 148

H bond, 152
Hard-soft acid base scale, 154
Heat
  of adsorption, 19, 127, 133, 135, 136,
    146, 160, 163
  isosteric, 17, 26
  treatment, 78, 82, 88, 98, 103, 111,
    112, 168
Henry's law, 26, 27, 45
  constant, 26, 30, 162
  theory of electrophoresis, 323

Heterogeneity
  chemical, 313
  macro-, 166
  molecular, 403
  surface, 14, 20–24, 159–163, 166,
    246
  type, 393
Homogeneity
  chemical, 158, 397
  of dispersion, 95
  of distribution, 364, 378
  of surfaces, 19, 25
  parameter, 108
  patches, 163
Humidity of carrier gas, 168
Hydrated silica, 355, 358
Hydrocarbons,
  adsorbed, 88
  adsorption of, 58, 157
  aromatic, 251, 254
  chains, 247
  halogenated, 33
Hydrodynamic
  chromatography, 216, 384
    wide-bore, 218
  diameter, 187, 229
  permeability, 301
  radius, 320
  thickness, 301, 320
  volume, 216, 378, 384, 390, 392,
    406
Hydrogen, 12, 24, 93
  as inert carrier gas, 46
  bond acidity and basicity, 155
  bonding, 43, 93, 162, 166, 385
  energy per, 35
  ion exchange, 355
  treatment, 25
    peroxide, 74
Hydrolysis, 274, 284, 293–295, 355
  controlled, 180
Hydrophilic
  eluent, 304
  gel, 187
  groups, 281, 283
  polymer, 68

Hydrophobic
  adsorbent, 232, 244
  agent, 133
  effect, 312
  factor, 228
  force, 146
  interaction, 152, 283, 284, 292, 385
  matrix, 68
  site, 138
  surface properties, 283
Hydroxyl groups, 274
  surface, 189, 278, 279, 284
  surface concentration of, 293

Ideal gas, 130
Illites, 95, 98
Infinite dilution,
  extrapolation at, 52
  measurement at, 44, 56, 76, 82, 126,
    128, 130, 147
In-plane mobility, 330, 344, 346, 384
Insertion phenomena, 157–159
Interaction(s)
  acid–base, 66, 68, 85, 127, 130, 135,
    146, 153
  adsorbate–adsorbate, 26, 45, 47, 57
  adsorbate–adsorbent, 6, 57, 60, 231
  between matrix and filler, 42, 67
  capacity of a solid, 163
  composite-blend, 113
  dispersive, 61, 126, 127, 130, 147
  donor–acceptor, 61, 126, 127, 133, 136
  double layer, 198
  electrostatic, 54, 247, 390
  energy, 146
  enthalpic, 389, 397
  enthalpy of, 155
  entropic, 389, 391, 397
  London potential of, 150, 155
  mechanism, 155
  molecular, 228, 231
  nonpolar, 185, 187
  nonspecific, 232
  nonsteric, 264, 270
  parameter, 64, 108, 112, 139, 291
  physicochemical, 146

[Interaction(s)]
  polar, 127
  polymer substrate, 384
  polymer–polymer, 45, 126
  polymer–solvent, 45
  probe–substrate, 51, 62, 136, 150
  solute–silica, 274, 282, 292
  specific parameter of, 66
  specific, 152, 229, 235, 246, 248, 254,
    391, 404
Interfacial
  adsorption energy, 337
  area, 380
  cake, 362
  compatibility, 77
  conformation, 333, 359, 379
  exchange, 338, 378, 384
  layer thickness, 134
  phenomena, 330
  reconformation, 359, 361, 373
  relaxation, 378, 380
  shear strength, 113, 139
  tension, 138
Intermolecular
  cohesive energy, 93
  hydrogen bonding, 93
  probe interaction, 149
Internal
  diameter of column, 132
  pore, 265, 289
  rotation, 257
  standard, 178, 216
Intrinsic viscosity, 363, 411, 419–422
Ion exchange, 264, 265
  chromatography, 312
Irreversible
  adsorption, 149, 181, 195, 199, 331,
    334, 337, 366, 397
  retention, 178

James–Martin coefficient, 149
Jamming limit, 337, 343, 361
Jump,
  frequency, 336, 344, 345, 347
  length, 336, 343, 344

Kaolinite, 95, 98
Kinetic coefficient of adsorption, 331–
    333, 337, 357, 361

Lamellar
  product, 166
  structure, 157
Laminar flow, 219, 233
Lateral resolution, 191
Layer thickness, 319–322
Lewis, 130, 153
  acid–base interaction, 358
  acid probe, 48, 52, 154
  base probe, 48
Lignin, 44, 74, 88, 92
Linear hydrocarbons, 157, 158
Linear semicrystalline structure, 92
Localized model of adsorption, 330
London,
  component of surface energy, 58, 84,
    98, 130, 147
  interaction potentiel, 150, 155

"Magic" agglomeration number, 208
Magnetic fluids, 186
Masking of silanol groups, 265, 267–271,
    308
Matrix
  acid number of, 68
  adhesion, 281
  chemical modification of, 69
  liquid, 113
  porous, 286, 290
  surface compatibility, 42, 95
Mechanical
  compression of carbon black, 88
  oscillator method, 395
  properties
    of composites, 42, 68, 82, 126, 132,
      136, 139
    of papers, 139, 140
    of particles, 148, 181
    of resins, 264, 300
Memory effect of the column, 195, 197
Metallic interaction, 153
Methanol-water mixture, 197

Methyl group, 251, 278
  energy of adsorption of, 56, 58, 130,
    151
  -substituted polyaromatic hydrocarbon,
    255
  surface area of, 51, 58, 72, 76, 130
Mica, 98, 148, 166
Micellar solution, 366, 369–372
Microcapillarity, 69
Micropores, 72, 289
Mobile phase, 178
  aqueous, 264
  binary, 240
  composition, 237, 240, 258, 274, 283
  definition, 126, 132, 331
  flow rate of, 129, 274
  ionic strength of, 284
  nonpolar, 231
  organic concentration in, 241, 242
  organic, 263
  polar, 234, 244
  properties of, 229
  viscosity of, 315
  water concentration in, 235, 241
Molar mass of probe, 163
Molecular
  area, 52, 163
  connectivity index, 228
  dimensions, 222, 380
  layer, 143
  level of solid, 150, 158, 160, 168
  method, 152
  model, 51
  modeling, 157
  orbital theory, 154
  organization, 385
  parameters, 150, 227, 228, 246
  probe, 126, 147, 157
  sieve, 85, 357
  size, 66, 166, 263, 284, 290, 291
  structure, 228, 255
  weight distribution, 300
  weight, 48, 57, 155, 264, 265, 286–
    290, 331, 338, 357, 363, 366, 378,
    384
Molecule geometry, 239

Monodisperse
  colloids, 191, 209, 212
  particle test, 208
  polymer, 286, 300, 393
Monomer concentration, 299, 300, 306,
    307, 314
Monomodal distribution function, 166
Morphology index, 157, 158
Multiple surface attachment, 294
Muscovite, 98, 146, 157, 166

Nanomorphology
  index, 158
  of surface, 148, 155–157
Negative peak, 185, 410
Nitrogen adsorption, 48
  liquid, 290
Nondispersive contribution to surface en-
    ergy, 42, 44, 56
Nonpolar interactions, 185
Nonpolar surface, 152
Nonspecific,
  adsorption, 264
  interactions, 150, 232, 265
  surface adsorption, 264

Optical trapping, 219
Ostwald ripening, 202, 203
Overadsorption, 343, 347, 359, 373, 376
Oxidation, 203
  anodic, 82, 111
  electrolytic, 78, 111
  nitric acid, 88
  surface, 43
Oxides
  mineral, 44, 95, 98, 103, 146, 147,
    154, 158, 174, 229
  organic, 48, 64, 74, 78, 264, 267, 282,
    305, 398, 427

Packed
  column, 46, 128, 132, 174, 263, 272,
    303, 307
Packing(s), 46, 175, 182, 187, 215, 216,
    263–265, 270
  ionization, 189

[Packing(s)]
  penetration, 187
  profile, 341
Partition coefficient
  bulk, 59
  equilibrium, 265
  surface, 57, 60, 61, 129, 149
Peak(s), 152, 349, 372
  adsorption, 369, 377
  area, 197, 212
  asymmetrical, 54, 55, 219
  broad, 212
  consecutive, 251
  convection, 219
  diffusion, 219
  Gaussian, 175
  height, 274
  initial, 343, 370, 372, 376
  monodisperse, 216
  negative, 185
  second, 166, 205, 209
  shape, 285
  shifting, 283
  single, 155, 161
  symmetrical, 54, 132, 219, 251
  third, 208
  width, 208
Permanent sorption of probe, 47
Phenolate group, 92
Photolysis, 185
Physical
  adsorption, 2, 6, 18, 20, 26
  characteristics
    of paper, 139
    of polymer, 147
    of sorbent, 228
  composition of bonded phase,
    281
  condition of chromatography,
    230
  interaction, 146
  model, 163
  nature of interaction, 42
  properties, 3, 33, 36
  surface imperfection, 24
  treatment, 77, 162

Physicochemical
  affinity, 60
  interaction between particles, 146
  model of retention, 231
  molecular parameter, 228, 230
  properties of sorbate, 228, 231
  surface characterization, 147
  surface properties, 148, 163
Physisorption
  of oligomers, 313
  of polyelectrolytes, 314
  of polymers, 313
Plasma treatment
  microwave, 88
  of graphite, 112
  of oxide, 98
  of pulp, 76, 108
Plate height, 176
Polar
  adsorbent, 229, 232
  compounds, 272, 406, 414
  copolymer, 418
  force, 166
  homopolymer, 419
  interaction, 127, 147
  mobile phase, 233–235, 241, 244, 248
  molecule, 52, 272
  probe, 47, 51, 61–66, 88, 127, 130,
    136, 148, 152, 155, 164
  stationary phase, 393, 418
  surface, 152, 229, 231
Polarizability
  molar deformation, 156
  molecular, 36
  of stationary phase, 155
  of surface atoms, 152
  probe, 150
Polyacrylonitrile-based fiber, 78, 110, 111
Polyamide fiber, 84
Polydispersity
  diameter, 181, 209, 212–214
  in polymer composition, 419
  in polymer molecular weight, 313,
    384
Polyelectrolyte, 199, 313, 379
  adsorption, 315, 318, 322, 366, 378

[Polyelectrolyte]
    layer, 318, 319, 320, 321, 326, 380
    model, 363
    characteristics, 378
    coacervation, 314
    desorption, 315, 384
    exchange, 378, 384
    immobilized, 314
    molecular dimension of, 380
    number of adsorbed, 365
    -protein interaction, 326
    reduced viscosity, 363
    retention, 321
    selectivity for protein separation, 314
    treated glass, 317
Polyester, 82
Polyethylene, 68
    as reference surface, 51, 152
    fiber, 85, 111, 126, 132, 133, 138
    glycol, 98, 281
    oxide, 282
    type polymer, 57
    untreated surface, 136, 140
Polyethyleneimine, 312
Polyethyleneterephthalate, 85
Polystyrene, 357, 358, 388, 395, 398,
    427, 429
    as synthetic matrix, 69
    block, 367, 368, 372
    gel, 174
    latex bead, 202, 359, 365
    modified fiber, 77, 135
    modified silica, 267, 289, 290
    -polyvinylpyridine copolymer, 361, 366
    resin, 263
    rubber-modified, 388
Pore(s)
    adsorption
        in, 85, 322, 389
        in micro-, 2
        in slit-like, 3, 8
    condensation in micro-, 24, 33
    contact area in micro-, 72
    cylindrical, 179, 180, 198, 302
    diffusion into, 174, 291
    micro-, 72

[Pore(s)]
    geometry, 319
    macro-, 265, 308
    radius, 301, 319
    random-plane, 179
    size, 176, 187, 269, 282, 302, 308,
        390, 402
    spherical, 179
    structure, 292
    volume, 189, 265, 269, 289, 389, 391
    wall, 300, 304
Precipitation/redissolution, 397, 403
Primary amino groups, 93
Printing ink pigment, 88, 105
Probe(s), 48, 126
    acid–base properties of, 52, 64
    electron donor–acceptor, 54, 66, 127
    enthalpy of vaporization of, 67
    heat of adsorption, 136, 163
    interaction of, 51
    mechanical retention of, 69
    model, 67
    n-alkane, 130, 147, 148, 151, 155, 156,
        158
    permanent sorption of, 47
    retention time of, 47, 52, 55, 59, 128,
        149, 157
    retention volume of, 61, 149
    surface area of, 51, 130, 163
        dispersive, 136
    surface partition coefficient of, 129
    surface tension, 131
    topology index of, 156
    volatile, 45, 168
Protein(s)
    adsorption, 312, 385
        model for, 334, 344
    binding, 314, 315, 317–319, 321,
        326
    binding site, 311, 315
    electrophoretic mobility of, 315, 322,
        324
    elution sequence of, 312
    molecular dimension of, 312
    net charge of, 324, 326
    retention of, 314

[Protein(s)]
  selective binding of, 317
  selectivity in separation of, 314
Pyrogenic silica, 148, 152, 158, 162,
  175

Q effect, 205

Radial and axial concentration distributions, 219
Raman spectroscopy, 43, 275
Random walk, 330, 335, 336, 343–345
Reconformation, 359, 361, 373, 379
Reference line, 59, 62, 66, 136
Relative retention, 36, 313
Relaxation process, 360, 370, 378, 380,
  385
Resolution, 326
  improved, 215, 223, 265, 270
  lateral, 191
  molecular weight, 282, 292, 305
  peak, 184, 185, 426, 428
  poor, 189, 269
  separation, 191, 251, 313, 414
Resolving power, 284
Retention, 313, 355
  colloid, 178
  copolymer, 398
  homopolymer, 398
  index, 36
  mechanism, 228
  model, 231–232
  polyelectrolyte, 321
  protein, 285, 292, 314
  solute, 290, 364
  sorbent, 228, 230, 233–255
Retention times, 2, 4–7, 12–17
  and adsorbent size, 24
  and adsorption isotherm, 161
  and calibration plot, 181
  and electrolyte concentration, 199, 202
  and flow rate, 69
  and internal standard, 178
  and negative peak, 185
  and particle diameter, 191, 208
  and sample concentration, 195

[Retention times]
  and solubility parameter, 399
  and surface properties, 45, 147, 157,
    291
  determination of, 54
  net, 149
  of potassium iodide, 192
  of probe, 47, 127, 128
  prediction of, 33
  relative, 36
Retention volume,
  and adsorption sites, 157
  and distribution coefficient, 389
  and free energy of adsorption, 61–65
  and hydrodynamic volume, 390
  and molecular weight, 393, 401
  and polymer interaction, 127
  and retention time, 149
  and surface energy of substrate, 72
  and temperature dependence, 136
  net, 56, 57, 59

Sample concentration effects, 195, 212,
  369, 370, 377, 419
Second gas–solid virial coefficient, 2, 3,
  6, 7, 10, 20, 24, 27
Secondary ion mass spectrometry, 43
Selective
  adsorption, 384
  detection, 211, 388, 419
  distribution, 217, 389
  protein binding, 326
  retardation, 339
  separation, 146, 166, 228, 232, 312,
    314, 352, 388
  solvent, 358
Selectivity
  mobile phase, 242
  stationary phase, 242, 248
Silanol group, 162, 164, 166, 185, 246,
  250, 285, 292, 293, 300, 308, 312,
  355
Silica
  bare, 182, 185, 187, 215
  -based resin, 265
  colloidal, 174

[Silica]
crystalline, 157
gel surface, 36, 229, 246–250, 284,
  394, 398, 403, 406
grafted, 98, 103
hydroxylated, 227, 251, 255, 294, 355,
  365
model surface, 158
modification, 266
particles, 95, 148, 178
polymer,
  coated, 268, 269, 270, 271–273,
    300, 312, 313, 316, 326
  grafted, 286, 289–292, 300, 302,
    306, 309, 313
  modification of, 313
precipitated, 158
pyrogenic, 152, 158, 162
silylation, 274, 275, 279, 281, 293, 294
wide pore, 181
Softness of acid and base, 154
Solvation energy, 155
Sorbate-cavity interaction, 26, 27, 30
Spatial configuration of probe, 51
Spreading pressure, 57, 129
Stabilizer, 180–183, 228
Stability
chemical, 273, 281, 300, 308
of colloids, 146, 198
of silica, 182
mechanical, 264, 300, 312
thermal, 264
Standard
adsorption state, 57
apparatus, 46
calibration plot, 268
colloids, 187, 216
enthalpy of adsorption, 129
entropy change, 60
entropy of adsorption, 129
free energy change, 60
free energy of adsorption, 129
internal, 178, 216
polymers, 189, 400
protein, 283
temperature, 13

[Standard]
variation in free energy of adsorption,
  150
Stationary phase, 14, 45, 65, 166
adsorption at, 174, 182, 188, 274, 331,
  385, 390, 397
composition, 176
packing, 132
partition coefficient, 59, 129, 389
surface energy of, 128
Statistical thermodynamic consideration,
  2, 4, 330
Stereochemistry of probe, 150
Steric hindrance, 275, 277–283
Structural information,
on adsorbate, 35, 36, 228–234, 246,
  313, 321, 359, 371, 385
on adsorbent, 2–6, 8, 10, 20, 23, 26,
  28–30, 33, 69–108, 157, 160,
  166, 169
Surfactant, 175, 217

Tacticity of macromolecule, 403
Tailing of peak, 14, 168, 182, 195
Talc, 95, 98, 148, 157, 166
Temperature determination, 5, 12–17, 47,
  132
Theoretical plates number, 148
Thermal treatment, 78, 82, 88, 98
Thermodynamic equilibrium, 331, 359
Transmission electron microscopy, 174,
  179
Two surface approach, 20–25, 32–34

Universal calibration, 175, 179, 266, 390

van der Waals interactions, 6, 8, 18, 24,
  36, 127, 146, 156, 176
Vapor pressure, 13, 17, 48, 57, 66, 129,
  148
Virial coefficient treatment, 2–6
Virus, 269

Wettability measurements, 78, 138, 146, 148
Wilhelmy-plate measurements, 48, 131, 138

Zeolite, 9, 10, 12, 26–31

Printed and bound by CPI Group (UK) Ltd, Croydon, CR0 4YY

24/10/2024

01778278-0003